IORA International Conference on Operations Research 2016

IOP Conference Series: Materials Science and Engineering
Volume 166

Bogor, Indonesia
27 August 2016

Editors:

Toni Bakhtiar
Fajar Delli Wihartiko
Sukono

ISBN: 978-1-5108-3670-9
ISSN: 1757-8981

Printed from e-media with permission by:

Curran Associates, Inc.
57 Morehouse Lane
Red Hook, NY 12571

CURRAN ASSOCIATES INC. proceedings.com

Some format issues inherent in the e-media version may also appear in this print version.

Printed by Curran Associates, Inc. (2017)

For permission requests, please contact the Institute of Physics at the address below.

Institute of Physics
Dirac House, Temple Back
Bristol BS1 6BE UK

Phone: 44 1 17 929 7481
Fax: 44 1 17 920 0979

techtracking@iop.org

Additional copies of this publication are available from:

Curran Associates, Inc.
57 Morehouse Lane
Red Hook, NY 12571 USA
Phone: 845-758-0400
Fax: 845-758-2633
Email: curran@proceedings.com
Web: www.proceedings.com

Table of contents

Volume 166

The Indonesian Operations Research Association (IORA) - International Conference on Operations Research 2016

27 August 2016, Bogor, Indonesia

Accepted papers received: 16 January 2017
Published online: 8 February 2017

Preface

Papers

FOREWORD
IORA International Conference on Operations Research 2016

Conference Chair:
Prof. Soewarto Hardhienata, *Universitas Pakuan, Indonesia*

The roots of operations research can be traced back several decades, when early efforts were made to use a scientific approach in the organization management. Originating from an interdisciplinary activity to solve complex problems in the military during World War II, the ideas advanced in military operations were then adapted to improve efficiency and productivity in the civilian sectors. Now operation research as well as management science (OR/MS) are viewed as a body of established mathematical models and techniques to solve complex management problems. It encompasses a growing number of technical areas arising in the direction and management of large systems of men, machines, materials and money in industry, business, government, defense, and other decision-making fields. According to the Operations Research Society of America (ORSA), OR/MS is concerned with scientifically deciding how to best design and operate man-machine systems, usually under conditions requiring the allocation of scarce resources.

OR/MS has had an impressive contribution on improving the efficiency of numerous organizations around the world by offering a best solution. In the process, OR/MS has made a significant support to increasing the productivity of the economies of various countries.

IORA International Conference on Operations Research 2016 is held in conjunction with the inauguration of Indonesian Operations Research Association (IORA). The conference initiates to bring together OR/MS researchers, academicians and practitioners, whose collective work has sustained continuing OR/MS contribution to decision-making in many fields of application. It can be considered as good platforms for the OR/MS community, particularly in Indonesia, to meet each other and to exchange ideas.

The primary objectives of the conference are:

1. to provide OR/MS researchers, academicians and practitioners an appropriate platform for sharing experiences, communication and networking with other experts within the nation and from around the world,

IOP Conf. Series: Materials Science and Engineering **166** (2017) 011001 doi:10.1088/1757-899X/166/1/011001

2. to promote and disseminate the scientific field of operations research in Indonesia through the establishment of the Indonesian Operations Research Association (IORA).

In the spirit of OR/MS has made a significant contribution to increasing the efficiency in various sectors, the theme of the conference will be

"*Better Decisions for Sustainable Development.*"

Date and Venue

Date: August 27, 2016 (Saturday), 08.00 – 17.00
Venue: IPB International Convention Center, Botani Square Building
Jl. Padjadjaran, Bogor 16127, Indonesia

IOP Conf. Series: Materials Science and Engineering **166** (2017) 011001 doi:10.1088/1757-899X/166/1/011001

2016

Better Decisions for Sustainable Development

IORA International Conference on Operations Research 2016

in conjunction with

The Inauguration of Indonesian Operations Research Association (IORA)

PROCEEDING

Organized by:

Department of Computer Sciences,
Pakuan University, Bogor, Indonesia

Supported by: Indonesian Operations Research Association (IORA), Bogor Agricultural University, Padjadjaran University, University of Indonesia, Atma Jaya Catholic University, Indonesia Open University, Kalimantan Institute of Technology, Binus University, Bandung Islamic University, UprightDecision

IOP Conf. Series: Materials Science and Engineering **166** (2017) 011001 doi:10.1088/1757-899X/166/1/011001

PROCEEDING

IORA International Conference on Operations Research 2016

in conjunction with

The Inauguration of Indonesian Operations Research Association (IORA)

Date:
August 27, 2016 (Saturday)

Venue:
IPB International Conference Center, Bogor, Indonesia

Theme:
Better Decisions for Sustainable Development

Editors:

Toni Bakhtiar, *Bogor Agricultural University (IPB)*, Indonesia
Fajar Delli Wihartiko, *Universitas Pakuan*, Indonesia
Sukono, *Universitas Padjadjaran*, Indonesia

WELCOME REMARKS

IORA International Conference on Operations Research 2016

From the President
Indonesian Operations Research Association (IORA)
Prof. Sudradjat Supian

Operations research (OR) has rich traditions from viewpoint of wide boundaries in terms of clients, broad range of methodological approaches used and awareness to multiple and often conflicting interests. Under this perspective both OR and development share some concerns related to the implementation, as well as the design, of planning strategies. Thus, the relevance of OR and development has increasingly engaged the attention of operational researchers in both the industrialized and less developed countries.

For this conference we choose the following theme for our stand of work: "Better Decisions for Sustainable Development." It is our conviction that OR can make significant contribution to the process of development in general and to developing countries in particular. By this conference we aim to promote the increase in the use of OR as a practical tool for problems in many aspects of development. The ability to analyze large and complicated problems with operations research techniques is expected to suggest better decisions such that saving of the cost to industry and government, asserting that OR has a broader role than just knowledge brokering or technology transfer.

This conference is in conjunction with the inauguration of Indonesian Operations Research Association (IORA). Establishment of IORA in 2014 is evidently intended to reinforce the abovementioned initiative. We hope IORA can be considered as good platforms for OR researchers, academicians and professionals in Indonesia to meet each other, exchange ideas and strengthen their collaboration.

We look forward to welcome you in IORA-ICOR 2016 on August 27, 2016, in Bogor, Indonesia.

IOP Conf. Series: Materials Science and Engineering **166** (2017) 011002 doi:10.1088/1757-899X/166/1/011002

WELCOME REMARKS

IORA International Conference on Operations Research 2016

From the Rector
Universita Pakuan
Dr. Bibin Rubini

Mayor of the City of Bogor, excellencies, distinguished guests, dear colleagues, ladies and gentlemen,

It is a great honor for me to welcome all of you to the Indonesian Operation Research Association - International Conference on Operations Research 2016, this time in Bogor. On behalf of Department of Computer Sciences, Universitas Pakuan, as the organizing committee supported by Indonesian Operation Research Association and other universities, I would like to express my heartfelt gratitude for your presence in this opening ceremony of the conference.

I am particularly thankful to Prof. Chung-Piaw Teo of National University Singapore for taking time off from your busy schedule to deliver lecture in this important conference. I also wish to thank Mr. Mohamad Risdaya Fadil who has traveled from Madrid to Bogor to share some excellent ideas as well as Prof. Abdul Talib bin Bon from Universiti Tun Hussein Onn, Malaysia.

As I know, this conference is in conjunction with the inauguration of the Indonesian Operation Research Association (IORA). Therefore, I am especially honored to welcome you again here, in this milestone event. I very much appreciate the initiative of Prof. Sudradjat Supian from Universitas Padjadjaran in establishing and coordinating the association. We know that the first step in any endeavor is always the hardest. But I do believe that in the near future IORA will be able to promote the increase of the use of operations research as a practical tool for problems in many aspects of development and make significant contribution to the process of development in general in Indonesia. I hope IORA can be considered as good platforms for researchers, academicians and professionals in Indonesia to meet each other, exchange ideas and strengthen their collaboration in operations research.

Finally, I wish all participants a most fruitful day with interesting and stimulating discussions and exchange of knowledge in this conference.

Thank you very much.

COMMITTEE

IORA International Conference on Operations Research 2016

Conference Chair: Prof. Soewarto Hardhienata (Universitas Pakuan)

Program Co-chairs: Dr. Sukono (UniversitaS Padjadjaran)

Dr. Toni Bakhtiar (Bogor Agricultural University)

Advisory Committee:

Prof. Irwin Yousept (Universität Duisburg-Essen, Germany)
Prof. Chung-Piaw Teo (National University Singapore, Singapore)
Mr. Mohamad Risdaya Fadil (Aeroelasticity Specialist, Madrid, Spain)
Prof. Ilias Mamat (Quest International University Perak, Malaysia)
Prof. Abdul Talib bin Bon (Universiti Tun Hussein Onn, Malaysia)
Prof. Ismail bin Mohd. (Univeristi Malaysia Perlis, Malaysia)
Prof. Mustafa Mamat (Universiti Sultan Zainal Abidin, Malaysia)
Dr. Admi Syarif (Lampung University)
Dr. Amril Aman (Bogor Agricultural University)
Dr. Bibin Rubini (Universitas Pakuan)
Prof. Asep Kuswandi Supriatna (Universitas Padjadjaran)
Prof. Edy Soewono (Institute of Technology Bandung)
Prof. Hadi Sutanto (Atma Jaya Catholic University)
Prof. Herman Mawengkang (North Sumatra University)
Dr. Sri Harijati (Indonesia Open University)
Prof. Sudradjat Supian (Universitas Padjadjaran)
Prof. Widodo (Gadjah Mada University)
Prof. Bermawi P. Iskandar (Institute of Technology Bandung)
Prof. Abdul Hakim Halim (Coordinator of Kopertis IV)
Prof. Djati Kerami (University of Indonesia)
Prof. Ronny Kastaman (Universitas Padjadjaran)

Scientific Committee:

Subchan (Kalimantan Institute of Technology)
Diah Chaerani (Universitas Padjadjaran)
Ratih Dyah Kusumawati (University of Indonesia)
Bib Paruhum Silalahi (Bogor Agricultural University)
Sri Setyaningsih (Universitas Pakuan)
Prihastuti Harsani (Universitas Pakuan)
Rieske Hadianti (Institute of Technology Bandung)

Isdaryanto Iskandar (Atma Jaya Catholic University)
Pramono Sidi (Indonesia Open University)

External Affairs and Publication Committee:

Defi Ekajaya Yuda (UprightDecision)
Sitta Alief Farihati (Indonesia Open University)
Rojali (BINUS University)

Finance Committee:

Arie Qurania (Universitas Pakuan)
Eman Lesmana (Universitas Padjadjaran)
Farida Hanum (Bogor Agricultural University)

Event Management Committee:

Eneng Tita Tosida (Universitas Pakuan)
Fajar Delli Wihartiko (Universitas Pakuan)
Iyan Mulyana (Universitas Pakuan)
Hagni Wijayanti (Universitas Pakuan)
Agung Prajuhana Putra (Universitas Pakuan)
Lita Karlitasari (Universitas Pakuan)
Mufarrih Ridhwan (Bandung Islamic University)
Prapto Tri Supriyo (Bogor Agricultural University)

IOP Conf. Series: Materials Science and Engineering **166** (2017) 011004 doi:10.1088/1757-899X/166/1/011004

Peer review statement

All papers published in this volume of *IOP Conference Series: Materials Science and Engineering* have been peer reviewed through processes administered by the proceedings Editors. Reviews were conducted by expert referees to the professional and scientific standards expected of a proceedings journal published by IOP Publishing.

IOP Conf. Series: Materials Science and Engineering **166** (2017) 012001 doi:10.1088/1757-899X/166/1/012001

Modelling of capital asset pricing by considering the lagged effects

Sukono[1*], Y. Hidayat[2], A. Talib bin Bon[3], S. Supian[4]

[1,4]Department of Mathematics, Universitas Padjadjaran, Indonesia

[2]Department of Statistics, Universitas Padjadjaran, Indonesia

[3]Department of Production and Operations, Universiti Tun Hussein Onn Malaysia, Malaysia.

[*]corresponding author: sukono@unpad.ac.id

Abstract. In this paper the problem of modelling the Capital Asset Pricing Model (CAPM) with the effect of the lagged is discussed. It is assumed that asset returns are analysed influenced by the market return and the return of risk-free assets. To analyse the relationship between asset returns, the market return, and the return of risk-free assets, it is conducted by using a regression equation of CAPM, and regression equation of lagged distributed CAPM. Associated with the regression equation lagged CAPM distributed, this paper also developed a regression equation of Koyck transformation CAPM. Results of development show that the regression equation of Koyck transformation CAPM has advantages, namely simple as it only requires three parameters, compared with regression equation of lagged distributed CAPM.

1. Introduction

Referring Allen & Bujang [1] and Franses & Oest [5] standard form of Capital Asset Pricing Model (CAPM) was first developed separately by Sharpe in 1964, Lintner in 1965 and Mossin in 1969, therefore, the model is often called the CAPM form Sharpe-Lentner-Mossin. Most of the making of CAPM is based on a series of assumptions, namely: (i) An evaluation of the portfolio based on the expected return and standard deviation of the portfolio over a specified period of time; (ii) investor action based solely on considerations of expected return and standard deviation of the portfolio; (iii) The assets of the individual (individual assets) can completely broke up the smallest (fully divisible). Based on asset, it can buy assets on the desired amount; (iv) There is an interest rate and a risk-free savings (risk free lending and borrowing rate).This rate applies to all investors; (v) There are no transaction costs and income taxes; (vi) Information can be obtained directly (instantly) and free for all investors; (vii) individual investor action can not affect the price of the asset. In contrast, the actions of all investors (together and in the same direction) may affect the price of assets in the market; (viii) Each investor has the same hope (homogeneous expectations) against expectations of return, standard deviation and covariance of assets; and (ix) All assets can be traded (marketable) [6], [10].

According to Allen & Bujang [1] and Kuehn *et al.* [7], CAPM regression Equality is used to analyze the effect of the market index return on asset returns. The regression equation of CAPM is to link the level of expected returns on risk assets with a risk of the asset in a balanced condition [11]. In the equilibrium conditions of the capital markets, the difference between the market return with the return of risk-free assets referred to as the market risk premium, while the difference between the returns of assets with a risk-free asset return is referred to as the risk premium asset. In the CAPM, the risk premium of assets was affected by the independent variable market risk premium in the same period [11].We estimate that the risk premium assets are not only influenced by the market risk premium in the same period, but it can be influenced also by the market risk premium in the period - previous period or periods lagged. So that the CAPM model like this, known as the lagged distributed CAPM model. The weakness of distributed lagged CAPM is no sign that is efficient to determine the length lagged [3].

Associated with the lagged distributed, according Franses & Oest [5], Koyck transformation model assumes that the lagged effect on the explanatory variables are unlimited (infinite), but the coefficient parameter (for the variables lagged) down geometrically. Transformation model of Koyck has advantages, namely simple as it only requires three parameters. Therefore, in this paper intends to develop equation of Koyck transformation CAPM. The goal is to get an alternative model CAPM analysis. To clarify the application of this CAPM models, in this paper is given a numerical illustration.

2. Model of Capital Asset Pricing Model (CAPM)

In this section we explore the regression equation of CAPM, distribution lagged CAPM, and the development of the model of Koyck transformation CAPM.

2.1. Regression Equations of CAPM

The basic equation of the standard CAPM is known that the balance of the capital market will be indicated by the asset markets, where the line connecting the investment portfolio of a risk-free opportunity to opportunity-risk investment portfolio [2]. This relationship applies to all assets, either efficient or inefficient. To determine the position of the market portfolio, it needs to be combined between risky assets [6], [7].If we let r_{ft} returns risk-free assets at time t, then the expectation of risk-free asset is $\mu_f = E(r_{ft})$, and the variance of the risk free asset is $\sigma_f^2 = Var(r_{ft}) = 0$. All investors are assumed to invest in the same portfolio, namely the market portfolio. This assumption is valid because the assumptions in the CAPM, which all investors using the same analysis, which uses Markowitz's method [9]. In a state of equilibrium, all risk assets should be on the market portfolio, because all investors would hold that portfolio [8].

If the portfolio is composed of all the assets in the market, and we let r_{mt} market return at time t, then the expected market return is $\mu_m = E(r_{mt})$ and the variance of the market return is $\sigma_m^2 = Var(r_{mt})$. The difference between the return expectations of the market with expectations of returns risk-free assets amounted to $[E(r_{mt}) - \mu_f]$ referred to as market risk premiums, and the ratio between the risk of a market premium to market risk σ_m, ie, $[E(r_{mt}) - \mu_f]/\sigma_m$ is the slope of the capital market line equation [9].If we let r_{pt} portfolio return of capital markets at the time T, then the portfolio return expectations of the capital market is $\mu_p = E(r_{pt})$, and the variance of portfolio return of capital markets is $\sigma_p^2 = Var(r_{pt})$. Line equation of capital market portfolios can be expressed as:

$$E(r_{pt}) = E(r_f) + \frac{[E(r_m) - E(r_f)}{\sigma_m}\sigma_p \tag{1}$$

Slope $[E(r_m) - \mu_f]/\sigma_m$ is the market price of the portfolio risk efficiency. The market price shows the additional return required by the market [11], [8].

Furthermore, suppose r_t assets return at time t, with expectations of asset returns $\mu_t = E(r_t)$ and variance $\sigma_t^2 = Var(r_t)$. Based on the concept of capital market portfolios line mentioned above, the relationship between $E(r_t)$, $E(r_{mt})$, and $E(r_{ft})$, can be expressed as:

$$E(r_t) - E(r_{ft}) = \beta\{E(r_{mt}) - E(r_{ft})\} \tag{2}$$

where β is the slope. The difference between the expected return of assets to the expected return of risk-free assets amounted to $[E(r_t) - E(r_{ft})]$ known as the risk premium for assets [2].

Equation (2) empirically cannot be tested statistically, since the equation (2) is an equality expectation. Expectations equation is a value that has not been observed. Therefore, in order regression equation of CAPM empirically testable shall be amended as follows:

$$r_t - r_{ft} = \beta_0 + \beta_1(r_{mt} - r_{ft}) + e_t \tag{3}$$

Therefore the risk-free asset returns have flats constant, it can be written as $\mu_f = E(r_{ft})$. Because it is a risk-free assets, then the variance is $\sigma_f^2 = Var(r_{ft}) = 0$ [11]. So the equation (3) can be expressed as:

$$r_t - \mu_f = \beta_0 + \beta_1(r_{mt} - \mu_f) + e_t \tag{4}$$

Where β_0 is the constant term, β_1 is the slope, and e_t is the residual. Sequence of residual $\{e_t\}$ are assumed white noise, which is normal distribution with the average 0 and variance σ_e^2 [5].

Estimation of equation (4) can be done with the least squares method.

2.2. *Distributed Lagged CAPM*

The assumptions in the making of the standard CAPM is still used in the making of a distributed lagged CAPM. An important difference that the distributed lagged CAPM can be accommodate a possible effect of the risk premium on some past period, while the standard CAPM cannot do it [2].

Let r_t be an asset returns in the period t, and r_{mt} market index return in the period t. It is known that the return of risk-free assets in the period t, r_{ft} has the average $\mu_f = E(r_{ft})$ constant and $\sigma_f^2 = Var(r_{ft}) = 0$. The regression equation of distributed lagged CAPM as follows:

$$r_t - \mu_f = \alpha_0 + \beta_0(r_{mt} - \mu_f) + \beta_1(r_{mt-1} - \mu_f) + ... + \beta_l(r_{mt-l} - \mu_f) + e_t \tag{5}$$

Where α_0 is constant term, β_i $(i = 1,...,l)$ is coefficient of market risk premium, with l is length lagged, and e_t is the residuals row of the regression equation of lagged distributed CAPM, which is assumed to be normally distributed white noise with the average 0 and variance σ_e^2. Constants α_0 and coefficients β_i of parameters where $i = 1,...,l$ can be estimated using the least squares method (least square). A difficulty in the CAPM model with the lagged is to determine the length lagged influential.

Estimation of length lagged with Ad-Hoc method. To overcome the difficulties of determining the length of the lagged effect, it can be done by looking at signs of stability coefficient parameters. The way that is done by looking at the changes in positive sign (+) or negative (-) of the coefficient parameters. Length lagged is selected at the time showed stability (continued positive or continued negative) if lagged changed in length. If lagged extended, then the sign of the coefficient parameter inconsistent, it means that lagged cannot be selected. The weakness of this method is there are no clues about long lagged. Another weakness especially respect to economic data (time series) often have a high correlation (Gujarati, 1978).

2.3. *Koyck Transformation CAPM*

Based on the shape of CAPM with lagged, in this paper can be developed into a form of Koyck transformation CAPM, which equation can be written as follows:

$$r_t - \mu_f = \alpha_0 + \beta_0(r_{mt} - \mu_f) + \beta_0\lambda(r_{mt-1} - \mu_f) + \beta_0\lambda^2(r_{mt-2} - \mu_f) + ... + \varepsilon_t \quad (6)$$

or can be written into a simpler form as follows:

$$r_t - \mu_f = \alpha_0 + \beta_0\sum_{l=0}^{\infty}\lambda^K(r_{mt-l} - \mu_f) + \varepsilon_t \quad (7)$$

The first lagged from equation (7) as follows:

$$r_{t-1} - \mu_f = \alpha_0 + \beta_0\sum_{l=0}^{\infty}\lambda^l(r_{mt-l-1} - \mu_f) + \varepsilon_{t-1} \quad (8)$$

If equation (8) multiple by λ, the obtained:

$$\lambda(r_{t-1} - \mu_f) = \lambda\alpha_0 + \beta_0\sum_{l=0}^{\infty}\lambda^{l+1}(r_{mt-l-1} - \mu_f) + \lambda\varepsilon_{t-1} \quad (9)$$

When the equation (7) is subtracted by equation (9) obtained:

$$(r_t - \mu_f) - \lambda(r_{t-1} - \mu_f) = \alpha_0 - \lambda\alpha_0 + \beta_0(r_{mt} - \mu_f) + \varepsilon_t - \lambda\varepsilon_{t-1} \quad (10)$$

or

$$(r_t - \mu_f) = \alpha_0(1-\lambda) + \beta_0(r_{mt} - \mu_f) + \lambda(r_{t-1} - \mu_f) + \varepsilon_t - \lambda\varepsilon_{t-1} \quad (11)$$

For example $v_t = \varepsilon_t - \lambda\varepsilon_{t-1}$, equation (11) can be written as follows:

$$r_t - \mu_f = \alpha_0(1-\lambda) + \phi_0(r_{mt} - \mu_f) + \lambda(r_{t-1} - \mu_f) + v_t \quad (12)$$

Compared to CAPM by lagged form, Koyck transformation CAPM simpler form only requires three parameters, namely: α_0, λ and β_0. In addition, there is no reason multikolinier occurrence. Noteworthy is that the initial form (equation (6)) begins with lagged, but the final form (equation (12)) ends with autoregresi. Referring Franses & Oest [5], equation (11) is a ARMAX model. Coefficient parameter equation (11) can be estimated using the method of instrumental variables or conditional *likelihhod*.

3. Illustration Numerical Results and Discussion

In this section we explore numerical illustration of estimation of CAPM regression, estimation of distributed lagged CAPM, and estimation of Koyck transformation CAPM.

3.1. Estimation of CAPM regression

Suppose that there are as many as 400 stocks return data S and the market return (ISHG). It is known that the risk-free asset returns are relatively constant, with an average of $\hat{\mu}_f$ =0.0026462 and variance $\hat{\sigma}_f^2 = 0$. Estimator CAPM regression such data is:

$$\underset{Stat-t}{r_t - 0.0026462} = \underset{(40.53)}{0.00412} + \underset{(33.04)}{0.256}(r_{mt} - 0.0026462) + \varepsilon_t$$

R^2 =73.30% F = 109.85 $P - Value$ = 0.000

- The verification test of the parameter estimator

  To constants estimator $\hat{\beta}_0$, the hypothesis being tested is $H_0 : \hat{\beta}_0 = 0$ with alternative $H_0 : \hat{\beta}_0 \neq 0$. Statistical $t_{rasio}(\hat{\beta}_1)$ = 33.04. While at significant level α = 0.05; critical value $t_{(0.05;925)}$ = 1.645. Show that $t_{ratio}(\hat{\beta}_1) > t_{(0.05;925)}$, hypothesis H_0 is rejected, it is mean that $\hat{\beta}_1$ = 0.256 significant.
- The suitability test of the model

The suitability test generated value determination R^2 =73.30%. It means that the market risk premium is correlated quite strongly with the risk premium stocks. To test the significance of the model, the hypothesis is $H_0: \hat{\beta}_0 = \hat{\beta}_1 = 0$ with alternative $H_1: \exists \hat{\beta}_i \neq 0 (i = 0,1)$. The regression model generates statistics $F =$ 109.85with probability $P - Value =$ 0.000; at the level of significance $\alpha =$ 0.05. It is clear that $P - Value < \alpha$, hypotehsis H_0 is rejected . it means regression model is significant. Residual normality test shows $\varepsilon_t \sim N(0,0.00000376)$, it means that *residual* ε_t *white noise*.

3.2. Estimation of distributed lagged CAPM

Suppose that there are as many as 400 stocks return data S and the market return (JCI). It is known that the risk-free asset returns are relatively constant, with an average of $\hat{\mu}_f$ =0.0026462 and variance $\hat{\sigma}_f^2 = 0$. Regression estimator of CAPM for such data is

$$r_t - \mu_f = \underset{(18.24)}{0.00388} + \underset{(27.32)}{0.436} (r_{mt} - \mu_f) + \underset{(16.26)}{0.270} (r_{mt-1} - \mu_f) + \underset{(17.20)}{0.201} (r_{mt-2} - \mu_f) + \underset{(8.23)}{0.132} (r_{mt-3} - \mu_f) + \varepsilon_t$$

Stat–t

R^2 =79.60% F = 381.99 $P - Value$ = 0.000

The length of the lag regression estimator is 3. The coefficient signs become inconsistent parameter if lag extended again. It means that positive sign turn negative sign.

- The verification test of the parameter estimator

  To constants estimator $\hat{\alpha}_0$,the hypothesis being tested is $H_0: \hat{\alpha}_0 = 0$ with alternative $H_0: \hat{\alpha}_0 \neq 0$. Statistical $t_{rasio}(\hat{\alpha}_0)$ =18.24; while at significant level $\alpha =$ 0.05; critical value $t_{(0.05;925)}$ = 1.645. Because $|t_{rasio}(\hat{\alpha}_0)| > t_{(0.05;925)}$ so that hypothesis H_0 is rejected. It means that $\hat{\alpha}_0 = 0.00388$ is significant. For parameter estimator $\hat{\beta}_0$, the hypothesis being tested is $H_0: \hat{\beta}_0 = 0$ with alternative $H_0: \hat{\beta}_0 \neq 0$. Statistical $t_{rasio}(\hat{\beta}_0)$ = 27.32; it shows $t_{rasio}(\hat{\beta}_0) > t_{(0.05;925)}$. So that hypothesis H_0 is rejected and $\hat{\beta}_0$ = 0.436 is significant. Parameter estimator $\hat{\beta}_1$, the hypothesis being tested is $H_0: \hat{\beta}_1 = 0$ with alternative $H_0: \hat{\beta}_1 \neq 0$. Statistical $t_{rasio}(\hat{\beta}_1)$ = 16.26; it shows $t_{rasio}(\hat{\beta}_1) > t_{(0.05;925)}$. So that hypothesis H_0 is rejected and $\hat{\beta}_1$ = 0.276 is significant. Parameter estimator $\hat{\beta}_2$, the hypothesis being tested is $H_0: \hat{\beta}_2 = 0$ with alternative $H_0: \hat{\beta}_2 \neq 0$. Statistical $t_{rasio}(\hat{\beta}_2)$ = 17.20; it shows $t_{rasio}(\hat{\beta}_2) > t_{(0.05;925)}$. So that hypothesis H_0 is rejected and $\hat{\beta}_2$ = 0.261 is significant. Parameter estimator $\hat{\beta}_3$, the hypothesis being tested is $H_0: \hat{\beta}_3 = 0$ with alternative $H_0: \hat{\beta}_3 \neq 0$. Statistical $t_{rasio}(\hat{\beta}_3)$ = 8.23; it shows $t_{rasio}(\hat{\beta}_3) > t_{(0.05;925)}$. So that hypothesis H_0 is rejected and $\hat{\beta}_3$ = 0.132 is significant.

- The suitability test of the model

  The suitability test generated value determination R^2 =79.60%. It means that the market risk premium is correlated quite strongly with the risk premium stocks. To test the significance of the model, the hypothesis is $H_0: \hat{\alpha}_0 = \hat{\beta}_i = 0$ with alternative $H_1: \exists \hat{\alpha}_0 \neq 0$ and or $\hat{\beta}_i \neq 0$, ($i = 0,1,2,3$). The regression model generates statistics F =381.99with probability $P - Value$ = 0.000; at the level of significance $\alpha =$ 0.05. It is clear that $P - Value < \alpha$, hypotehsis H_0 is rejected. It means regression model is significant. Residual normality test shows $\varepsilon_t \sim N(0, 0.00001668)$, it means that *residual* ε_t *white noise*.

3.3. Estimation of Koyck transformation CAPM

Koyck models assume that the lagged effect on the explanatory variables are unlimited (infinite), but the coefficient parameter (for the variables lagged) down geometrically. Transformation model of Koyck has advantages, namely simple as it only requires three parameters, namely: α_0, λ and β_0.

Furthermore, suppose that $Y_{it} = S_{it} - \hat{\mu}_f$ in the risk premium return stock i ($i = 1, ..., N$ and N number of shares) at the time t ($t = 1, ..., T$ and T the number of data), and W market risk premium return on time. Transformation Koyck CAPM form refer ARMAX model of equation (12). Because regression Koyck Y_{it} to I_{mt} produces small coefficient of determination R^2. Variable Y_{it} is transformed to $M_{it} = \ln\{(1+Y_{it})/(1-Y_{it})\}$. Koyck regression equation to be estimated subsequently in the form $M_{it} = c_i + \beta_i I_{mt} + \lambda M_{it-1} + u_{it} - \lambda u_{it-1}$. Constant estimation $c_i = \alpha_i(1-\lambda_i)$, parameter coefficient λ_i and *multiplier* β_i conducted using the method of instrumental variables referring to equation (12) or conditional *likelihood* estimator. For example, the data will be analyzed regression of stock returns based Koyck transformation CAPM. Based on the results of the regression analysis, obtained estimator constants $\hat{c}_1$ = -0.0155 has a statistical value $t_{rasio}(\hat{c}_1)$ = -5.40 with probability P = 0.000. The hypothesis tested is H_0: $\hat{c}_1$ = 0alternatively: H_1: $\hat{c}_1 \neq$ 0.Using a significance level of α = 0.05was obtained critical value statistic $t_{(0.05;824)}$ = 1.962847. Because $t_{rasio}(\hat{c}_1) > t_{(0.05;824)}$, the hypothesis H_0 is rejected, which means significant estimator constants $\hat{c}_1$ = -0.0155.

To test the hypothesis estimator coefficient $\hat{\lambda}_1$ and *multiplier* $\hat{\beta}_1$ conducted in the same manner as a test of hypothesis estimator constants. The test results concluded that coefficient $\hat{\beta}_1$ = 0.713 and $\hat{\lambda}_1$ = 0.321 is significant.

The above regression coefficient of deterministic R^2 = 67.80%. Simultaneous test on statistical regression models F = 862.50 with probability P = 0.000.The hypothesis tested is H_0: $\hat{c}_1 = \hat{\beta}_1 = \hat{\lambda}_1$ = 0alternatively: H_1: $\exists\, \hat{c}_1 \neq 0$, $\hat{\beta}_1 \neq 0$ and or $\hat{\lambda}_1 \neq 0$. For significant level α = 0.05was obtained critical value statistic $F_{(0.05;3;824)}$ = 2.615708.Because$F > F_{(0.05;3;824)}$ and *P-Value*$< \alpha$, so that hypothesis H_0 is that means regression model M_{1t} = -0.0155 + 0.713 I_{mt} +0.321 $M_{1t-1} + u_{1t}$ -0.321 u_{1t-1} or M_{1t} = -0.0155 +0.713 $(r_{mt} - \hat{\mu}_f)$ +0.321 $(S_{1t-1} - \hat{\mu}_f)$ $+u_{1t}$ -0.321 u_{1t-1} is significant. Testing the assumption of normality*residula* u_t, results show that u_t normal distribution with an average $\hat{\mu}_{u1}$ = $-4.4479\times 10^{-17} \approx 0$ and standard deviation $\hat{\sigma}_{u1}$ = 0.07880 or variance σ^2_{u1} = 0.00621. Thus the above regression model accordingly.

4. Conclusions

In this paper discussed about modelling of Capital Asset Pricing Model (CAPM) with considering the delay effects. In the discussion of the CAPM includes the standard CAPM regression, lagged distributed CAPM regression, and CAPM regression Koyck transformation. Sharpe-Lentner-Mossin has introduced a regression equation of the standard CAPM. Regression standard CAPM does have a simple shape, but regression standard CAPM is not able to capture the effects of market risk premium in the past. Then the CAPM regression equation developed by lagged distributed, which is able to capture the effects of market risk premium in the past. The difficulty is not easy to determine the length lagged. In this paper developed CAPM regression Koyck transformation. Regression of CAPM Koyck transformation is a form ARMAX simple, and capable of capturing the effects of market risk premium in the past.

Acknowledgment
We would like thanks the Rector and Dean of Faculty of Mathematics and Natural Sciences Universitas Padjadjaran, which have provided a grant program, the Academic Leadership Grant (ALG), under the coordination of Prof. Dr. Sudradjat, M.S., which is a means to increase research activities and publications to researchers at the Universitas Padjadjaran.

References

[1] Allen, D.E. & Bujang, I., Conditional Beta Capital Asset Pricing Model (CAPM) and Duration Dependece Test, *Working Paper,* 18th World IMACS/MODSIM Congress, Cains, Australia 13-17 July 2009, [internet], 2009. [updated 2013 Jan 15; cited 2015 Apr 10]. Available from: http://www.mssanz.org.au/modsim09.

[2] Choudhary, K. & Choudhary, S. 2010. Testing Capital Asset Pricing Model: Empirical Evidences from Indian Equity Market. *Eurasian Journal of Business and Economics 2010*, 3 (6), 127-138.

[3] Demetriou, I.C. & Vassiliou, E.E., A Distributed Lag Estimator with Piecewise Monotonic Coefficients, *Proceedings of the Wold Congress on Enggineering,* 2008 Vol. II, WCE 2008, July 2-4, [internet] 2008, London, U.K. [updated 2013 Jan 15; cited 2015 Apr 10]. Available from: www.iaeng.org/publication/ WCE2008/WCE2008_pp1088-1094.pdf.

[4] Draper, N.R. & Smith H. 1998. *Applied Regression Analysis.* Third Edition. New York: John Wiley & Sons, Inc.

[5] Franses, P.H. & Van Oest, R., On the Econometrics of the Koyck Model, *Econometric Institute Report 2004-07,* retrived fromhttp://www.publishing.eur.nlfir/asset/1190/ei200407.pdf, on December 2009).

[6] Krause, A. 2001. *An Overview of Asset Pricing Models.* Preliminary Version. University of Bath School of Management.

[7] Kuehn, L.A., Simutin, M. & Wang,J.J., 2014, A Labor Capital Asset Pricing Model, *Working Paper,* Tepper School of Business Carnegie Mellon University, [internet], 2014. [updated 2015 Feb 15; cited 2015 Apr 10]. Available from: ww.andrew.cmu.edu/user/.../WangJessie_CV.pd...

[8] Shamim, M.A., Abid, Y. and Shaikh, E.A. 2014. Validity of Capital Asset ricing Model in Pakistan's Capital Market (Karachi Stock Exchange). *Journal of Emerging Issues in Economics, Finance and Banking* (JEIEFB), *An Online International Research Journal* (ISSN: 2306-367X) 2014 Vol: 3 Issue 4.

[9] Suartini, N.K.A & Mertha, I.M. Perbandingan CAPM dengan APT dalam Memrediksi Return Saham. Kertas Kerja. Fakultas Ekonomi Universitas Udayana (Unud), Bali, Indonesia.

[10] Tadelilin, E., *Portofolio Investasi*: Teori dan Aplikasi, Edisi Pertama. Yogyakarta: Penerbit Kanisius, 2010

[11] Wakyiku, D. 2010. Testing the Capital Asset Pricing Model (CAPM) on the Uganda Stock Exchange. *Working Paper.* African Institute for Mathematical Sciences, South Africa. E-mail: davidw@aims.ac.za.

IOP Conf. Series: Materials Science and Engineering **166** (2017) 012002 doi:10.1088/1757-899X/166/1/012002

Supply and demand analysis for flood insurance by using logistic regression model: case study at Citarum watershed in South Bandung, West Java, Indonesia

P. Sidi[1*], M. Mamat[2], Sukono[3*], S. Supian[4]

[1*]Department of Mathematics, Universitas Terbuka, INDONESIA
[2]Graduate School, Universiti Sultan Zainal Abidin, MALAYSIA
[3,4]Department of Mathematics, Universitas Padjadjaran, INDONESIA

E-mails: pram@ecampus.ut.ac.id; sukono@unpad.ac.id; sudradjat@unpad.ac.id

Abstract. Floods have always occurred in the Citarum river basin. The adverse effects caused by floods can cover all their property, including the destruction of houses. The impact due to damage to residential buildings is usually not small. Indeed, each of flooding, the government and several social organizations providing funds to repair the building. But the donations are given very limited, so it cannot cover the entire cost of repair was necessary. The presence of insurance products for property damage caused by the floods is considered very important. However, if its presence is also considered necessary by the public or not? In this paper, the factors that affect the supply and demand of insurance product for damaged building due to floods are analyzed. The method used in this analysis is the ordinal logistic regression. Based on the analysis that the factors that affect the supply and demand of insurance product for damaged building due to floods, it is included: age, economic circumstances, family situations, insurance motivations, and lifestyle. Simultaneously that the factors affecting supply and demand of insurance product for damaged building due to floods mounted to 65.7%.

1. Introduction

Flooding has always been a natural disaster that always happens every year in the Citarum river basin in the South Bandung West Java. The disadvantage of this disaster is diverse, ranging from the loss of: property, a car, people, until the damage of houses [4]. Various flood disaster management efforts have been made by the government, such as evacuating victims and provide financial aid for home repairs. However, the donation cannot cover the losses suffered by the community. Therefore, public participation needs to be improved so that they participate in preparation for the floods that continue to occur. Insurance damage of houses is a form of risk management that is effective in overcoming the loss [2].

Flood insurance is a form of cooperation between people who want to minimize the risk of damage uncertain. Flood insurance provides protection and guarantees if there is a risk of damage due to flood. Because of the risk of damage due to flood, it raised awareness of a person to bestow such risks to the insurance company by filing a flood insurance demand [5]. The demand for flood insurance encourages insurance companies to make efforts in attracting customers. Various insurance companies race to create a services product that can meet the needs of customers [10]. Understanding of customer behavior is an important task for the company to determine the factors that encourage customers to follow the life insurance. In order to know these factors, then the method can be used is the logistic regression analysis [1], [3].

This paper intends to conduct research, with the aim of analyzing the supply and demand of insurance caused by flood damage of houses Citarum river. The analyzed data is obtained by distributing questionnaires to gather public perception in the Citarum river basin in South Bandung regency. The analysis is done in stages as follows: (i) to analyze deals flood insurance products, (ii) the analysis of demand for insurance products, and (iii) analyze the factors that influence the demand for insurance products flooding.

2. Logistic Regression Model

Logistic regression is a model used to describe the relationship between independent variables and response variables with ordinal scale. In this logistic regression, the variable response with ordinal scale more than two categories, and each category can be rated. The model for logistic regression is *cumulative logit models*. According Fahrmeir [6], the cumulative probability model for k categories is:

$$P(Y \le r \mid x) = \frac{\exp(\theta_r + x'\beta)}{1+\exp(\theta_r + x'\beta)} \quad (1)$$

while the cumulative logit model as follows:

$$\log\left[\frac{P(Y \le r \mid x)}{P(Y > r \mid x)}\right] = \log\exp(\theta_r + \boldsymbol{x}'\boldsymbol{\beta}) = \theta_r + \boldsymbol{x}'\boldsymbol{\beta} \quad (2)$$

where $r = 1,\dots,k-1$, θ_r =constants to -*r*, $\boldsymbol{\beta} = (\beta_1, \beta_2, \beta_3, \dots, \beta_p)$ is vector coefficients of free variable, and $\boldsymbol{x}'=$ vector of free variable. Method of estimation parameters used in the logistic regression model is the Maximum Likelihood Estimation (MLE). This parameter estimator is obtained by maximizing the likelihood function respect to parameters. The likelihood functions for three categories as follows [7], [12]:

$$l(\boldsymbol{\beta}) = \prod_{i=1}^{n}\left\{\left[\frac{e^{\theta_1+x'\beta}}{1+e^{\theta_1+x'\beta}}\right]^{y_{1i}}\left[\frac{e^{\theta_2+x'\beta}-e^{\theta_1+x'\beta}}{\left(1+e^{\theta_2+x'\beta}\right)\left(1+e^{\theta_1+x'\beta}\right)}\right]^{y_{2i}}\left[\frac{1}{1+e^{\theta_2+x'\beta}}\right]^{y_{3i}}\right\} \quad (3)$$

In a logistic regression model, is necessary to test the significance of the regression coefficients simultaneously and partially. Testing the significance of the regression coefficients simultaneously is performed by using a likelihood estimation ratio test symbolized by *G*. According to Hosmer and Lemeshow [7], [9], this is defined as follows:

$$G = -2\ln\left[\frac{L_0}{L_1}\right] \quad (4)$$

with L_0 =likelihood in the condition H_0, L_1 =likelihood in the condition H_1. The hypothesis of this test is: $H_0 : \beta_1 = \beta_2 = \ldots = \beta_p = 0$ against alternative H_1: at least there is one $\beta_p \neq 0$. Criteria test the null hypothesis is rejected if $P\left[\chi^2(v) > G\right] < \alpha$, the level of significance was set.

Statistics G will follow Chi-Square distribution with degrees of freedom v. While testing the significance of regression coefficient partially is done by using the *Wald* test is defined as:

$$W = \frac{\hat{\beta}_l}{SE\left(\hat{\beta}_l\right)} \tag{5}$$

The hypothesis of this test is: $H_0 : \beta_i = 0$ against alternative H_1: $\beta_i \neq 0, 1 = 1, 2, \ldots, p$. Criteria test the null hypothesis is rejected if $P[|Z| > W] < \alpha$, the level of significance was set. Where, Z is a random variable that follows a standard normal distribution [7].

In addition, it is also testing the suitability of the model used to check the model obtained is appropriate or not in accordance with the observed data. The test is performed using Pearson test and Deviance test [6], [8]. Pearson statistic is defined as:

$$\chi^2 = \sum_{i=1}^{g} \chi^2{}_p\left(y_i, \hat{\pi}_i\right) \text{ with } \chi^2{}_p\left(y_i, \hat{\pi}_i\right) = n_i \sum_{j=i}^{k} \frac{\left(y_{ij} - \hat{\pi}_{ij}\right)^2}{\hat{\pi}_{ij}} \tag{6}$$

Deviance statistic is defined as:

$$D = 2\sum_{i=1}^{g} \chi_D^2\left(y_i, \hat{\pi}_i\right) \text{ with } \chi_D^2\left(y_i, \hat{\pi}_i\right) = n_i \sum_{j=1}^{k} y_{ij} \log\left(\frac{y_{ij}}{\hat{\pi}_{ij}}\right) \tag{7}$$

The hypothesis of this test is H_0: model is appropriate, against the alternative H_1: model is not appropriate. Pearson and Deviance statistics was spread by Chi-Square distribution with degrees of freedom $g(k-1)-p$. Decisions reject the null hypothesis. Criteria test the null hypothesis is rejected if: χ^2 and $D > \chi^2_{(g(k-1)-p)}$ or $P\left[\chi_v^2 > \chi^2_{pearson}\right]$ and $P\left[\chi_v^2 > D\right]$ less than α, the level of significance was set. If a null hypothesis is rejected, then the conclusion is that the model obtained is not appropriate. Conversely, if the opportunities generated is greater than opportunities desired or alpha, then H_0 accepted so that the model is appropriate [11], [12].

3. Methodology

The data used in this study are primary data that obtained by distributing questionnaires to customers of an insurance company at branch office, Bandung on 15-26 February 2016 as many as 98 people. Response variable used in this study is a request of flood insurance, while the independent variables were age (X_1), the economy condition (X_2), the family situation (X_3), insurance motivation (X_4), and lifestyle (X_5). The steps are performed in this study: the first is to test the assumption of multicollinearity. This test uses correlation matrix. Furthermore, the univariable of logistic regression model is analyzed. This analysis is used to sort out the variables that will be used in a multivariable of logistic regression model, and to generate a parameter estimator. After that, in this analysis was also performed coefficient test of logistic regression simultaneously by using a likelihood ratio test. While, the coefficient test of logistic regression partially was done by using *Wald* test. After getting the best logistic regression model, further it is performed testing for the suitability of the model.

4. Results and Discussion

From the results of instrument test showed that variable X_2, X_3, X_4, and X_5 are valid and reliable. Therefore, all of the study variables are eligible for use in this research. Before performing univariable logistic regression analysis, performed first assumption test of multicollinearity. Based on the examination results of multikolinieritas is obtained that correlation values between X_4 with X_3 for 0.627 is greater than the correlation between X_4 and Y is 0.374. It indicates the occurrence of multicollinearity, but it is not excluded for allegedly influencing the dependent variable. The next step is to perform univariable logistic regression analysis. In this analysis, first performed is the likelihood ratio test for univariable logistic regression model. Tests carried out by using Equation (4), and the results are given in Table 1.

Table 1. Results of Likelihood Ratio Test for Univariable Logistic Regression Model

Variables	Statistic of *G*	*P*-Value
X_1	0.573	0.574
X_2	12.453	0.000
X_3	53.363	0.000
X_4	25.421	0.000
X_5	57.244	0.000

Based on the results in Table 1, it can be seen that the p-value of the estimation ratio likelihood G of four independent variables, namely the economy condition (X_2), the family situation (X_3), insurance motivation (X_4), and lifestyle (X_5) is less than = 5 %. Therefore it can be concluded that the four significant variables that are not all the parameters in the model is zero. After that, the partial test of the value of the parameter estimator for univariable logistic regression model. Testing is done by using equation (5), and the results are given in Table 2.

Table 2. Result of Parameter Estimation, Value-Z of Univariable Logistic Regression Model

Variables	Predictors	Coefficients	Statistic of *Z*	*P*-Values
Age	C_1	-3.23312	-2.67321	0.00511
	C_2	-2.41325	-1.34320	0.13124
	C_3	0.67924	0.75627	0.34152
	X_1	0.02115	0.56489	0.48636
Economy condition	C_1	2.67413	2.11631	0.03547
	C_2	4.12536	3.22953	0.00251
	C_3	5.98525	4.77351	0.00001
	X_2	-0.43574	-3.76835	0.00000
Family situation	C_1	2.32356	3.24362	0.01784
	C_2	3.24798	4.85796	0.00002
	C_3	5.67595	7.00345	0.00000
	X_3	-1.38462	-5.17364	0.00000
Insurance motivation	C_1	2.47463	2.63352	0.02301
	C_2	4.26738	4.01732	0.00002
	C_3	5.98584	4.87642	0.00000
	X_4	-0.70185	-5.21364	0.00000
Lifestyle	C_1	4.32911	4.53722	0.00002
	C_2	5.88678	5.75691	0.00000
	C_3	10.1325	9.00231	0.00001
	X_5	-0.89150	-5.43510	0.00000

Based on Table 2, it is known that the statistical results of the value of Z and P-value on the age variable is not significant in the model, because $P[|Z| > 0.56489] = 0.48636$ more than $\alpha = 5\%$. The economy condition variable is significant in the model, because $P[|Z| > 3.76835] = 0.00000$ less than $\alpha = 5\%$. The family situation variable is significant in the model, because $P[|Z| > 5.17364] = 0.00000$ less than $\alpha = 5\%$. The insurance motivation variable is significant in the model, because $P[|Z| > 5.21364] = 0.00000$ less than $\alpha = 5\%$. The lifestyle variable is significant in the model, because $P[|Z| > 5.43510] = 0.00000$ less than $\alpha = 5\%$. Therefore, the independent variables that should be in the logistic regression models is the economy condition (X_2), the family situation (X_3), insurance motivation (X_4), and lifestyle (X_5).

Furthermore, logistic regression analysis will be done by using X_2, X_3, X_4, and X_5 so that it obtained results in Table 3.

Table 3. Results of Parameter Estimation, Standard Errors of Logistic Regression Model

Predictors	Coefficients	*SE*-Coefficients
C_1	8.97139	2.13838
C_2	13.27131	2.28379
C_3	15.57133	2.73726
X_2	-0.426424	0.13617
X_3	-0.724789	0.28345
X_4	-0.089617	0.15675
X_5	-0.827016	0.25379

Based on the estimation results of logistic regression in Table 3, it is obtained three logit models, namely:

Logit regression model 1:

$$\log[P(Y \le 1)] = 8.97139 - 0.348356X_2 - 0.724789X_3 - 0.089617X_4 - 0.827016X_5;$$

Logit regression model 2:

$$\log[P(Y \le 2)] = 13.27131 - 0.348356X_2 - 0.724789X_3 - 0.089617X_4 - 0.827016X_5;$$

Logit regression model 3:

$$\log[P(Y \le 3)] = 15.57133 - 0.348356X_2 - 0.724789X_3 - 0.089617X_4 - 0.827016X_5.$$

After that, testing the logistic regression coefficient simultaneously by using estimation ratio test likelihood and it is obtained the value $G = 74\ 543$ with a P-value $= 0.000 < \alpha = 5\%$. It can be concluded that at least one of the independent variables significantly influence the response variable. Then, the logistic regression coefficient test partially using the *Wald* test, refer to the equation (5), and the results are given in Table 4.

Table 4. Results of Wald Test

Variables	Statistic of *Z*	*P*-Value
X_2	-3.74	0.004
X_3	-4.37	0.002
X_4	-1.68	0.437
X_5	-5.64	0.000

On the results of the Wald test is known that there are three independent variables are variables X_2, X_3 and X_5 which has a smaller chance of α. This means that the independent variable is significant at the level 5%. In addition, there is one independent variable is the variable X_4 which have greater opportunities than the level 5%. It can be concluded that the factor of economic conditions, the family situation, and lifestyle have a significant influence on demand for flood insurance. Furthermore, the suitability tests of the model by using Pearson and Deviance test. Tests were done according to Equation (6) and (7). Based on the results of testing the suitability of the model obtained *p-value* for the Pearson method was not significant because $P[\chi^2_{216} > 216.455]$ =0.437 more than $\alpha = 5\%$. In addition, *p-value* for the method of Deviance is not significant because $P[\chi^2_{216} > 148.132] =$ 1.000 more than $\alpha = 5\%$. This means accepting H_0, so it can be concluded that the model obtained in accordance with observed data. Based on the results of *Pseudo R-Square* is obtained Nagelkerke value of 65.7%. This means that 65.7% variability of the dependent variable is able to be explained by variable X_2, X_3, X_4, and X_5 while the remaining 34.3% is explained by other variables.

Thus the interpretation of the best logistic regression for logit model 1, it can be concluded that the chances of economic condition variable (X_2) affect demand for flood insurance for the first category were lower by 0.426424 as compared to the second category. Chances of the family situation variable (X_3) affect demand for flood insurance for the first category were lower by 0.724789 as compared to the second category. Chances of insurance motivation variable (X_4) affect demand for flood insurance for the first category were lower by 0.089617 as compared to the second category. Chances of lifestyle variable (X_5) affect demand for flood insurance for the first category were lower by 0.827016 as compared to the second category.

Interpretation of the logit models 2, it can be concluded that the chances of economic condition variable (X_2) affect demand variables flood insurance for second categories was lower by 0.426424 compared to the third category. Chances of the family situation variable (X_3) affect demand for flood insurance for both categories was lower by 0.724789 compared to the third category. Chances of insurance motivation variable (X_4) affect demand for life insurance for both categories was lower by 0.089617 compared to the third category. Chances of lifestyle variable (X_5) affect insurance demand for both categories was lower by 0.827016 compared to the third category.

Interpretation of the logit models 3, it can be concluded that the chances of economic condition variable (X_2) affect demand for the third category of flood insurance was lower by 0.426424 compared to the fourth category. Chances of the family situation variable (X_3) affect demand for flood insurance for the third category was lower by 0.724789 compared to the fourth category. Chances of insurance motivation variable (X_4) affect demand for life insurance for the third category was lower by 0.089617 compared to the fourth category. Chances of lifestyle variable (X_5) affect insurance demand for the third category was lower by 0.827016 compared to the fourth category.

5. Conclusion

In this paper has carried out analysis of the factors affecting demand and supply flood insurance, with a case study in the Citarum river basin Bandung regency, West Java. Results of analysis showed that the supply and demand for insurance of damage to buildings caused by floods, influenced by factors: age, economy condition, family situation, insurance motivation, and lifestyle. From the results of data analysis that has been done, there are three best ordinal logistic regression models on the factors that affect the supply and demand for insurance of damage to buildings caused by floods significantly. Based on Pseudo R-Square, insurance demand of damage to buildings caused by floods able to be explained by the logistic regression model amounted to 65.7%, while the rest of 34.3% is explained by other factors. After knowing the factors that significantly influence demand for flood insurance, it is expected that an insurance company branch office in Bandung can further increase the sales of products, especially products of flood insurance.

Acknowledgment

We would like thanks the academic leadership grant (ALG), Faculty of Mathematics and Natural Sciences, University Padjadjran, which has provided facilities to conduct research and publication.

References

[1] R. Baker, C. Weinand, J. Jeng, H. Hoeksema, S. Monstrey, S. Pae, R. Spence and D. Wilson. 2009. Using Ordinal Logistic Regression to Evaluate the Performance of Laser-Doppler Predictions of Burn-Haling Time. *BMC Medical Research Methodology,* (online), 9-11, (www.biomedcentral.com), diakses 5 Mei 2013.

[2] J. Czajkowski, H. Kunreuther, and E. Michel-Kerjan. 2013. Catastrophe Model Based Quantification of Riverine and Storm Surge Flood Risk in Texas. *Working Paper* # 2013-01. The Wharton School, University of Pennsylvania.

[3] S. Das and R. Rahman. 2011. Application of Ordinal Logistic Regression Analysis in Determining Risk Factors of Child Malnutrition in Bangladesh. *Nutrition Journal, (online),* 10:124, (www.nutritionj.com), diakses 5 Mei 2013.

[4] V.A. Dei-Tutu. 2002. Flood Hazards, Insurance, and House Prices-A Hedonic Property Price Analysis. *M.S. Research Paper.* Department of Economics, College of Arts and Sciences, East Carolina University.

[5] R.E. Dumm, C. Nyce, D.L. Eckles, and J. Volkman-Wise. 2015. Demand for Catastrophe Insurance and the Representative Heuristic. *Working Paper.* Florida State University.

[6] L. Fahrmeir and G. Tutz. 1994. *Multivariate Statistical Modelling Based on Generalized Linear Models.* New York: Springer Verlag.

[7] D.W. Hosmer and S. Lemeshow. 2000. *Applied Logistic Regression.* Second Edition. Canada: John Wiley & Sons, Inc.

[8] S. Javali and P. Pandit. 2010. A Comparison of Ordinal Regression Models in an Analysis of Factors Associated with Periodontal Disease. *Journal of Indian Society of Periodontology,* (online), 14(3):155-159, (www.jisponline.com), diakses 5 Mei 2013.

[9] A. Luca. 2011. Ordinal Logistic Regression for the Estimate of the Response Functions in the Conjoint Analysis. *iBusiness,* (online), 3:383-389, (http://www.SciRP.org/journal/ib), diakses 2 Mei 2013.

[10] Y. Paudel, W. J. W. Botzen, and J. C. J. H. Aerts. Estimation of Insurance Premiums for Coverage Against Natural Disaster Risk: an Application of Bayesian Inference. Nat. Hazards Earth Syst. Sci., 13, 737–754, 2013. Published by Copernicus Publications on behalf of the European Geosciences Union.

[11] R.T. Ramadhayanti, I.N. Parta, and H. Permadi. 1912. Implementasi Regresi Logistik Ordinal pada Faktor-Faktor yang Mempengaruhi Permintaan Asuransi Jiwa: (Studi Kasus di AJB Bumiputera 1912 Cabang Malang Dieng). *Paper.* Universitas Negeri Malang.

[12] C. Trihendradi. 2007. *Kupas Tuntas Analisis Regresi, Strategi Jitu Melakukan Analisis Hubungan Causal.* Yogyakarta: ANDI.

IOP Conf. Series: Materials Science and Engineering **166** (2017) 012003 doi:10.1088/1757-899X/166/1/012003

Mean-variance portfolio optimization by using time series approaches based on logarithmic utility function

E. Soeryana[1*], N. Fadhlina[2], Sukono[3], E. Rusyaman[4], S. Supian[5]

[1,3,4,5]Department of Mathematics, Universitas Padjadjaran, Indonesia
[2]Department of Mathematics, FST, Universiti Malaysia Terengganu, Malaysia

[*]Email: endangsoeryana@yahoo.co.id

Abstract. Investments in stocks investors are also faced with the issue of risk, due to daily price of stock also fluctuate. For minimize the level of risk, investors usually forming an investment portfolio. Establishment of a portfolio consisting of several stocks are intended to get the optimal composition of the investment portfolio. This paper discussed about optimizing investment portfolio of Mean-Variance to stocks by using mean and volatility is not constant based on logarithmic utility function. Non constant mean analysed using models Autoregressive Moving Average (ARMA), while non constant volatility models are analysed using the Generalized Autoregressive Conditional heteroscedastic (GARCH). Optimization process is performed by using the Lagrangian multiplier technique. As a numerical illustration, the method is used to analyse some Islamic stocks in Indonesia. The expected result is to get the proportion of investment in each Islamic stock analysed.

1. Introduction

Investment is basically invest some capital into some form of instrument (asset), can be either fixed assets or financial assets. Investing in financial assets can generally be done by buying shares in the stock market. Investing in stocks, investors will be exposed to the risk that the magnitude of the problem along with the magnitude of the expected return [4]. The greater the expected return, generally the greater the risk to be faced. Investment risk is describing rise and fall stock price changes at any time can be measured by the value of variance [9].

The strategy is often used by investors in the face of the risks of investing is to form an investment portfolio. Establishment of an investment portfolio is essentially allocates capital in a few selected stocks, or often referred to diversify investments [5]. The purpose of the establishment of the investment portfolio is to get a certain return with minimum risk levels, or to get maximum returns with limited risk. To achieve these objectives, the investor is deemed necessary to conduct analysis of optimal portfolio selection. Analysis of portfolio selection can be done with optimum investment portfolio optimization techniques [8].

Therefore, this paper studied the paper on portfolio optimization model of Mean-Variance, where the average (mean) and volatility (variance) assumed the value is not constant, which is analyzed using time series model approach (time series). Non constant mean analyzed using models Autoregressive Moving Average (ARMA), whereas non constant volatility analyzed using models of the Generalized Autoregressive Conditional Heteroscedasticity (GARCH) [8]. Methods such analysis is then used to analyze a stock in Indonesia. The purpose of this analysis is to obtain the proportion of investment capital allocation in some stocks are analyzed, which can provide a maximum return with a certain level of risk.

2. Methodology

In this section will discuss the stages of analysis includes the calculation of stock returns, mean modeling, volatility modeling and portfolio optimization.

2.1. *Stocks Return*

Suppose P_{it} stock price i at time t, and r_{it} stock return i at time t. The value of r_{it} can be calculated using the following equation.

$$r_{it} = \ln\left(\frac{P_{it}}{P_{it-1}}\right) \tag{1}$$

where $i = 1,...,N$ with N number of stocks that were analyzed, and $t = 1,...,T$ with T the number of stock price data observed ([9] , [10]).

2.2. *Mean Models*

Suppose $\{r_{it}\}$ is stock return i at time t are stationary, will follow the model of ARMA(p,q) if for every t have the following equation:

$$r_{it} - \phi_1 r_{it-1} - ... - \phi_p r_{it-p} = a_{it} + \theta_1 a_{it-1} + ... + \theta_q a_{it-q},$$

or can be written as

$$r_{it} = \phi_1 r_{it-1} + ... + \phi_p r_{it-p} + a_{it} + \theta_1 a_{it-1} + ... + \theta_q a_{it-q}, \tag{2}$$

where $\{a_{it}\}$~WN$(0, \sigma_i^2)$, which means sequence of residual $\{a_{it}\}$ normally distributed white noise with mean 0 and variance σ_i^2. Sequence $\{r_{it}\}$ is a model ARMA(p,q) with mean μ_{it}, if $\{r_{ti} - \mu_{it}\}$ is a model ARMA(p,q) ([3], [7]).

Stages of the process modeling the mean include: (i) identification of the model, (ii) parameter estimation, (iii) diagnostic test, and (iv) prediction [10].

2.3. *Volatility Models*

Volatility models in time series data in general can be analyzed using GARCH models. Suppose $\{r_{it}\}$ is Islamic stock returns i at time t is stationary, the residuals of the mean model for Islamic stock i at time t is $a_{it} = r_{tt} - \mu_{it}$. Residual sequence $\{a_{it}\}$ follows the model GARCH(g,s) when for each has the following equation:

$$a_{it} = \sigma_{it}\varepsilon_{it},\ \sigma_{it}^2 = \alpha_{i0} + \sum_{k=1}^{g} \alpha_{ik} a_{it-k}^2 + \sum_{j=1}^{s} \beta_{ij}\sigma_{it-j}^2 + \varepsilon_{it}, \tag{3}$$

with $\{\varepsilon_{it}\}$ is a sequence of residual volatility models, namely the sequence of random variables are independent and identically distributed (IID) with mean 0 and variance 1. Parameter coefficients satisfy the property that $\alpha_{i0} > 0$, $\alpha_{ik} \geq 0$, $\beta_{ij} \geq 0$, and $\sum_{k=1}^{\max(g,s)} \alpha_{ik} + \beta_{ij} < 1$ ([8], [10]).

Volatility modeling process steps include: (i) the estimated mean model, (ii) test of ARCH effects, (iii) identification of the model, (iv) the estimated volatility models, (v) test of diagnosis, and (vi) prediction [10].

2.4. *Prediction of l –Step Ahead*

Using the mean and volatility models, aiming to calculate the prediction of mean $\hat{\mu}_{it} = \hat{r}_{ih}(l)$ and volatility $\hat{\sigma}_{it}^2 = \hat{\sigma}_{ih}^2(l)$, for l-period ahead of the starting point prediction h ([1], [10]). The prediction results of mean $\hat{\mu}_t = \hat{r}_{ih}(l)$ and volatility $\hat{\sigma}_{it}^2 = \hat{\sigma}_{ih}^2(l)$, will then be used for portfolio optimization calculations below.

2.5. Risk Tolerance

Each individual investor generally has a curve equation or utility functions that vary according to the preferences of its attitude towards an investment risk. Based on the utility function which is owned by an investor, we can determine the risk aversion function and risk tolerance function. Here is discussed a form of utility function in order to determine the function risk aversion and risk tolerance function. Suppose W stated the assets (funds) held in order to make an investment. Suppose that one investor has a logarithmic utility function form, the equation is as follows (Norstad, 2011):

$$U(W) = \log_e^{(k+W)}\text{; with the provision } W > k \text{ and constant } k > 0.$$

The first and second derivatives of the function of this utility are as follows:

$$U'(W) = \frac{1}{k+W} = (k+W)^{-1} > 0 \text{ and } U''(W) = -\frac{1}{(k+W)^2} < 0.$$

Thus the function of risk aversion R_R can be determined as follows:

$$R_R(W) = -\frac{U''(W)}{U'(W)} = -\frac{(k+W)^{-1}}{-\frac{1}{(k+W)^2}} = \frac{1}{k+W}$$

and the function of risk tolerance τ is a constant function shaped as:

$$\tau(W) = \frac{1}{R_R(W)} = \frac{1}{\frac{1}{k+W}} = k+W. \quad (4)$$

2.6. Portfolio Optimization Model

Suppose r_{it} Islamic stock return i at time t, where $i = 1,...,N$ with N the number of stocks that were analyzed, and $t = 1,...,T$ with T the number of Islamic stock price data observed. Suppose also $\mathbf{w}' = (w_1,..., w_N)$ weight vector, $\mathbf{r}' = (r_{1t},..., r_{Nt})$ vector stock returns, and $\mathbf{e}' = (1,...,1)$ unit vector. Portfolio return can be expressed as $r_p = \mathbf{w}'\mathbf{r}$ with $\mathbf{w}'\mathbf{e} = 1$ ([5], [12]). Suppose $\boldsymbol{\mu}' = (\mu_{1t},..., \mu_{Nt})$, expectations of portfolio μ_p can be expressed as:

$$\mu_p = E[r_p] = \mathbf{w}'\boldsymbol{\mu}. \quad (5)$$

Suppose given covariance matrix $\boldsymbol{\Sigma} = (\sigma_{ij})_{i,j=1,...,N}$, where $\sigma_{ij} = Cov(r_{it}, r_{jt})$. Variance of the portfolio return can be expressed as follows:

$$\sigma_p^2 = \mathbf{w}'\boldsymbol{\Sigma}\mathbf{w}. \quad (6)$$

Definition 1. *A portfolio* p^* *called* (mean-variance) *efficient if there is no portfolio* p *with* $\mu_p \geq \mu_{p^*}$ and $\sigma_p^2 < \sigma_{p^*}^2$ [5].

To get efficient portfolio, typically using an objective function to maximize

$$2\tau\mu_p - \sigma_p^2,\ \tau \geq 0$$

where the parameters of the investor's risk tolerance. Means, for investors with risk tolerance τ ($\tau \geq 0$) need to resolve the problem of portfolio

$$\text{Maximize}\{2\tau\mathbf{w}'\boldsymbol{\mu} - \mathbf{w}'\boldsymbol{\Sigma}\mathbf{w}\} \qquad (7)$$

The condition $\mathbf{w}'\mathbf{e} = 1$

Please note that the completion of (6), for all $\tau \in [0, \infty)$ form a complete set of efficient portfolios. Set of all points in the diagram-(μ_p, σ_p^2) related to efficient portfolio so-called efficient surface (efficient frontier) ([2], [11]).

Equation (6) is the optimization problem of quadratic convex [5]. Lagrangian multiplier function of the portfolio optimization problem is given by

$$L(\mathbf{w}, \lambda) = 2\tau\mathbf{w}'\boldsymbol{\mu} - \mathbf{w}'\boldsymbol{\Sigma}\mathbf{w} + \lambda(\mathbf{w}'\mathbf{e} - 1) \qquad (8)$$

Based on the Kuhn-Tucker theorem, the optimality condition of equation (7) is $\frac{\partial L}{\partial \mathbf{w}} = 0$ and $\frac{\partial L}{\partial \lambda} = 0$. Completed two conditions of optimality mentioned equation, the equation would be the optimal portfolio weights as follows:

$$\mathbf{w}^* = \frac{1}{\mathbf{e}'\boldsymbol{\Sigma}^{-1}\mathbf{e}}\boldsymbol{\Sigma}^{-1}\mathbf{e} + \tau\{\boldsymbol{\Sigma}^{-1}\boldsymbol{\mu} - \frac{\mathbf{e}'\boldsymbol{\Sigma}^{-1}\boldsymbol{\mu}}{\mathbf{e}'\boldsymbol{\Sigma}^{-1}\mathbf{e}}\boldsymbol{\Sigma}^{-1}\mathbf{e}\} \qquad (9)$$

Furthermore, with substituting $\mathbf{w}^*$ into equation (4) and (5), respectively obtained the values of the expectation and variance of the portfolio [9]. As a numerical illustration, will be analyzed some stocks as the following.

3. Illustrations

In this section will discuss the application of the method and the results of the analysis stage of the observation that includes stocks data, the calculation of Islamic stock returns, modelling the mean of stocks, volatility modelling, prediction of the mean and variance values, and the process of optimization. In this section we explore numerical illustration of estimation of CAPM regression, estimation of distributed lagged CAPM, and estimation of Koyck transformation CAPM.

3.1. Stocks Data

The data used in this study is secondary data, in the form of time series data (time series) of some of the daily stock price, which includes the names of stocks: AKRA, CPIN, ITMG, MYOR, and TLKM. Stock price data, the data used is the closing stock price, for over 1360 days starting from January 1, 2010 till June 30, 2015 were downloaded from www.finance.yahoo.com [6]. Stock prices data will then be processed by using statistical software of Eviews-6 and Matlab.

3.2. Stocks Return and Stationarity

Stock returns of five firms in this study were calculated using equation (1). Figure-1 below indicates the five stock chart returns are analyzed.

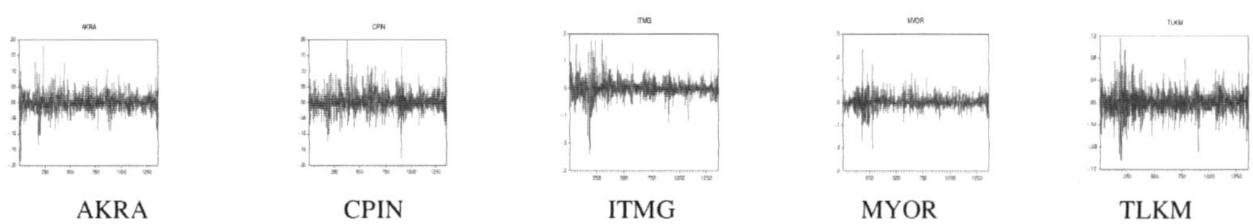

AKRA CPIN ITMG MYOR TLKM

Figure-1 FiveStockCharts

It can be seen in Figure-1 that the five stock return data were analyzed have stationary. For stationary testing is done using the ADF test. Statistic results respectively values are -34.24848; -33.79008; -30.20451; -40.04979; and -28.36974. Further, if the specified level of significance $\alpha =$ 5%, can be obtained by a standard normal distribution critical value is -2.863461. It is clear that the value of the test statistic for all ADF of stocks are analyzed located in the rejection region, so that everything is stationary.

3.3. Mean Modelling of Stocks Return

Stock return data will be used to estimate mean model using the software of Eviews-6. First, the identification and estimation of mean model carried out through the sample autocorrelation function (ACF) and partial autocorrelation function (PACF). Based on the pattern of ACF and PACF, tentative models can be specified from each stock returns. The estimation results indicate that the best models are respectively ARMA(1,0), ARMA(1,0), ARMA(1,0), ARMA(7,0), and ARMA(2,0).Second, the significance test for parameters and significance tests for models indicate that the mean model for all stocks analyzed have significance. Third, diagnostic tests for these models are done by using the data residual correlogram and Ljung-Box test hypotheses. The test results show the residuals of the models are white noise. Results residual normality test showed normal distribution. Therefore residuals for all stock has been analyzed are white noise.

Equations of the results mean modeling for the five stocks will be written simultaneously with the volatility models each of stocks, which will be estimated following.

3.4. Volatility Modelling of Stocks Return

Modeling of volatility in this section is done by using statistical software of Eviews-6. First, carried out the detection elements of autoregressive conditional heteroscedasticity (ARCH) to the residual a_t, using the ARCH-LM test statistic. Statistical value of the results obtained χ^2 (obs*R-Square) each of stock returns AKRA, CPIN, ITMG, MYOR, and TLKM respectively are: 31.76757; 76.75431; 40.55526; 48.58576; 9.270781, and 125.2410 by probability each of 12:050.0000 or smaller, which means that there are elements of ARCH.

Second, the identification and estimation of volatility models. This study used models of generalized autoregressive conditional heteroscedasticity (GARCH) which refers to equation (3). Based on squared residuals correlogram a_t^2, the ACF and PACF graphs of each selected model of volatility that might be tentative. Volatility model estimation each of stock return performed simultaneously (synchronously) with mean models. After going through tests of significance for parameters and significance tests for models, all equations are written below have been significant. The best model are respectively:

- Stock of AKRA follow the model ARMA(1,0)-GARCH(1,1) with equation:
  $r_t = 0.073891\ r_{t-1} + a_t$
  and
  $\sigma_t^2 = 0.000014 + 0.040015\,\alpha_{t-1}^2 + 0.9404318\,\sigma_{t-1}^2 + \varepsilon_t$
- Stock of CPIN follow the model ARMA(1,0)-GARCH(1,1) with equation:
  $r_t = 0.089639\ r_{t-1} + a_t$
  and
  $\sigma_t^2 = 0.000052 + 0.134049\,\alpha_{t-1}^2 + 0.820716\,\sigma_{t-1}^2 + \varepsilon_t$
- Stock of ITMG follow the model ARMA(1,0)-GARCH(1,1) with equation:
  $r_t = 0.193825\ r_{t-1} + a_t$

and

$$\sigma_t^2 = 0.000012 + 0.066024\,\alpha_{t-1}^2 + 0.923108\,\sigma_{t-1}^2 + \varepsilon_t$$

- Stock of MYOR follow the model ARMA(7,0)-GARCH(1,1) with equation:

$$r_t = 0.102007\,r_{t-7} + a_t$$

and

$$\sigma_t^2 = 0.000009 + 0.044332\,\alpha_{t-1}^2 + 0.945801\,\sigma_{t-1}^2 + \varepsilon_t$$

- Stock of TLKM follow the model ARMA(2,0)-GARCH(1,1) with equation:

$$r_t = -0.084289 r_{t-2} + a_t$$

and

$$\sigma_t^2 = 0.000019 + 0.139166\alpha_{t-1}^2 + 0.824540\sigma_{t-1}^2 + \varepsilon_t$$

Based on the ARCH-LM test statistics, the residuals of the models for stock AKRA, CPIN, IMTG, MYOR, and TLKM there is no element of ARCH, and also has white noise. Mean and volatility models are then used to calculate the values $\hat{\mu}_t = \hat{r}_t(l)$ and $\hat{\sigma}_t^2 = \sigma_t^2(l)$ recursively.

3.5. Prediction of Mean and Variance Values

Models of mean and volatility estimation results of five stocks in the previous stage, is then used to perform one step ahead prediction for the mean and variance values. Results predicted mean and volatility values are given in Table 1 below.

Table 1. Predictive Values of Mean and Variance One Period Ahead for Each Stocks

Islamic Stocks	Model of Mean-Volatility	Mean ($\hat{\mu}_t$)	Variance ($\hat{\sigma}_t^2$)
AKRA	ARMA(1,0)-GARCH(1,1)	0.001407	0.001200
CPIN	ARMA(1,0)-GARCH(1,1)	0.004919	0.001840
IMTG	ARMA(1,0)-GARCH(1,1)	0.014656	0.001078
MYOR	ARMA(7,0)-GARCH(1,1)	0.004940	0.000956
TLKM	ARMA(2,0)-GARCH(1,1)	0.004406	0.001399

3.6. Mean-Variance Portfolio Optimization

In this part we did the process of portfolio optimization calculations. Portfolio optimization is done with referring to the equation (6). The data used for process optimization are the values of the mean and variance are given in Table 2. Using the values of mean in Table 2, column $\hat{\mu}_t$, used to form the mean vector as: $\boldsymbol{\mu}^T = (0.001407\;\; 0.004919\;\; 0.014656\;\; 0.004940\;\; 0.004406)$, amount of the stock that were analyzed.

Furthermore, by using the values of variance in Table-2, column $\hat{\sigma}_t^2$, and together with the values of the covariance between stocks, used to form the covariance matrix $\boldsymbol{\Sigma}$ and the inverse matrix $\boldsymbol{\Sigma}^{-1}$ as follows:

$$\boldsymbol{\Sigma} = \begin{pmatrix} 0.001200 & 0.000136 & 0.000251 & 0.000113 & 0.000401 \\ 0.000136 & 0.001840 & 0.000092 & 0.000315 & 0.000225 \\ 0.000251 & 0.000092 & 0.001078 & 0.000512 & 0.000133 \\ 0.000113 & 0.000315 & 0.000512 & 0.000956 & 0.000075 \\ 0.000401 & 0.000225 & 0.000133 & 0.000075 & 0.001399 \end{pmatrix}$$

and

$$\Sigma^{-1} = 10^3 \times \begin{pmatrix} 0.9613 & -0.0345 & -0.2020 & 0.0257 & -0.2522 \\ -0.0345 & 0.5904 & 0.0738 & -0.2237 & -0.0801 \\ -0.2020 & 0.0738 & 1.3037 & -0.6955 & -0.0406 \\ 0.0257 & -0.2237 & -0.6955 & 1.4880 & 0.0150 \\ -0.2522 & -0.0801 & -0.0406 & 0.0150 & 0.8030 \end{pmatrix}$$

Optimization done in order to determine the composition of the portfolio weights, and thus the portfolio weight vector is determined by using equation (8). The weight vector calculation process, the values of risk tolerance τ determined by the simulation begins value $\tau = 0.000$ with an increase of 0.001. If it is assumed that short sales are not allowed, then the simulation is stopped when the value of τ=0.036, because it has resulted in a portfolio weight at least there is a negative value. The portfolio weights calculation results are given in Table 2.

Table 2. Process of Mean-Variance Portfolio Optimization

τ	w_1	w_2	w_3	w_4	w_5	$\mathbf{w}^T\mathbf{e}$	$\hat{\mu}_p$	$\hat{\sigma}_p^2$	$\hat{\mu}_p - \hat{\sigma}_p^2$	$\hat{\mu}_p / \hat{\sigma}_p^2$
	AKRA	CPIN	IMTG	MYOR	TLKM	Sum	Mean	Variance	Maximum	Ratio
0.000	0.2150	0.1406	0.1895	0.2629	0.1920	1	0.0059	0.00043136	0.00546864	13.7161
0.001	0.2093	0.1411	0.2025	0.2555	0.1916	1	0.0061	0.00043151	0.00566849	14.0507
0.002	0.2036	0.1417	0.2155	0.2480	0.1913	1	0.0062	0.00043195	0.00576805	14.6893
0.003	0.1978	0.1422	0.2285	0.2406	0.1909	1	0.0064	0.00043268	0.00596732	14.6893
0.004	0.1921	0.1428	0.2414	0.2331	0.1905	1	0.0065	0.00043370	0.00606630	14.9921
0.005	0.1864	0.1433	0.2544	0.2257	0.1902	1	0.0066	0.00043502	0.00616498	15.2832
0.006	0.1807	0.1439	0.2674	0.2182	0.1898	1	0.0068	0.00043663	0.00636337	15.5621
0.007	0.1750	0.1444	0.2803	0.2108	0.1894	1	0.0069	0.00043853	0.00646147	15.8284
0.008	0.1693	0.1450	0.2933	0.2033	0.1891	1	0.0071	0.00044073	0.00665927	16.0817
0.009	0.1636	0.1455	0.3063	0.1959	0.1887	1	0.0072	0.00044322	0.00675678	16.3217
0.010	0.1579	0.1461	0.3193	0.1884	0.1883	1	0.0074	0.00044600	0.00695400	16.5481
0.011	0.1522	0.1466	0.3322	0.1810	0.1880	1	0.0075	0.00044907	0.00705093	16.7608
0.012	0.1465	0.1472	0.3452	0.1736	0.1876	1	0.0077	0.00045344	0.00724656	16.9597
0.013	0.1407	0.1477	0.3582	0.1661	0.1872	1	0.0078	0.00045610	0.00734390	17.1445
0.014	0.1350	0.1483	0.3711	0.1587	0.1587	1	0.0080	0.00046005	0.00753995	17.3154
0.015	0.1293	0.1488	0.3841	0.1512	0.1865	1	0.0081	0.00046430	0.00763570	17.4724
0.016	0.1236	0.1494	0.3971	0.1438	0.1461	1	0.0083	0.00046884	0.00783116	17.6155
0.017	0.1179	0.1499	0.4101	0.1363	0.1858	1	0.0084	0.00047367	0.00792633	17.7449
0.018	0.1122	0.1505	0.4230	0.1289	0.1854	1	0.0086	0.00047879	0.00812121	17.8608
0.019	0.1065	0.1510	0.4360	0.1214	0.1850	1	0.0087	0.00048421	0.00821579	17.9633
0.020	0.1008	0.1516	0.4490	0.1140	0.1847	1	0.0088	0.00048992	0.00831008	18.0528
0.021	0.0951	0.1521	0.4619	0.1065	0.1843	1	0.0090	0.00049592	0.00850408	18.1295
0.022	0.0893	0.1527	0.4749	0.0991	0.1840	1	0.0091	0.00050221	0.00859779	18.1937
0.023	0.0836	0.1533	0.4879	0.0916	0.1836	1	0.0093	0.00050880	0.00879120	18.2459
0.024	0.0779	0.1538	0.5009	0.0842	0.1832	1	0.0094	0.00051568	0.00888432	18.2863
0.025	0.0722	0.1544	0.5138	0.0767	0.1829	1	0.0096	0.00052285	0.00907715	18.3154
0.026	0.0665	0.1549	0.5268	0.0693	0.1825	1	0.0097	0.00053032	0.00916968	18.3336
0.027	0.0608	0.1555	0.5398	0.0619	0.1821	1	0.0099	0.00053808	0.00936192	18.3413
0.028	0.0551	0.1560	0.5527	0.0544	0.1818	1	0.0100	0.00054613	0.00945387	18.3390
0.029	0.0494	0.1566	0.5657	0.0470	0.1814	1	0.0102	0.00055447	0.00964553	18.3270
0.030	0.0437	0.1571	0.5787	0.0395	0.1810	1	0.0103	0.00056311	0.00973689	18.3059
0.031	0.0380	0.1577	0.5917	0.0321	0.1807	1	0.0105	0.00057204	0.00992796	18.2760
0.032	0.0322	0.1582	0.6046	0.0246	0.1803	1	0.0106	0.00058126	0.01001874	18.2379
0.033	0.0265	0.1588	0.6176	0.0172	0.1799	1	0.0107	0.00059078	0.01010922	18.1919
0.034	0.0208	0.1593	0.6306	0.0097	0.1796	1	0.0109	0.00060059	0.01029941	18.1386
0.035	0.0151	0.1599	0.6435	0.0023	0.1792	1	0.0110	0.00061069	0.01038931	18.0783
0.036	0.0094	0.1604	0.6565	-0.0052	0.1788	1	0.0112	0.00062108	0.01057892	18.0115

Based on the results of the optimization process are given in Table 2, the pair of points ($\hat{\mu}_p$, $\hat{\sigma}_p^2$) efficient portfolio can be formed or the so-called efficient frontier as given in Figure 2a. This graph shows the efficient frontier decent area for investors with different levels of risk tolerance to make an investment. Also by using the optimization process results in Table 2, it can be calculated ratio value $\hat{\mu}_p$ towards $\hat{\sigma}_p^2$ for each level of risk tolerance. The ratio calculation results can be shown as inFigure2.b. This ratio shows the relationship between the optimum portfolio return expected with variance as a measure of risk.

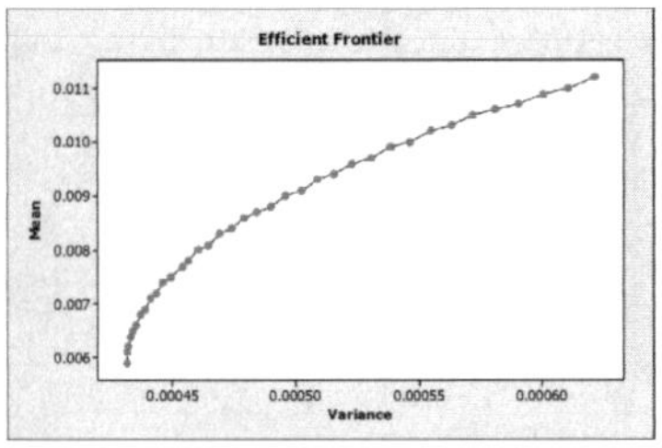

Figure 2a. Efficient Frontier

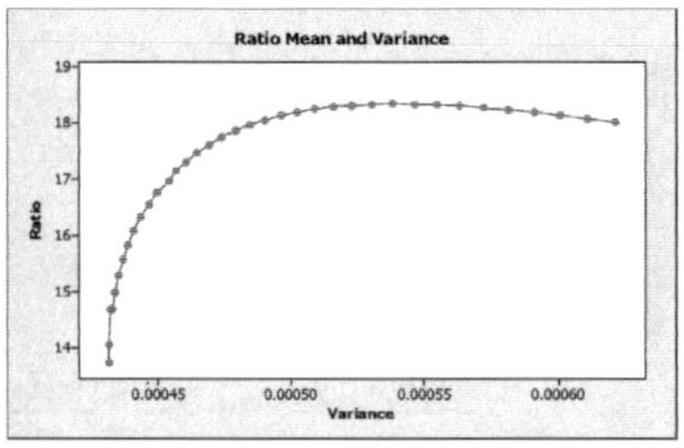

Figure 2b. Ratio of Mean-Variance

Based on the results of the calculation of portfolio optimization, the optimum value is achieved when the value of the portfolio's risk tolerance $\tau = 0.027$. The portfolio produces mean value of $\hat{\mu}_p$ = 0.0099 with the value of risk as the variance $\hat{\sigma}_p^2$ = 0.00053808.

Composition weight of the maximum portfolio respectively are: 0.0608, 0.1555, 0.5398, 0.0619, and 0.1821. This provides reference to investors that invest in stocks of AKRA, CPIN, ITMG, MYOR, and TLKM, in order to achieve the maximum value of the portfolio, the composition of the portfolio weights are as mentioned above.

4. Conclusions

In this paper we analyzed the Mean-Variance portfolio optimization on some stocks by using non-constant mean and volatility models approaches in some stocks traded in the Islamic capital market in Indonesia. The analysis showed that some of stocks which analyzed follow the ARMA(p,q)-GARCH(g,s) models. Whereas, based on the results of the calculation of portfolio optimization, it is produced that the optimum is achieved when the composition of the portfolio investment weights in stocks of AKRA, CPIN, ITMG, MYOR, and TLKM, respectively are 0.0608, 0.1555, 0.5398, 0.0619, and 0.1821. The composition of the portfolio weights thereby will produce a portfolio with mean value of 0.0099 and the value of risk, measured as the variance of 0.00053808.

Acknowledgment

We would like thanks the Rector and Dean of Faculty of Mathematics and Natural Sciences Universitas Padjadjaran, which have provided a grant program, the Academic Leadership Grant (ALG), under the coordination of Prof. Dr. Sudradjat, M.S., which is a means to increase research activities and publications to researchers at the UniversitasPadjadjaran.

References

[1] E. Febrian & A. Herwany, Volatility Forecasting Models and Market Co-Integration: A Study on South-East Asian Markets, *Working Paper in Economics and Development* Studies, Department of Ekonomics, Padjadjaran University, 2009.

[2] S. Goto & Y.Xu, On Mean Variance Portfolio Optimization: Improving Performance Through Better Use of Hedging Relations, *Working* Paper, Moore School of Business, University of

South Carolina, 2012.

[3] D.N. Gujarati, *Basic Econometrics*, Fourth Edition, The McGraw–Hill Companies, Arizona, 2004.

[4] A. Kheirollah & O. Bjarnbo. A Quantitative Risk Optimization of Markowitz Model: An Empirical Investigation on Swedish Large Cap List, *Master Thesis,* in Mathematics/Applied Mathematics, University Sweden, Department of Mathematics and Physics, 2007. www.mdh.se/polopoly_fs/ 1.16205!MasterTheses.pdf.

[5] H. H. E. Panjer, et al., Financial Economics: With Applicationsto Investments, Insurance, and Pensions, Schaumburg, Ill.: The Actuarial Foundation, 1998.

[6] R.A. Rifqiawan, AnalisisPerbedaan Volume PerdaganganSaham-Saham yang Optimal Pada Jakarta Islamic Index (JII) di Bursa Efek Indonesia (BEI), Program Studi Magister Sains Akuntansi, Program Pascasarjana, Universitas Diponegoro, Semarang, 2008.

[7] Shewhart, A., Walter & S. S. Wilks, *Applied Econometric Time Series*, John Wiley & Sons, Inc, United States of America, 2004.

[8] Shi-Jie Deng, Heavy-Tailed GARCH models: Pricing and Risk Management Applications in Power Market, *IMA Control & Pricing in Communication & Power Networks* (2004),7-17 Mar http://www.ima.umn.*edu/talks/.../deng/power_ workshop_ ima032004-deng.pdf.*

[9] Sukono, Subanar& D. Rosadi,Pengukuran*VaR*DenganVolatilitasTakKonstandan*Efek Long Memory,Disertasi*,PrtogramStudi S3 Statistika, Jurusan Matematika, Fakultas Matematika dan Ilmu Pengetahuan Alam, Universitas Gajah Mada, Yogyakarta, 2011.

[10] R. S. Tsay, *Analysis of Financial Time Series,* Second Edition, USA: John Wiley & Sons, Inc, 2005.

[11] A. Yoshimoto, The Mean-Variance Approach To Portfolio Optimization Subject To Transaction Costs. *Journal of the Operations Research Society of Japan*, Vol. 39, No. 1, March 1996

[12] D. Zhang, Portfolio Optimization with Liquidity Impact, *Working Paper*, Center for Computational Finance and Economic Agents, University of Essex, 2006. www.orfe.princeton.edu/ oxfordprinceton5/slides/yu.pdf.

Causal relationship model between variables using linear regression to improve professional commitment of lecturer

S. Setyaningsih
Department of Computer Science and Doctoral Program of Education Management, Pakuan University, INDONESIA
Email: sri_setya@yahoo.com

Abstract. The main element to build a leading university requires lecturer commitment in a professional manner. Commitment is measured through willpower, loyalty, pride, loyalty, and integrity as a professional lecturer. A total of 135 from 337 university lecturers were sampled to collect data. Data were analyzed using validity and reliability test and multiple linear regression. Many studies have found a link on the commitment of lecturers, but the basic cause of the causal relationship is generally neglected. These results indicate that the professional commitment of lecturers affected by variables empowerment, academic culture, and trust. The relationship model between variables is composed of three substructures. The first substructure consists of endogenous variables professional commitment and exogenous three variables, namely the academic culture, empowerment and trust, as well as residue variable ε_y. The second substructure consists of one endogenous variable that is trust and two exogenous variables, namely empowerment and academic culture and the residue variable ε_3. The third substructure consists of one endogenous variable, namely the academic culture and exogenous variables, namely empowerment as well as residue variable ε_2. Multiple linear regression was used in the path model for each substructure. The results showed that the hypothesis has been proved and these findings provide empirical evidence that increasing the variables will have an impact on increasing the professional commitment of the lecturers.

1. Introduction

An understanding of the lecturers as executors of University Tridharma need to be supported by a commitment to the profession. Lecturers are expected to serve as experts in their respective sectors in accordance with the progress of science and technology and development of society. Lecturers who have high professional commitment to be easily powered by its leaders. In addition, lecturers should be able to improve the qualifications and educational strata, applying academic culture, and willing to work to improve the welfare of lecturers as well as activities in the community. The parameters of the development of lecturers in carrying out its duties and functions not only can be seen from the productivity of the implementation of University Tridharma, but normatively can be seen from the level of education and the functional career of the lecturers. Lecturers who have high trust towards the organization will accept the values and goals of the organization, willing to work in earnest on behalf of the organization, and have a strong desire to maintain membership in the organization.

Determination of the sample has been proportionally based on the amount of scattered lecturer at each faculty. Target respondents were conducted with a random using proportional random sampling technique from Cochran formula that involves the determination of the sample size or enter the characteristics contained in the population. Therefore, the minimal sample size will be able to reflect the actual conditions of the population. Cochran formula in determining the sample size is not only expensive but also enter the alpha level of the characteristics contained in the population. Calibration of the instrument in this study include instrument validity test and reliability test. Research instruments to be valid, therefore, before further research is held, each item of each variable instrument should be analyzed to see the relationship between the scores of each item with the total score or correlation in a variable. The type of validity used is internal validity or the validity of which was obtained using Pearson Product Moment Correlation. Reliability of the items to a valid instrument variables were analyzed through Cronbach Alpha.

2. Linear Regression

In this section, the hypotheses generated from the research framework defined by the model Path. In connection with this, the data analysis technique used is path analysis with correlation analysis method is done partially or jointly. Tests done after the data is collected using multiple linear regression analysis, the analysis used to determine the influence of empowerment, cultural, academic and trust to the professional commitment.

Model of the relationship between variables is composed of three substructures, namely:
Substructure-1: $Y= \beta_{y1}X_1 + \beta_{y2}X_2 + \beta_{y3}X_3 + \varepsilon_y$,
Substructure-2: $X_3 =\beta_{31}X_1+\beta_{32}X_2+\varepsilon_3$
Substructure-3: $X_2 = \beta_{21}X_1+\varepsilon_2$.

As for testing the assumptions used tests for normality and linearity test. Normality test aims to determine whether a data error estimates normal distribution or not. Good data is the data had normal or nearly normal distribution through the test Liliefors (L).

3. Statistical Hypothesis

Statistical hypothesis in this study is based on testing of empirical data obtained through a survey of respondents. Statistical hypothesis is formulated as follows:

1. H_0 : $\beta_{y1} \leq 0$ 2. H_0 : $\beta_{y2} \leq 0$ 3. H_0 : $\beta_{y3} \leq 0$
H_1 : $\beta_{y1} >$ H_1 : $\beta_{y2} > 0$ H_1 : $\beta_{y3} > 0$
4. H_0 : $\beta_{31} \leq 0$ 5. H_0 : $\beta_{32} \leq 0$ 6. H_0 : $\beta_{21} \leq 0$
H_1 : $\beta_{31} > 0$ H_1 : $\beta_{32} > 0$ H_1 : $\beta_{21} > 0$

4. Data Analysis and Discussion

The data in this study using primary data which obtained through questionnaires distributed directly to the lecturer as respondents in every faculty in Pakuan University. Questionnaire distributed is as much as 135 questionnaires. Using the formula for the proportion of the sample, the determination of sample size proportional to each work unit is determined as follows:

Table 1 The distribution of lecturer samples per-faculty

No.	Faculty	Academic level of Lecturer			Sum
		Expert assistant	Lector	Associate Professor	
1	Law	3	10	3	16
2	Economy	8	15	5	28
3	Math. & Nat. Sciences	13	15	7	35
4	Fac. Of Education	7	9	7	23
5	Letters	3	8	3	14
6	Technic	2	10	7	19
	Total	36	10	7	19

5. Calibration of Instruments

5.1. Validity Test

Research instruments to be valid, therefore, before being held for further research, every grain of variable instruments should be analyzed the relationship between the scores of each item with the total score (correlation in the variable). The validity of the type used is the internal validity or the validity of which was obtained using Pearson Product Moment correlation, using the test criteria.

From these calculations generated that grains are valid and not-valid (dropped). With a significance level of 5% and n = 30 with 0.361 correlation value. The validity of the instrument is determined by comparing with $r_{calculated}$ with r_{table}.

a) If $r_{calculated} > r_{table}$, those items are declared as invalid, and will be used for data collection.
b) If $r_{calculated} \leq r_{table}$, those items are declared as not valid.

5.2. Reliability Test

Test for reliability against the grain of the instrument from a valid professional commitment was analyzed using Alpha Cronbach, which determines the values of the variance of each item and then calculating Alpha Cronbach reliability coefficient. Instrument reliability coefficient calculation performed after an invalid item (drop) is not used in this calculation. By using references from Anderson and Hair, if the value of Alpha Cronbah coefficients greater than 0.70 (r_{11}> 0.7) means that reliability is sufficient (sufficient reliability).

5.3. Testing of Requirements Analysis

Before implementation the path analysis, data must meet several requirements for statistical tests, namely:

5.3.1 Normality Test for Errors

For the normality of the distribution of the error used Liliefors Test. Testing for normality error is performed to determine that the observed sample derived from normal distributed population. Conditions in the test of error is when H_0 is accepted in error estimates (Y- $\hat{Y}$) normal distribution, and if H_0 is rejected because of an error of estimate (Y- $\hat{Y}$) is not normal.

Testing for normality Liliefors (L), with the testing criteria at significance level α = 0.05 as follows: H_0 accepted if $L_{calculated} < L_{table}$ and H_0 will be rejected if $L_{calculated} > L_{table}$.

Based on the results of the overall calculation error normality test, in this study can be seen in the summary in Table 2 below:

Table 2. Error Normality Test Summary

No	Error Estimation	n	$L_{calculated}$	L_{table}		Decission
				α = 0,05	α = 0,01	
1	Y to X_1	135	0.06	0.08	0.09	Normal
2	Y to X_2	135	0.07	0.08	0.09	Normal
3	Y to X_3	135	0.06	0.08	0.09	Normal
4	X_3 to X_1	135	0.04	0.08	0.09	Normal
5	X_3 to X_2	135	0.05	0.08	0.09	Normal
6	X_2 to X_1	135	0.07	0.08	0.09	Normal

5.3.2 Significance test and Linearity of Regression Coefficients

The last test is required prior to analysis is the significance test and linearity of regression models. Linearity test aims to test whether the regression line of the independent variable on the dependent variable has a linear relationship or not.

If two mentioned data is linearly related, it is predicted that these two variables have a unidirectional relationship. To test the linearity of the data used F-test, with the testing criteria as follows:

H_0: Regression Model is figured as linear, if $F_{calculated} \leq F_{table}$
H_1: Regression Model is figured as non linear, $F_{calculated} > F_{table}$

Testing the hypothesis of linearity test performed as follows:

Table 3 Results Summary of Linearity Test of Regresion Model

No	Relationship Model between Variable	F- calculated	F- table (α =0,05)	F-able (α = 0,01)	Test results of Linearity Pattern
1	Y to X_1	0.92	1.49	1.77	Linear
2	Y to X_2	1.69	1.49	1.77	Linear
3	Y to X_3	1.04	1.49	1.77	Linear
4	X_3 to X_1	1.45	1.49	1.77	Linear
5	X_3 to X2	0.91	1.49	1.77	Linear
6	X_2 to X_1	1.65	1.49	1.77	Linear

6. Model Analysis

The model of the relationship between variables in the substructure-1 consists of an endogenous variable professional commitment (Y) and three exogenous variables, trust (X3), academic culture (X2) and empowerment (X1), and one variable residue that ε_y. Based on statistical analysis using SPSS version 17.0 was obtained the following results:

Table 4. Value of Path Coeffisien at Substructure-1

Model	Unstandardized Coefficients		Standardized Coeficients	T	Sig
	ß	Std Error	Beta		
(constant)	39.324	6.875		5.720	0.000
Empowerment (X_1)	0.261	0.060	0.343	4.353	0.000
Academic Culture (X_2)	0.138	0.069	0.171	2.003	0.047
Trust (X_3)	0.157	0.074	0.185	2.124	0.036

Multiple linear regression analysis is used to determine the effect of independent variables, namely empowerment, cultural, academic and trust on the dependent variable professional commitment. The results of multiple linear regression analysis model of the relationship between variables in the substructure-1 of this study are shown in Table 4.

The results of the data analysis are used to create multiple linear regression equation and its interpretation as follows: multiple linear regression equations on the substructure -1 that has been done so that we get the regression equation Y = 0,343X1 + 0.171X2 + 0.185X3. In the above table shows that all path coefficients are significant. The results of the analysis proved that for all path coefficients are significant, the model does not need to be fixed. Interpretation of multiple regression equation that occur can be interpreted as follows: a = 39.324 is constant, meaning that if empowerment, academic culture, and trust is equal to zero, then the professional commitment of lecturers amounted to 39.324. Beta = 0.343 is the regression coefficient of empowerment, meaning that with the increasing empowerment of the lecturer, it will further enhance the professional commitment of lecturers with the assumption that academic culture and trust are held as constant.

Beta = 0.185 is the regression coefficient of trusts, meaning that with the increasing trust of a lecturer will further enhance the professional commitment of lecturers with assuming that empowerment and academic culture variables are held as constant.

The relationship model between variables in the substructure-2 consists of the endogenous variables, namely trust (Y) and two exogenous variables, namely academic culture (X2) and empowerment (X1), as well as a residual variable ε_4. Based on this relationship, then the path model on the substructure-2 are as follows: X3 = $X_3 = \beta_{31}X_1 + \beta_{32}X_2 + \varepsilon_3$.

The result of the calculation through SPSS-17 obtained by the path coefficients or standardized coefficients (beta) on the substructure -2 can be written in Table 5

Table 5 The value of path coeffisien on Substructure-1

Model	Unstandardized Coefficients		Standartized Coeficientsf	T	Sig
	ß	Std Error	Beta		
(constant)	43,724	7.141		6.123	0.000
Empowerment (X_1)	0.350	0.062	0.391	5.615	0.000
Academic Culture (X_2)	0.472	0.066	0.496	7.127	0.000

From Table 5 above it can be shown that the path model on substructure-2 is $\hat{X}_3 = 0.39X_1 + 0.50X_2$. The coefficient of determination (R2), which is an analysis used to determine how large the contribution of independent variables between empowerment, academic culture, and trust simultaneously to the professional commitment of lecturers, shown in percentages. The value of the coefficient of determination R^2_{X321} amounted to 0.67 (67%).

Table 6. Result Summary of Empirical Model on Substructural-2

Model 1	R	R square	Adjustd R Square	Std Error of the Estimate
	0.817[a]	0.668	0.663	10.39020

a. Predictors: (constant), empowerment, academic culture
b. Dependent variable: trust.

The magnitude of the effect of other variables outside X_1, X_2 to X_3 are:

	X_1	X_2
	1	0,693879
R =	0,693879	1

Inverse Matrix:

$$R^{-1} = \begin{bmatrix} C_{11} & C_{12} \\ C_{21} & C_{22} \end{bmatrix} = \begin{bmatrix} 1{,}928519 & -1{,}133816 \\ -1{,}33816 & 1{,}928519 \end{bmatrix} \begin{bmatrix} r_{41} \\ r_{42} \end{bmatrix} = \begin{bmatrix} \rho_{41} \\ \rho_{42} \end{bmatrix}$$

$$\rho_{yi} X_i = (R^{-1})(r_{yixk}) = \begin{bmatrix} \rho_{41} \\ \rho_{42} \end{bmatrix} = \begin{bmatrix} 1{,}928519 & -1{,}133816 \\ -1{,}33816 & 1{,}928519 \end{bmatrix} = \begin{bmatrix} 0{,}735232 \\ 0{,}767444 \end{bmatrix} = \begin{bmatrix} 0{,}390 \\ 0{,}496 \end{bmatrix}$$

$$\varepsilon_i = \sqrt{1 - R^2_{yixk}} \text{, so } \varepsilon_3 = \sqrt{1 - 0{,}668} = 0{,}576$$

The relationship model between variables in the substructure-3 consists of the endogenous variable, namely academic culture and the exogenous variable, namely empowerment (X1), as well as a residual variable namely ε_2. Based on this relationship, the path model on substructure-3 is as follows: $X_2 = \beta_{21}X_1 + \varepsilon_2$. The result of the calculation through SPSS-17 obtained by the path coefficients or standardized coefficients (beta) on the substructure-3 are presented in Table 7.

Table 7. Value of Path Coeffisient on Substructure-3

Model	Unstandardized Coefficients		Standardized Coefficients	T	Sig
	ß	Std Error	Beta		
(constant)	48.268	8.368		5.768	0.000
Empowerment (X_1)	0.654	0.059	0.694	11.113	0.000

Path model on substructure-3, $\widehat{X}_2 = 0.69\ X_1$. The coefficient value of determination R^2_{X21} amount 0.48% .

Table 8. Summary of Empirical Model Results on Substructure-3

Model 1	R	R Square	Adjust R Square	Std Error of the Estimation
	0.694	0.481	0.478	13.614

Dependent variable: academic culture

The magnitude of the effect of other variables outside X_1, X_2 to X_3 are:

$$R = \begin{matrix} X_1 & X_2 \end{matrix} \\ R = \begin{bmatrix} 1 & 0{,}693879 \\ 0{,}693879 & 1 \end{bmatrix}$$

$$\rho_{21} = r_{21} = 0{,}693 \text{ dan } R^2 = 0{,}481467$$

$$\varepsilon_i = \sqrt{1 - R^2_{yixk}} \text{ , so } \varepsilon_2 = \sqrt{1 - 0{,}4811467} = 0{,}72$$

Path Diagram of the Research:

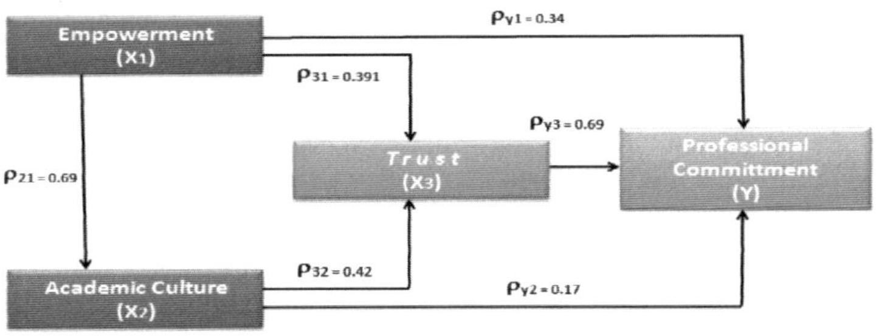

Figure 1. Path Diagram of the Research

7. Hypothesis Test

After analysis of the structural model has been done, the final result obtained is used to test the hypothesis in order to determine the direct and indirect effect between variables. The proposed hypothesis inferred through path coefficient value calculation, recapitulation hypothesis testing, and significance for each stream studied.

Level of significance = 0,05 or 5%, degrees of freedom (dk)= n-1-k. Table Value = $t_{\frac{\alpha}{2}}(n-1-k)$ = (0,05/2;34-1-3) = 0,025; 30=1,97.

Table 9. Recapitulation of Hypothesis Test

Variable	Path Coefficient	$T_{calc.}$	t_{table}	Test Decision
X_1 to Y	$\rho_{y1} = 0.34$	4.35	1.97	H_0 rejected, H_1 accepted. There is a direct positive influence between professional commitment and empowerment
X_2 to Y	ρ_{y2} = 0,17	2.00	1.97	H_0 rejected,H_1 accepted. There is a direct positive influence between academic culture and professional committment
X_3 to Y	ρ_{y4} = 0,19	2.12	1.97	H_0 rejected, H_1 accepted. There is a direct positive influence between trust and professional committment
X_1 to X_3	ρ_{41} = 0,39	5.62	1.97	H_0 rejected, H_1 accepted. There is a direct positive influence between empowerment and trust
X_2 to X_3	ρ_{42} = 0,50	7.13	1.97	H_0 rejected, H_1 accepted. There is a direct positive influence between academic culture and trust
X_1 to X_2	ρ_{21} = 0,69	5.77	1.97	H_0 rejected, H_1 accepted. There is a direct positive influence between academic culture and professional committment

8. **Conclusions**

Variables: empowerment, academic culture, and trust have a positive direct impact toward professional commitment of lecturers, this result proved through the following test:

1. Multiple linear regression analysis through three models of the substructure. The results of multiple regression analysis equation is: Substructure-1 $\hat{Y}$ = 0,343X_1 + 0.171X_2 + 0.185X_3. Substructure-2 $\hat{X}_3$ = 0.39X_1 + 0.50X_2. substructure-3, $\hat{X}_2 = \beta_{21}X_1$. Multiple linear regression above proves that the variables empowerment, academic culture, and trust has a positive direct effect toward professional commitment, among the most dominant variable is empowerment.
2. T-test is a simple test of significance correlation between variables, indicate that variables empowerment, academic culture, and trust has a significant influence toward professional commitment because of the value t_{count} is greater than the value t_{table}.
3. F_{test} is testing the significance of regression between variables. Results of the F_{test} proved that empowerment, academic culture, and trust affect the professional commitment of lecturer since F_{count} is greater than the value of F_{table}.
4. The results of the test coefficient of determination (R^2) proves that empowerment, academic culture, and trusts have positive influence toward professional commitment of 67.4%, and the remaining 32.6% is influenced by the other variables that are not used in this study, such as career development for lecturers and the performance appraisal system of lecturers.

References

[1]. Adrian, R, Cosma, M. (2013), "*The International Annual Scientific Session Strategies*" Bucharest National Defence University.
[2]. Brick, J. (2011), *Academic Culture: A Student's Guide to Studying at University*, SecondEdition, Australia: Macmillan Publisher.

[3]. Chegini, MG, Roohoallah K. (2013), "*the Relationship between Empowerment and Organizational Commitment'*, International Research Journal of Applied and Basic sciences.
[4]. Cochran, William G. (2010), *Teknik Penarikan Sampel.*Jakarta: UI-Press.
[5]. George Jennifer M. George and Gareth R. Jones.(2012), *Understanding and Managing Organizational Behavior,* Sixth Edition, New Jersey: Pearson Edu., Inc.
[6]. Blau, J. (2011), *The measurement and prediction of career commitment*" Journal of Occupational Psychology. The British Psychological Society". Article first published online.
[7]. Judeh M. (2012), *An Analysis of the Relationship between trust and Employee empowerment" International Business Management6(2).* ISSN; 1993-5250.
[8]. Kannan, R. and S. Muthu Perumal Pillai, *An Examination on the Professional Commitment of Engineering College Teachers,* Medwell Journal: International Business Management 2 (6), 200
[9]. Metin, HK, Suleyman, S. (2009), *The Impact of Empowerment and Quality Culture on Job Satisfaction,* Journal of Global Strategic Management.
[10]. Shockley-Zalabak, PS., Sherwyn PM, dan Michael, Z. (2010), *Building the High-Trust Organization: Strategies for Supporting Five Key Dimensions of Trust,* San Fransisco: Jossey-Bass.
[11]. Winch, C and Gingell, J. (2008), *Philosophy of Education: The Key Concepts,*Second Edition, New York: Routledge.

Solving portfolio selection problems with minimum transaction lots based on conditional-value-at-risk

E.P. Setiawan and D. Rosadi

Mathematics Department, Universitas Gadjah Mada, INDONESIA

E-mails: e2r4.ps@gmail.com, dedirosadi@gadjahmada.edu

Abstract. Portfolio selection problems conventionally means 'minimizing the risk, given the certain level of returns' from some financial assets. This problem is frequently solved with quadratic or linear programming methods, depending on the risk measure that used in the objective function. However, the solutions obtained by these method are in real numbers, which may give some problem in real application because each asset usually has its minimum transaction lots. In the classical approach considering minimum transaction lots were developed based on linear Mean Absolute Deviation (MAD), variance (like Markowitz's model), and semi-variance as risk measure. In this paper we investigated the portfolio selection methods with minimum transaction lots with conditional value at risk (CVaR) as risk measure. The mean-CVaR methodology only involves the part of the tail of the distribution that contributed to high losses. This approach looks better when we work with non-symmetric return probability distribution. Solution of this method can be found with Genetic Algorithm (GA) methods. We provide real examples using stocks from Indonesia stocks market.

1. Introduction

In the investment world, portfolio selection problem conventionally means 'minimizing the risk, given the certain level of returns' from some financial assets. Since Markowitz's seminal paper on 1952, this problem has found a mathematical formulation, and became a popular and interesting topic in financial mathematics. Markowitz introduced mean-variance portfolio optimization model, a quadratic programming problem with variance as risk measurement [1]. This model became a milestone in finance theory, but has high computational complexity and too many parameters that must be estimated from the data [2]. This models also rely on assumption that investor are risk averse and the return of assets are normally distributed [3]. Some research suggest that these assumptions are rarely satisfied in practical use.

Many researchs had done to simplify the portfolio selection problems and to certain that these assumptions are realistic. By change the risk measurement, one can convert the portfolio selection problems from mean-variance quadratic programming to a linear programming, which were simpler and easier for being solved than a quadratic programming. Some risk measurement that can be used i.e mean absolute deviation, as introduced by Konno [4], and Conditional Value-at-Risk or CVaR that has been developed by Rockafellar and Uryasev [5]. Pflug [6] noted that CVaR is a coherent risk measure, having the following properties: translation-equivariant, positively homogeneous, convex, monotonic with respect to stochastic dominance of order 1, and monotonic with respect to monotonic dominance of order 2. These properties make CVaR looks attractive as a risk measure. Another approach that using CVaR as constraint had been studied by Uryasev [7] and Palmquist *et al.* [8].

On the other hand, the solution of the portfolio selection problem that obtained by both quadratic and linear programming method are in real numbers, which may generate some problem in real application. Each asset usually has its minimum transaction lots, which are different in many stock markets around the world. A transaction lot, also simply called lot or round, represents a standardized quantity associated with a specific asset, set by the regulatory body in the stock exchange. A lot represents the basis for a transaction on the financial asset. The number of lots must be integer, so the solution from the portfolio selection problem cannot be used immediately.

Some classical portfolio selection considering minimum transaction lots were developed based on Mean Absolute Deviation (MAD), variance (like Markowitz's model), and semi-variance as risk measure [2,3]. Regarding the attractive properties of Conditional Value-at-Risk (CVaR), in this paper we investigated the portfolio selection method considering minimum transaction lots with CVaR as risk measure. This method only involves the part of the tail of the distribution of the investment return that contributed to high losses. The CVaR also looks better when the probability distribution of the assets return is not symmetric. Different from the method that used by Angelelli *et al.* [10], in this paper we use genetic algorithm method to solve the optimization problem.

This paper organized as follows. In the section 2, we discuss the Conditional Value-at-Risk as risk measure and its properties. The related portfolio selection method, considering minimum transaction lots, discussed in the section 3. Empirical studies with stocks from Indonesia Stock Exchange (IDX) are presented in the section 4. The last, section 5 devoted for some conclusion.

2. Conditional Value-at-Risk

The definition of Conditional Value-at-Risk cannot be separated from the Value-at-Risk or VaR. With respect to a specified probability level β, the β-VaR of a portfolio is the lowest amount α in order to with probability β, the loss will not exceed α. In mathematical expression, we can write

$$VaR_\beta\left(R_p\right)=\min\left\{R \mid P\left(-R_p \geq R\right) \leq 1-\beta\right\}$$

where P denotes the probability function. Typical values for β are 0.90, 0.95, and 0.99. Although VaR is a well-known risk measure that used by most financial institutions around the world, this measure has several undesirable properties: VaR is not subadditive, lack of convexity, and does not take the magnitude or amount of the losses beyond the VaR value into account. These undesirable features motivated the development of Conditional Value-at-Risk.

Conditional Value-at-Risk (CVaR), also known as mean excess loss, mean shortfall, tail VaR, and expected tail loss (ETL), measures the expected amount of losses in the tail of the distribution of possible portfolio losses, beyond the VaR. Taking the portfolio return R_p as a random variable, we can define the CVaR as

$$CVaR_\beta\left(R_p\right)=E\left[-R_p \mid -R_p \geq VaR_\beta\left(R_p\right)\right] \quad (1)$$

To describe some mathematical properties of the CVaR measure, we introduced some notations below. Let us denote by $\mathbf{w}$, the n-dimensional portfolio vector such that each component w_i represent the amount of shares held in financial asset-i. Then, we denote by $\mathbf{y}$ in $\mathbb{R}^m$ a random vector describing the uncertain outcomes of the economy, e.g. market parameters. We define the function $f(\mathbf{w},\mathbf{y})$ as *loss function*, represent the loss associated with the portfolio vector $\mathbf{w}$. Note that for each $\mathbf{w}$, the loss function $f(\mathbf{w},\mathbf{y})$ is a one-dimensional random variable in $\mathbb{R}$. Also we can assumed that the underlying probability distribution of **y** in $\mathbb{R}^m$ have density, denote by $p(\mathbf{y})$.

Following Rockafellar and Uryasev [5], since CVaR of the losses of portfolio **w** is the expected value of the losses conditioned on the losses being in excess of VaR, assuming that all random values are discrete, we have

$$CVaR_\beta(\mathbf{w}) = E\left[f(\mathbf{w},\mathbf{y}) \mid f(\mathbf{w},\mathbf{y}) \geq VaR_\beta(\mathbf{w})\right] = \frac{\sum_{\{\mathbf{y} \mid f(\mathbf{w},\mathbf{y}) > VaR_\beta(\mathbf{w})\}} p(\mathbf{y}) f(\mathbf{w},\mathbf{y})}{\sum_{\{\mathbf{y} \mid f(\mathbf{w},\mathbf{y}) > VaR_\beta(\mathbf{w})\}} p(\mathbf{y})} \tag{2}$$

In the continuous case, we get the equivalent form

$$CVaR_\beta(\mathbf{w}) = E\left[f(\mathbf{w},\mathbf{y}) \mid f(\mathbf{w},\mathbf{y}) \geq VaR_\beta(\mathbf{w})\right] = (1-\beta)^{-1} \int_{f(\mathbf{w},\mathbf{y}) > VaR_\beta(\mathbf{w})} p(\mathbf{y}) f(\mathbf{w},\mathbf{y})\, dy \tag{3}$$

It can be proved that the CVaR is always at least as large as VaR. But, the most important property of CVaR is that the CVaR is a coherent risk measure, a property which was defined by Artzner *et al.* [9]. It means that (1) if there only positive returns, then the CVaR should be non-positive; (2) the risk of a portfolio of two assets be less than or equal to the sum of the risks of the individual assets; (3) if the portfolio is increased c times, the risk becomes c times larger,and (4) cash or another risk-free asset does not contribute to the portfolio risk. Another important property of CVaR, proved by Pflug [6] is that CVaR is convex in the following sense: For arbitrary (possibly dependent) random variables $\mathbf{y}_1$ and $\mathbf{y}_2$, and $0 < \lambda < 1$,

$$CVaR_\beta(\lambda \mathbf{y}_1 + (1-\lambda)\mathbf{y}_2) \leq \lambda CVaR_\beta(\mathbf{y}_1) + (1-\lambda) CVaR_\beta(\mathbf{y}_2) \tag{4}$$

3. Minimum Lots Portfolio Selection with Conditional Value-at-Risk

The minimum lots portfolio selection models with conditional Value-at-Risk can be established from the Conditional Value-at-Risk portfolio optimization model.

3.1 CVaR Optimization

From the above formula, we need to determine the VaR before we calculate the CVaR. This problem cause CVaR optimization turn out to be somewhat tricky in practice. Fortunately, Rockafellar and Uryasev [5] discovering simpler approach. The key in this approach is a characterization of VaR(**w**) and CVaR(**w**) in terms of the function F_β on $X \times \mathbb{R}$, defined by

$$F_\beta(\mathbf{w},\alpha) = \alpha + (1-\beta)^{-1} \int_{y \in \mathbb{R}^m} \left[f(\mathbf{w},\mathbf{y}) - \alpha\right]^+ p(\mathbf{y})\, dy \tag{5}$$

where $[t]^+ = \max\{t, 0\}$, that can be used instead of CVaR. There are three important properties of that function: $F_\beta(\mathbf{w},\alpha)$ is a convex and continously differentiable function with respect to α, the $\text{VaR}_\beta(\mathbf{w})$ is a minimizer of $F_\beta(\mathbf{w},\alpha)$, and the minimum value of $F_\beta(\mathbf{w},\alpha)$ is $\text{CVaR}_\beta(\mathbf{w})$.

Let y_j be the return on j-th instrument, and $\mathbf{y} = (y_1, y_2, \ldots, y_n)$ is a random vector. Also, let $p(\mathbf{y})$ denote the distribution of **y**, a joint distribution of the return of each instrument. Assume p(**y**) is independent of **w**. Using notations above, the return on the portfolio is the sum of the returns on each individual instruments in the portfolio, weighted by the proportions of each asset in the portfolio. The loss of this portfolio is the negative of the return, that is

$$f(\mathbf{w},\mathbf{y}) = -[w_1 y_1 + \ldots + w_n y_n] = -\mathbf{w}^T\mathbf{y} \tag{6}$$

Substituting equation (6) to equation (5), we get

$$F_\beta(\mathbf{w},\alpha) = \alpha + (1-\beta)^{-1} \int_{y\in\mathbb{R}^m} \left[-\mathbf{w}^T\mathbf{y} - \alpha\right]^+ p(\mathbf{y})\,dy \tag{7}$$

The integral in equation (7) above needs to be approximated, since this method is proposed to use when **y** doesn't follow a special distribution, such as normal distribution, student-t distribution, etc. One of the approximation can be done by sampling the probability distribution of y according to its density $p(\mathbf{y})$ by simulation. Suppose that we have t data as sample set, $\mathbf{y}_1, \mathbf{y}_2, \ldots, \mathbf{y}_t$, Uryasev [7] suggest to use this approximate function

$$\tilde{F}_\beta(\mathbf{w},\alpha) = \alpha + \frac{1}{(1-\beta)q}\sum_{k=1}^{t}\left[-\mathbf{w}^T\mathbf{y}_k - \alpha\right] \tag{8}$$

The minimization problem above, minimizing $\tilde{F}_\beta$ to get the approximate solution for the minimization of F_β, is equivalent to

$$\text{minimize } \alpha + \frac{1}{q(1-\beta)}\sum_{k=1}^{t} u_k \tag{9}$$

subject to linear constraint

$$u_k \geq 0 \tag{10}$$

$$\mathbf{w}^T\mathbf{y}_k \geq -\alpha - u_k \tag{11}$$

$$\mathbf{w}^T\mathbf{1} = \sum_{j=1}^{n} w_j = 1 \tag{12}$$

Both the objective function and the constraint above are linear, so this optimization problem can be solved by standard linear programming techniques.

3.2 Additional Minimum Transaction Lots

A transaction lot, also simply called lot or round, represents a standardized quantity associated with a specific asset, set by the regulatory body in the stock exchange. A lot represents the basis for a transaction on the financial asset, or minimum quantity of an asset that can be traded. For example, in Indonesian Stock Exchange regular transaction, the minimum transaction lot for stocks and for option contract is 100 units and 1 unit, respectively. It means that an investor should buy or sell stocks in multiplication of 100 unit, but can buy only one or two option contracts.

The number of lot assets that can be bought or sold by an investor must be integer. It means that there are minimum amount of capital that can be invested in that assets. The total capital invested in that assets has to be expressed as a multiple of the minimum amount. For example, if the Stock Exchange Board establishes that security A has to be bought in multiples of 100 unit, and the price is IDR 11,500, this implies that the security A has minimum round of IDR 1,150,000.

Let N_j be the number of units of security j required as minimum quantity on transaction (lot), and p_j be the market price for security j at the date of the purchase of the portfolio. We denote by c_j the purchasing price for the minimum lot of each security j, expressed in terms of money and is equivalent to $c_j = N_j p_j$. Trivially, when asset j traded without minimum lot, we take $N_j = 1$, so $c_j = p_j$. Next, we define the integer variable x_j, $j = 1, 2, ..., m$, represents the number of minimum lots for each security j which will make part of the optimal portfolio. So, the quantity $c_j x_j$ represents the part of the total available budget that the investor decides to put to the security j. With these notations, we can define the portfolio selection problem with minimum transaction lot based on CVaR as minimizing

$$\min \; CVaR_{\beta}\left(R_p\right) \tag{13}$$

subject to

$$\sum_{j=1}^{m} x_j c_j \leq b \tag{14}$$

$$x_j \in \mathbb{Z} \tag{15}$$

As it can be observed above, equation (13) is the objective function which determines the CVaR as a risk measurement to be minimized. Equation (14) considered budget (b) that available for the investment, so that the total amount of investment is not greater than the available budget. Last, the (15) ensure that the number of lots should be integer.

3.3 Solution

The CVaR (13) can be linearized by substituting the linear function (7) and its approximation (8) or by using the equivalent form (9) and the additional constraint (11), become

$$\min \; \alpha + \frac{1}{q(1-\beta)} \sum_{k=1}^{t} u_k$$

subject to

$$\sum_{j=1}^{m} x_j c_j \leq b$$

$$\mathbf{x}^T \mathbf{y}_k \geq -\alpha - u_k$$

$$x_j \in \mathbb{Z}$$

It can be understood that the portfolio optimization model based on CVaR considering transaction lots is a mixed integer linear program, because only the x_j is restricted to integer values. There are some methods to obtain solution from a mixed integer linear program, for example by branch-and-bound method and cutting plane method [11]. Unfortunately, neither of the two methods can be claimed for being uniformly effective in solving the integer linear program. Mansini, Ogryczak, dan Speranza [12] also noted that integer linear programming models with CVaR measure are harder to solve than the corresponding models with mean absolute deviation (MAD) measure. An increasing number of assets also makes the model harder to solve.

Another approach that can be used to solving complex optimization problem is heuristic method, for example genetic algorithm. Proposed by Holland [13] in 1975, this method were developed from the Darwin's principal of natural selection and genetic theory. In genetic algorithms, an initial population consist of several solution, represented as chromosomes, is selected randomly as the first generation. Two chromosomes as "parents" were selected among them, and produce "child" by cross over and mutation operators. A fitness function, which is originated from the objective function of the optimization problem, used to evaluate the desirability of each solution represented by each chromosome. Chang *et al.* [3] explained basic steps of genetic algorithm as

Generate an initial population
Evaluate fitness of individuals in the population
Repeat
  Select parents from the population
  Recombine (mate) parents to produce children using cross over and mutation operators
  Evaluate fitness of the children
  Replace some or all of the population by the children
Until a satisfactory solution has been found

Since the study by Arnone *et al.*[14], genetic algorithm had used to solving portfolio selection problems. Many studies have showed that genetic algorithm can efficiently find near optimal or even the optimal solutions for optimization problems. Specially, the use of genetic algorithm in optimizing portfolio based on the Conditional Value at Risk (CVaR) has been described by Lin and Ohnishi [15]. This study can be seen as a development from their study, as in this study there is additional constraint i.e. minimum transaction lots.

4. Empirical Study

In this empirical study, we choose ten stocks from LQ-45 index, an blue-chips stock list published by Indonesian Stock Exchange (IDX) for period July 2015 to July 2016. We compute the continuously compounded return or log return, based on definition by Tsay [16]. The descriptive statistics of log return for each stock are presented in Table 1.

In Table 1 we can see that some stocks have large excess kurtosis, for example ADHI and SRIL. There are also relatively large difference between skewness of each stock return. It means that standard deviation (or variance) is not enough to describe the risk, so we use the conditional value at risk (CVaR). Here, the CVaR were calculated by historical method, so it is not influenced by the probability distribution of the loss.

Assume that our transaction are happened in Indonesian Stock Exchange (IDX), with regularly minimum lots equivalent to 100 unit stock. The opening price of each stock as given in Table 2.

Table 1. Descriptive statistics and 5% Conditional VaR of log return for some stocks

Stock Code	Mean	Standard deviation	Skewness	Excess Kurtosis	CVaR 5 %
PTPP	0.00056020	0.0193677	0.3554030	2.554045	0.03888579
ADHI	0.00166336	0.0300079	0.0929876	4.492607	0.06302641
PTBA	0.00027220	0.0337169	0.7406433	1.504704	0.06038276
LPPF	0.00111219	0.0282435	0.5336275	1.162816	0.05238701
SRIL	0.00024344	0.0435830	1.0833290	6.817095	0.08953712
BBNI	0.00014429	0.0229853	0.5762866	1.397740	0.04318361
BMTR	-0.00004335	0.0346544	0.3274879	0.746279	0.06991387
ICBP	0.00142647	0.0199368	0.8724702	2.460618	0.03476379
JSMR	-0.00002437	0.0221391	0.4233325	2.990921	0.04902375
TLKM	0.00136858	0.0167517	-0.2146708	0.523264	0.03657948

Table 2. Price of each stocks (in Indonesian Rupiah, IDR)

Stock Code	Price/unit	Price/lot	Stock Code	Price/unit	Price/lots
PTPP	3,870.00	387,000.00	BBNI	5,400.00	540,000.00
ADHI	2,830.00	283,000.00	BMTR	1,045.00	104,500.00
PTBA	9,875.00	987,500.00	ICBP	8,700.00	870,000.00
LPPF	20,000.00	2,000,000.00	JSMR	5,400.00	540,000.00
SRIL	268.00	26,800.00	TLKM	4,350.00	435,000.00

4.1 CVaR Portfolio Optimization without Transaction Lots

Suppose that an investor wants to invest IDR 1,000,000,000 into a portfolio consist of ten stocks above. To obtain the weights of each assets in the portfolio, we use genetic algorithm method. Parameters of the proposed genetic algorithm are set according to Hopgood [17], Lin and Liu [18], and Chang et al. [3] as follows. All computation were conducted on a notebook with an Intel i5 2.8 GHz processor for about 10 minutes.

Table 3. Some method and parameters of the Genetic Algorithm

Method/Parameters	Value
Chromosome length	10 (equal to the number of assets in this portfolio)
Representation of genes	real number between 0 and 1
Number of population	50
Crossover method	Uniform crossover
Crossover probability	0.7
Mutation method	Random mutation
Mutation probability	0.1
Selection method	Tournament (roulette-wheel) selection
Elitism	3

Without regarding the minimum transaction lots, the objective function is just minimizing the CVaR of the portfolio. On other side, genetic algorithm's goal is to get the chromosome with largest fitness, so we state the 'minimizing CVaR' as 'maximizing the negative of CVaR'. The result of this algorithm is stated in Table 4 below.

Table 4. Number of lots of each assets on the optimal portfolio *without* regarding the transaction lots, $\beta = 0.95$

Stock Code	Iteration 100x	Iteration 250x	Iteration 500x	Iteration 1000x
PTPP	120	233	227	233
ADHI	113	53	6	8
PTBA	31	25	21	22
LPPF	30	5	5	0
SRIL	918	577	1023	1069
BBNI	463	299	259	237
BMTR	170	88	63	14
ICBP	344	500	540	553
JSMR	52	70	92	120
TLKM	476	457	421	410
Total Investment	995,823,400	997,559,600	996,748,900	996,501,200

From result on table 4 above, we can conclude that different number of generations in genetic algorithm can yield slightly different number of lots on each assets. The limitation to only buy assets in regular lot transaction leads to some difference between the total money spent to bought the asset and the total available money for investment purpose.

4.2 CVaR Portfolio Optimization with Transaction Lots

In this sub section, we look for portfolio weight with CVaR as a risk measure and minimum transaction lots as additional constraint. First, we calculate the weight of each assets in the portfolio. The number of lots were calculated by multiply the weight of each assets by the total investment and divide it by each price per lot asset. Last, we measure the 'true' weight, as the number of lots, multiply by the price, and divide by the total assets. The portfolio's CVaR which will be minimized were calculated from this 'true' weight. This more complex fitness function needs more time to execute when implemented in the genetic algorithm.

For the convenience, the parameters of Genetic Algorithm is same as that used in sub section 4.1 above, i.e. that mentioned in table 3. The result of portfolio optimization with minimum transaction lots are shown in table 5, 6, and 7 below for different β of CVaR.

Table 5. Number of lots of each assets on the optimal portfolio regarding the transaction lots, with CVaR $\beta = 0.90$

Stock Code	Iteration 100x	Iteration 250x	Iteration 500x	Iteration 1000x
PTPP	355	409	405	360
ADHI	113	41	54	63
PTBA	50	48	55	52
LPPF	18	8	3	3
SRIL	623	483	522	739
BBNI	156	160	164	136
BMTR	234	143	87	35
ICBP	313	364	388	414
JSMR	159	148	126	153
TLKM	592	576	563	552
Total Investment	995,793,400	994,709,900	994,448,100	994,295,700

Table 6. Number of lots of each assets on the optimal portfolio regarding the transaction lots, with CVaR $\beta = 0.95$

Stock Code	Iteration 100x	Iteration 250x	Iteration 500x	Iteration 1000x
PTPP	245	196	216	247
ADHI	121	37	35	18
PTBA	40	30	27	21
LPPF	27	10	3	3
SRIL	675	629	962	959
BBNI	325	300	308	250
BMTR	169	65	48	73
ICBP	376	480	483	511
JSMR	118	88	90	105
TLKM	394	480	467	454
Total Investment	996,018,500	995,502,700	995,218,600	994,499,700

Table 7. Number of lots of each assets on the optimal portfolio regarding the transaction lots, with CVaR $\beta = 0.99$

Stock Code	Iteration 100x	Iteration 250x	Iteration 500x	Iteration 1000x
PTPP	120	270	339	302
ADHI	79	88	63	78
PTBA	74	67	73	60
LPPF	14	9	1	5
SRIL	757	499	312	547
BBNI	442	266	292	257
BMTR	156	40	27	17
ICBP	442	501	500	530
JSMR	119	122	78	84
TLKM	236	274	295	287
Total Investment	996,564,600	995,656,200	997,381,100	994,689,100

From Tables 5, 6, and 7 above, it can be seen that both the β value of CVaR and the number of generation influence the number of lots of each assets on the portfolio. However, the number of generation seems not related with the total investment.

4.3 Portfolio performance

From practical point of view, after an investment portfolio had constructed, it is need to be monitored and evaluated periodically. The purpose of monitoring and evaluating are to maintain the portfolio return and prevent the larger risk. In this case, performance of each portfolio above measured daily for 30 market days after. Figure 1 stated the LQ45 index movement for this period.

From figure 1 above, we could say that the market of stock in LQ45 index is bullish. As we could compute the daily continuously compounded return or log return, we could compute the continuously compounded multiperiod return as the sum of daily log return [17]. The portfolio total return then calculated as the sum of 100 times number of lots for each assets times return of each assets.

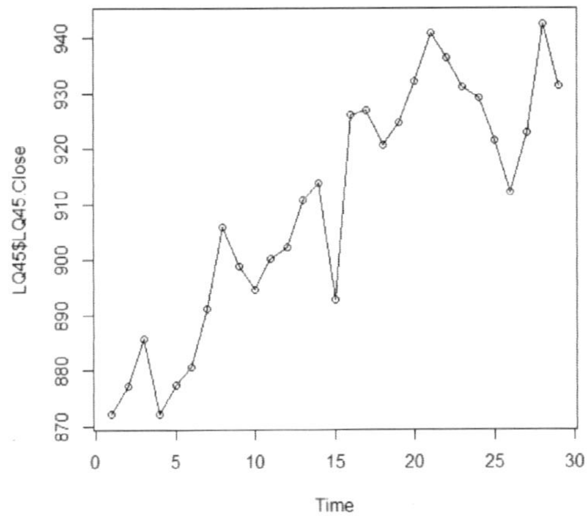

Figure 1. Movement of LQ45 index in 30 market days after portfolio selection

It is look necessary to know, how the additional constraint about minimum lot transaction affect the portfolio performance, i.e. the return and CVaR of the portfolio. Comparison between two portfolios with same β but different additional constraint can be examined from figure 2 below.

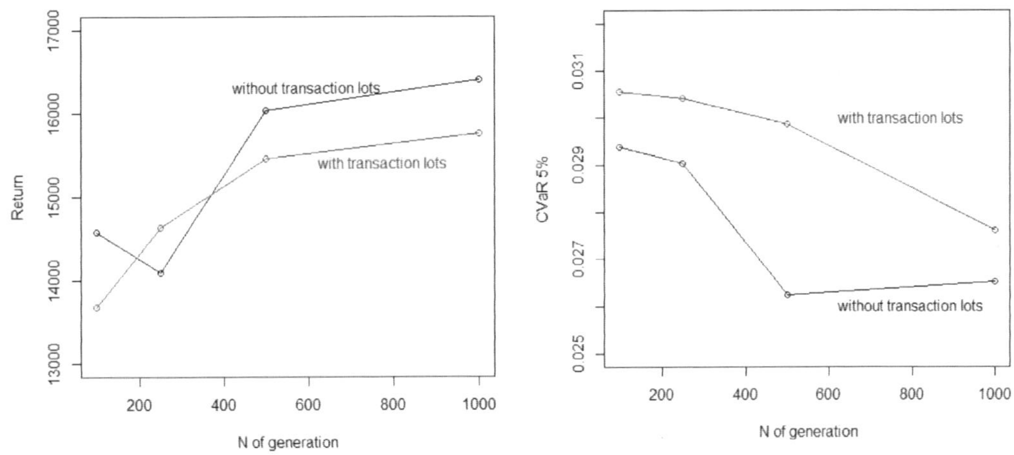

Figure 2. Portfolio return and CVaR for 30 market days for CVaR optimization with and without minimum transaction lots

Figure 2 above shows that the additional minimum transaction lots in the portfolio optimization give higher CVaR than portfolio without considering minimum transaction lots. However, for larger number of generations, it seems that both portfolio will give approximately same value of CVaR. For the return, its can looks that when the number of generation equal to 250, both of the portfolio give same return.

From figure 3 below, we could see the relationship between confidence level (β) of CVaR to the return and CVaR of the portfolio. For small number of generation, different confidence level yield different portfolio return. However, it will converge to approximately same return when the number of generation become larger, more than 1000. For larger number of generation, different beta CVaR also yielding portfolio with different risk, as measured by CVaR on the right side.

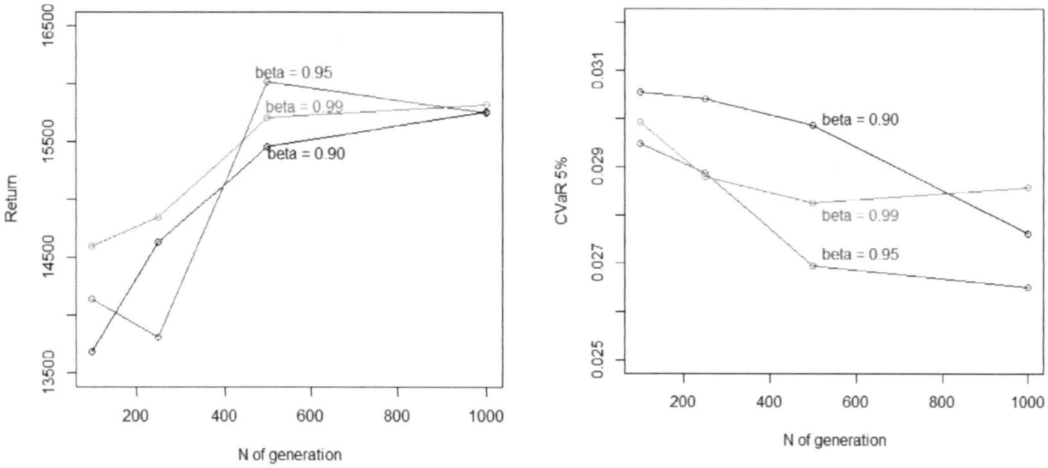

Figure 3. Portfolio return and CVaR for 30 market days for different β-CVaR optimization with minimum transaction lots

As explained in Table 3, there are some parameters in the genetic algorithm approach, i.e. cross over probability and mutation probability. Both of them also analyzed to find whether they affect the return and/or the risk of the optimal portfolio.

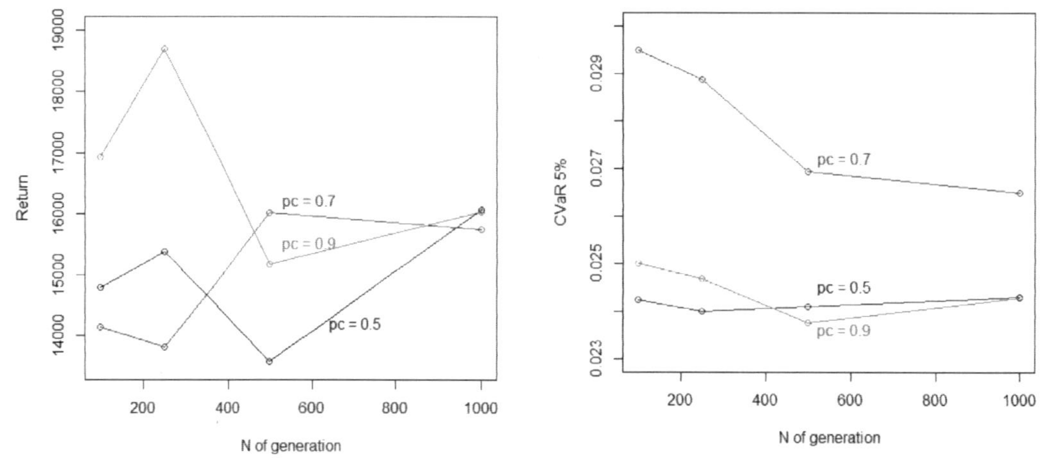

Figure 4. Portfolio return and CVaR for 30 market days for β-CVaR optimization with minimum transaction lots, with different crossover probability

From Figures 4 and 5, it looks that the probability of mutation and probability of crossing over of genetic algorithm affect the return and the CVaR of the portfolio. For the CVaR, which wanted to be minimized, too small mutation probability cause the CVaR down slowly, and may difficult to reach the global minima. By compared figure 4 to figure 2 above, we could carefully select the parameters of the genetic algorithm so the portfolio optimization problem constrained with minimum transaction lots will yield higher return than portfolio without minimum transaction lots.

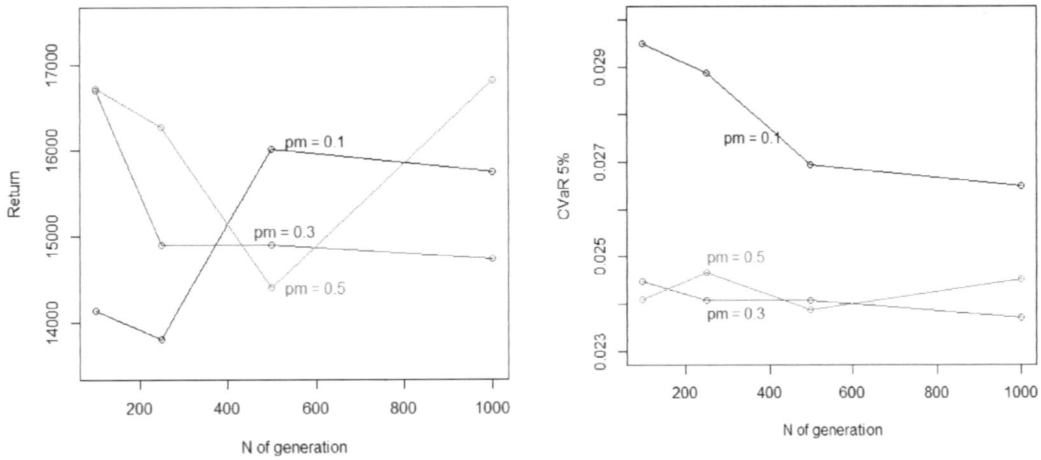

Figure 5. Portfolio return and CVaR for 30 market days for β-CVaR optimization with minimum transaction lots, with different mutation probability

Table 8. Reward-to-variability (Sharpe) Ratio of some optimum portfolio with minimum transaction lots and CVaR as risk measure. The stock description is refer to Table 3, unless mentioned.

Stock Description	Iteration 100x	Iteration 250x	Iteration 500x	Iteration 1000x
Varying significance level				
β = 0.90	3.269	3.853	4.156	3.998
β = 0.95	3.702	3.806	3.788	3.699
β = 0.99	3.933	4.771	5.362	4.977
Varying the mutation probability				
pm = 0.1	3.794	3.725	3.725	4.223
pm = 0.3	3.702	3.806	3.788	3.699
pm = 0.5	4.639	3.394	3.893	3.754
Varying the crossover probability				
pc = 0.5	3.946	4.116	3.989	3.943
pc = 0.7	3.702	3.806	3.788	3.699
pc = 0.9	3.031	3.797	4.004	3.620

Last, to compare all models above simultaneously, we compute the reward-to-variability ratio, which popular known as Sharpe ratio (See Elton *et al.* [19]). Here, reward is measured by difference between mean return of the portfolio and the risk-free-rate. The variability measured by variance of the portfolio. Note that the larger the ratio, the better the performance of the portfolio. As a rule of thumb, portfolio with Sharpe ratio > 1 considered good, Sharpe ratio > 2 considered very good, and Sharpe ratio > 3 considered excellent. The result is provided in the table 8.

From the reward-to-variability ratio above, we conclude that the number of generation or iteration should be more than 100. Using higher confidence level of CVaR, reward or excess return that will received by the investor also higher. Consistent to explanation above, the mutation and cross over probability also influence both the return and the variance.

IOP Conf. Series: Materials Science and Engineering **166** (2017) 012005 doi:10.1088/1757-899X/166/1/012005

5. Conclusion and Future Research

Portfolio optimization using Conditional Value at Risk (CVaR) as risk measurement has attractive property, i.e. the coherent property of CVaR. To build a portfolio optimization based on CVaR with minimal transaction lots as an additional constraint, genetic algorithm become an useful tool to solving the model and get the amount of asset that should be bought by the investor. However, the properties and method of genetic algorithm may influence the solution that obtained, and the portfolio can't well compared each other. In the future, more research is needed to get better solution, using other heuristic method and/or adding more constraint such as cardinality constraint and asset weight constraint.

References

[1] Markowitz HM 1952 Portfolio selection *J. of Finance* **7** 77-91
[2] Mansini R and Speranza MG 1999 Heuristic algorithms for the portfolio selection problem with minimum transaction lots *Eur. J. of Oper. Res.* **114** 219-233
[3] Chang T-J, Yang S-C and Chang K-J 2009 Portfolio optimization problems in different risk measures using genetic algorithm *Expert Sys. with App.* **36** 10529-10537
[4] Konno H and Yamazaki H 1991 Mean-absolute deviation portfolio optimization model and its application to Tokyo Stock Market *Manag. Sci.* **7** 519-531
[5] Rockafellar RT and Uryasev S 2000 Optimization of conditional value-at-risk *J. of Risk* **2** 21-41
[6] Pflug G 2000 Some remarks on the value-at-risk and the conditional value-at-risk *Probabilistic Constrained Optimization: Methodology and Applications* ed Uryasev S (Kluwer Academic Publishers) chapter 1 pp 1-11
[7] Uryasev S 2000 Conditional value-at-risk: optimization algorithms and applications *Financial Engineering news* **14** 1-5
[8] Palmquist J, Uryasev S and Krokhmal P 2002 Portfolio optimization with conditional value-at-risk objective and constraints *J. of Risk* **4** 11-27
[9] Artzner P, Delbaen F, Eber JM and Heath D 1999 Coherent Measures of Risk *Math. Finance* **9** 203-228
[10] Angelelli E, Mansini R and Speranza MG 2008 A comparison of MAD and CVaR models with real features *Journal of Banking and Finance* **32** 1188-1197
[11] Taha, HA 1992 *Operation Research, An Introduction Fifth Edition* Macmillan Publishing Company New York
[12] Mansini R, Ogryczak W and Speranza, MG 2015 *Linear and Mixed Integer Programming for Portfolio Optimization* Springer London
[13] Holland JH 1975 *Adaptation in Natural and Artificial Systems* University of Michigan Press
[14] Arnone S, Loraschi A and Tettamanzi A 1993 A genetic approach to portfolio selection *Neural Network World* **6** 597-604
[15] Lin S and Ohnishi M 2006 Optimal Portfolio Selection by CVaR based Sharpe Ratio *Scientiae Math. Japonicae Online* 1229-1251
[16] Tsay R S 2010 *Analysis of Financial Time Series Third Edition* John Wiley and Sons Inc. New Jersey
[17] Lin C-C and Liu Y-T 2008 Genetic algorithms for portfolio selection problems with minimum transaction lots **185** 393-404
[18] Hopgood AA 2001 *Intelligent Systems for Engineers and Scientist,* CRC Press
[19] Elton EJ, Gruber MJ, Brown SJ and Goetzmann WN 2014 *Modern Portfolio Theory and Investment Analysis* 9th edition John Wiley and Sons Inc. New York

Decision support system in an international-voice-services business company

R. Hadianti[1], S. Uttunggadewa[2], M. Syamsuddin[3] and E. Soewono[4]

[1,2,3,4]Research Consortium of Financial Modeling, Optimization and Simulation, Center for Mathematical Modeling and Simulation. Gedung Labtek III Lt. 1 Jalan Ganesha 10 Bandung 40132, INDONESIA.

E-mail: hadianti@math.itb.ac.id, s_uttunggadewa@math.itb.ac.id, msyamsuddin8@gmail.com, esoewono@bdg.centrin.net.id

Abstract. We consider a problem facing by an international telecommunication services company in maximizing its profit. From voice services by controlling cost and business partnership. The competitiveness in this industry is very high, so that any efficiency from controlling cost and business partnership can help the company to survive in the very high competitiveness situation. The company trades voice traffic with a large number of business partners. There are four trading schemes that can be chosen by this company, namely, flat rate, class tiering, volume commitment, and revenue capped. Each scheme has a specific characteristic on the rate and volume deal, where the last three schemes are regarded as strategic schemes to be offered to business partner to ensure incoming traffic volume for both parties. This company and each business partner need to choose an optimal agreement in a certain period of time that can maximize the company's profit. In this agreement, both parties agree to use a certain trading scheme, rate and rate/volume/revenue deal. A decision support system is then needed in order to give a comprehensive information to the sales officers to deal with the business partners. This paper discusses the mathematical model of the optimal decision for incoming traffic volume control, which is a part of the analysis needed to build the decision support system. The mathematical model is built by first performing data analysis to see how elastic the incoming traffic volume is. As the level of elasticity is obtained, we then derive a mathematical modelling that can simulate the impact of any decision on trading to the revenue of the company. The optimal decision can be obtained from these simulations results. To evaluate the performance of the proposed method we implement our decision model to the historical data. A software tool incorporating our methodology is currently in construction.

Keywords: decision support, international telecommunication, revenue maximization

1. Introduction

Telecommunication industry is continuously and simultaneously growing as the result of the various innovations of today's telecommunication technology. Nowadays, customers of this industry get some benefits from the growth of over-the-top (OTT) players in this industry that continuously expanding the scope of their offerings, the competitiveness in this industry is very high which leads to lower and lower customer cost of communication. Without any investment in faster network capabilities, cost control, business efficiencies and partnerships, any telecommunication company cannot be survive in the current situation.

We consider a problem facing by an international telecommunication services company which provides voice services, data services, and network provider, in the very high competitive situation. The voice services of this company are classified as wholesale services, international hubbing, extension voice service, and international direct dial. We focus on profit maximization from wholesale services, in which this company trades voice traffic with more than seventy business partners. There are four trading schemes that can be chosen by this company, namely, flat rate, class tiering, volume commitment, and revenue capped. Each scheme has a specific characteristic on company revenue assurance, company potential loss, and attractive or not for business partners to transfer significant traffic volume. Each scheme has a number of scheme parameters, such as rate, volume deal or revenue deal, where its values are determined by both parties. In section 2 we will describe more about the schemes and we also discuss that the last three schemes are regarded as strategic schemes to be offered to business partner to ensure incoming traffic volume for both parties. This company and each business partner need to choose an optimal agreement in a certain period of time that can maximize the company's profit. In this agreement, both parties agree to use a certain trading scheme, rate and volume/revenue deal. In the past, the agreements were usually decided by the sales officers without any decision support system. Therefore, not all agreements yield optimal profits. A decision support system is then needed in order to give a comprehensive information to the marketing officers to deal with the business partners.

The need of the decision support system is triggered by the fact that the company's revenue has a decreasing trend, contrasts to the traffic volume that has an increasing trend. Figure 1 below shows this phenomenon. The factor that causes this phenomenon is shown in Figure 2, which tells us the average rates to the six classes of destinations. The average rates have decreasing trends, and these cause the company's revenue is decreasing.

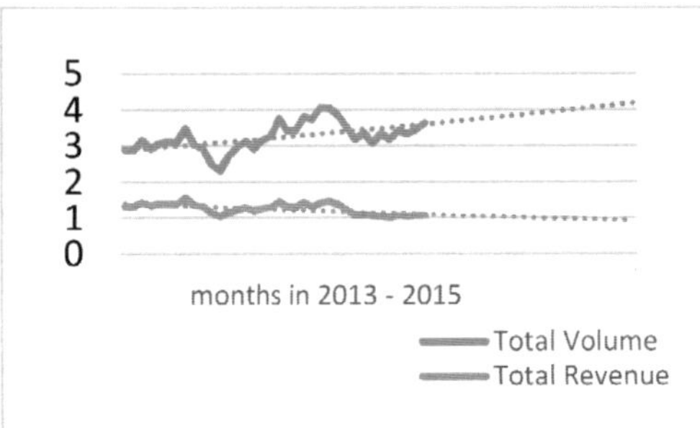

Figure 1. The incoming traffic volume (in 1000 million minutes) and the revenue (in US $ 10 million)

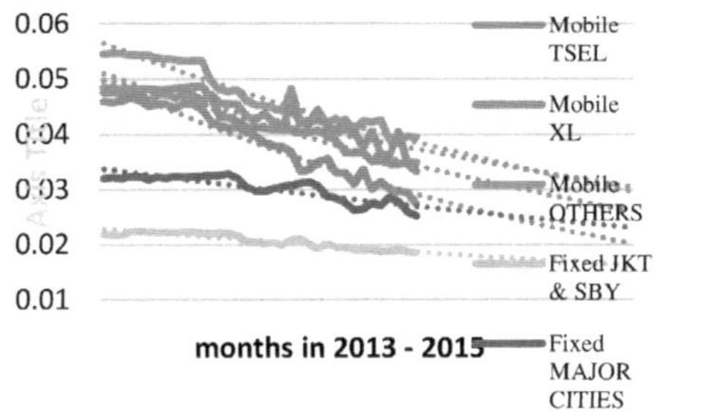

Figure 2. The average rates (US$/min and its extrapolations

If the rates are decreasing monotonically, then nothing can be done to control or to maximize revenue. But we can see from Figure 2 that the rates are not monotonically decreasing, which tells us that we need to explore what factors that influence the rates. Our elaboration on the company business process gave a conclusion that the rate is determined by negotiation between the company sales and business partners sales, and we may find some careless negotiations which leads to very low effective rates or unexpected business partner responds that cause un-optimal revenues. The decision support system is expected to give a guidance to sales officers in order to obtain win-win agreements to both the company and its business partners.

This paper studies a mathematical modeling that can simulate the impact of any decision on trading to the revenue of the company. In this paper, the study is focused on the incoming traffic analysis, so in the remaining of this paper we just consider the company's revenue since the company's profit can be studied by analyzing both incoming and outgoing traffic. The mathematical model is derived by performing following steps.

1. Analyzing the potential loss that can be imposed by a trading scheme.
2. Historical data analysis.
3. Modeling the expected revenue as a function of volume realization, rate, and others trading scheme parameters,
4. Modeling the optimal decision.

To evaluate the performance of the proposed method we implement our decision model to the historical data. We obtained that by performing our decision strategy, some unexpected phenomenon can be avoided so that the company's revenue can be optimized. The decision strategy is implemented into a software that can be used by sales officers. This software is on the trial stage, and is expected to be operated soon.

This paper is organized as follows. After this introduction, in section 2 we describes more the four trading schemes that can be chosen by sales officers. We also discuss the potential losses that can be imposed by the trading schemes. In Section 3, we present the results of our historical data analysis concerning on the price elasticity. By the analysis data we want to know at what level of rate reduction that will yield an increasing traffic volume. Besides that, we also want to know what the impact is if we increase the rate. After that, in Section 4 we derive a mathematical model for the expected revenue due to a decision. The results in Sections 2, 3 and 4 then is combined to derive an optimal decision model, which we present in Section 5. In Section 6 we give some examples on the decision model implementation. We end this paper by presenting some discussions in Section 7.

2. Trading schemes and the potential of loss

In each negotiation process, any sales officer and the business partner will choose one of four trading schemes, namely flat rate, class tiering, volume commitment, and revenue capped, for the next bilateral agreement. The description of all trading schemes are given in the following table.

Table 1. The description of the trading schemes

Trading scheme	Description
Flat rate	This trading scheme is a normal scheme, where both parties agree to deliver their traffic volumes at the specific rates, regardless how much volume of the traffic they deliver.
Class Tiering	By this trading scheme, both parties agree to set tier classes of volume in which each class is associated to a specific rate. In general, the rate for the lowest tier class is higher than the rate for the second tier class, and the rate for the second tier class is higher than the rate for the third tier class (if any). This setting is aimed to create an attracting scheme, so that the company and the business partner are eager to deliver more significant traffic volume.
Volume Commitment	By this trading scheme, the company commits to deliver a minimum traffic volume (which is called the first volume deal) during the period of the agreement at a certain rate (which is called the first rate deal). If the total volume exceeds the first volume deal, then the rate for the volume after the first volume deal follows the second rate deal. If during the period of the agreement the company could not fulfill the commitment, then the volume shortfall cost have to be paid in the grace period. On the other hand, the business partner has the same obligation in delivering its traffic volume, with respect to its volume deals and rate deals.
Revenue Capped	By this trading scheme, the company commit to pay a certain amount (which is called the revenue deal) to the business partner as the cost for terminating traffic volume delivered to the business partner to the end users, regardless how much volume of the traffic the company deliver.

Table 2 below shows examples of the trading schemes and its parameters involved.

Table 2. Examples of trading schemes.

Trading	Volume deal	Rate (US	Revenue deal
Flat rate	-	0.0273	
Class Tiering	100,000	0.0390	
	> 100,000	0.0360	
Volume Commitment	100,000	0.0400	
	> 100,000	0.0290	
Revenue Capped	-	-	1,415,000.00

It is clear that if we have a significant volume of traffic, the class tiering and the volume commitment trading schemes are preferable than the flat rate trading scheme since with an appropriate the first volume deal it can generate a lower effective rate, which can be obtained from the application of the second rate deal (and the third rate deal, if any). It should be noted that if the first volume deal is not appropriate, in the sense it is to high compare to our volume of traffic realization, by the volume commitment trading scheme will impose the shortfall loss that is the cost that has to be paid for unfulfillment of volume commitment. But this kind of loss will not be imposed by the class tiering trading scheme. The revenue capped trading scheme also can generate a lower effective rate if the traffic volume is very high (the effective rate is equal to the revenue deal divided by the traffic volume).

From the company point of view, the business partner revenue deal in the revenue capped trading scheme or the business partner first volume deal in the volume commitment trading scheme assure the company revenue. The opportunity for the business partner to gain a lower rate can be regarded as probability for the company to gain loss. But we note that the probability the company to gain loss in the revenue capped trading scheme is higher that the probability in the class tiering or the volume commitment trading schemes since in the last two trading schemes the minimum effective rate is bounded below by the minimum rate deal, meanwhile in the revenue capped trading scheme the higher traffic volume delivered the lower effective rate obtained. We summarize our analysis in the following table.

Table 3. Characteristics of trading schemes.

Trading scheme	Revenue assurance for the company	Opportunity for business partner to gain a lower	Probability of the company loss
Flat Rate	×	×	none
Class Tiering	×	√	low
Revenue Capped	√	√	high
Volume Commitment	√	√	low

IOP Conf. Series: Materials Science and Engineering **166** (2017) 012006 doi:10.1088/1757-899X/166/1/012006

3. Historical data analysis

We perform data analysis in order to validate the assumption we got that the voice product is price sensitive, in the sense that business partner will send more incoming traffic volume if the price is decreased. Contrary, business partner will not interested to send a significant incoming traffic volume if the rate is considered too high. This assumption is a well-known assumption, as written also in Meisner and Stefan [3] and Scott, *et al.* [7]. The assumption can easily understood by following figures 3 – 5 below. Figure 3 shows the historical data of the incoming traffic volume, the effective rate, and the average rate (obtained by averaging all rates applied to the others business partners to the same destinationtion). We can see that during two years the rate was never changed, and was always higher than the average rate. The higher rate made the business partner not very eager to deliver more and more traffic volume, causes a deacreasing trend in the traffic volume which leads to a deacresing trend of revenue.

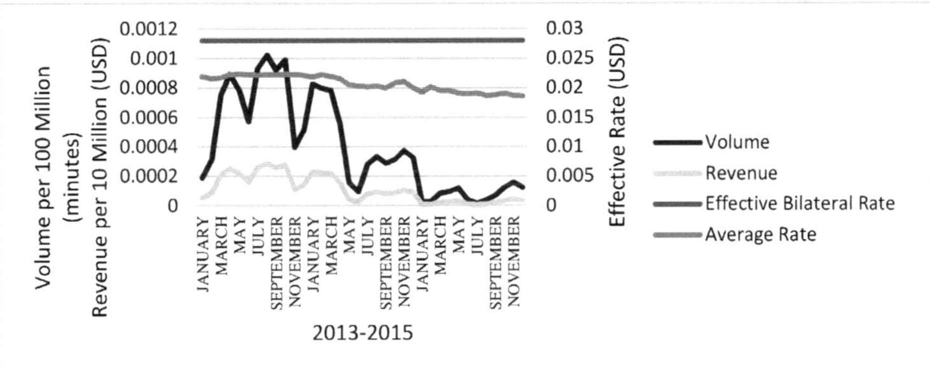

Figure 3. Analysis of Incoming Traffic Volume
Destination: Indonesia Fixed Jakarta & Surabaya - From Partner X

Figure 4 shows a very interesting phenomenon where the business partner seems do not want to deliver traffic anymore when the rate is increased.

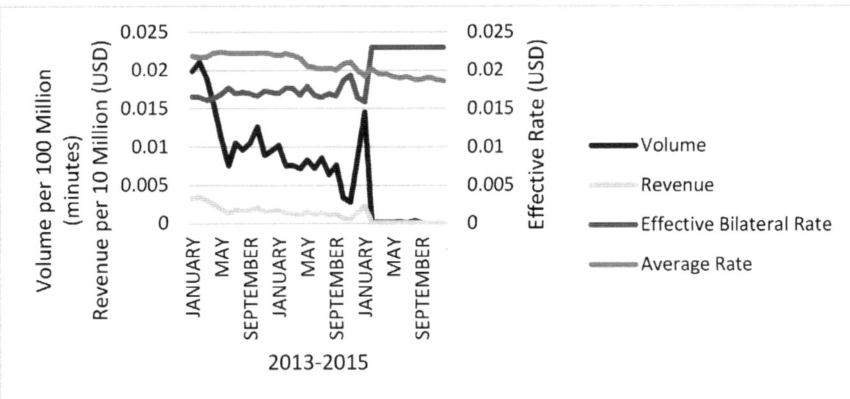

Figure 4. Analysis of Incoming Traffic Volume
Indonesia Fixed Jakarta & Surabaya – from partner Y

Figure 5 shows another interesting phenomenon where the business partner eagers to deliver a significant traffic volume when the rate is decreased quite significantly, and the business partner does not eager to deliver a significant traffic volume when the rate reduction is not significant.

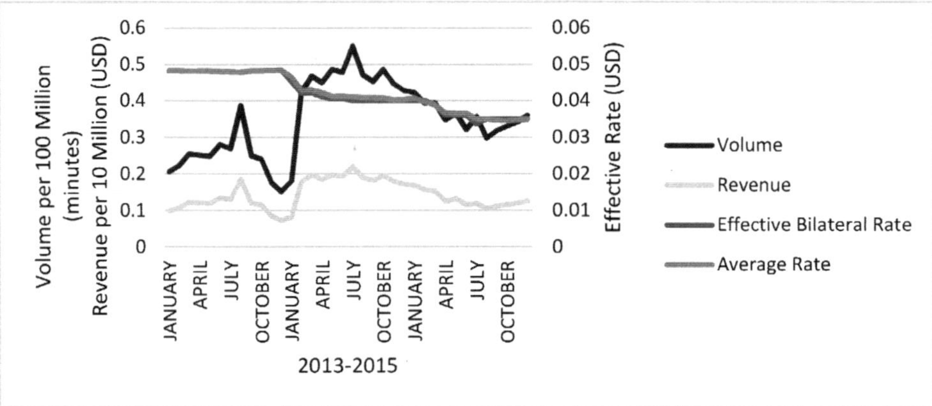

Figure 5. Analysis of Incoming Traffic Volume
Indonesia Mobile XL – from partner Z

We perform data analysis to obtain the statistics of these phenomena, and the results are written in the following table.

Table 4. Actions on rate change and its impacts

Action	Impact
Increasing the rate	Incoming traffic volume will be deacreased
Decreasing the rate < 7.5% the last rate	The trend of incoming traffic volume will be preserved
Decreasing the rate ≥ 7.5% the last rate	With probability 0.83 the incoming traffic volume will be positively shock with a certain increment

The results bring us to an assumption we make in our mathematical model that with probability 0.83 business partner will deliver more traffic volume if we decrease the rate at least 7.5% from the last rate, and business partner will deliver less traffic volume if we increase the rate. But we cannot decrease the rate freely since the rate have to greater or equal to so called $critical_{rate}$ the rate which reflects the operational cost.

4. Modeling the expected revenue

The decision support system is created to control the company revenue. We can control the company revenue by controlling the rate. Our discussion in the previous section tells that we just can keep the last rate or decrease the rate at least 0.75% of the last rate. There are two possible behavior:

- **Behavior 1**. If we keep the last rate or if we decrease the rate 0.75% of the last rate, then in the future the traffic volume will continue to behave as before.
- **Behavior 2**. If we decrease the rate by 0.75% of the last rate then in a certain period of time (of average three months) the traffic volume will increase with a certain slope that depends on the business partner, after that the traffic volume will back to its behavior before the rate is decreased. We formulate these two types of future traffic volume behaviors in the following, where we model the traffic volume for six following months.

Let:

V_0 be the traffic volume at the last month,

t be the months, $1 \le t \le 6$,

d_1 be the average increasing slopes experienced by the historical traffic volume,

d_2 be a slope parameter of the last six months traffic volume model (which leads to **Behavior 1**)

$$v(t) = Ae^{d_2 t}, t = 1,2,\dots,6 \quad (1)$$

ER_1 be the expected revenue in the following six months if the last rate is preserved, where the traffic volume follows model (1),

ER_{opt} be the expected revenue in the following six months if the rate is decreased by 7.5% of the last rate, under the condition that the volume follows **Behavior 2**,

$$V(t) = \begin{cases} -\frac{d_1}{3}t^2 + 2d_1 t + V_0, & 1 \le t \le 3 \\ (V_0 + 3d_1)e^{-d_2(t-3)}, & 4 \le t \le 6. \end{cases} \quad (2)$$

ER_2 be be the expected revenue in the following six months if the rate is decreased by 7.5% of the last rate.

$rate_{end}$ be the rate in the last month.

Then,

$$ER_1 = rate_{end} * \sum_{t=1}^{6} Ae^{d_2 t} \quad (3)$$

$$ER_{opt} = (rate_{end} - 7.5\%\, rate_{end})\left(\sum_{i=1}^{3}\left(-\frac{d_1}{3}i^2\right) + 2d_1 i + V_0 + \sum_{i=1}^{3}(V_0 + 3d_1)e^{-d_2 i}\right) \quad (4)$$

$$ER_2 = 0.83ER_{opt} + 0.17ER_1 \quad (5)$$

From (1) we obtain the expected traffic volume in the month is $EV = V_0 e^{d_2}$. These models are used in **Simulation 1**, which is used to know the impact of decresing the rate to the expected revenue. This simulation will be used as a subroutine in **Algorithm 1** that will be discussed in the next section.

Simulation 1

Input

- the last trading scheme and its parameters (rate deal/volume deal/revenue deal)
- the traffic volume in the last six months

Steps

1. Evaluate A, d_2 by using model (1).
2. Evaluate er (the effective rate in the last month).
3. Evaluate ER_1, ER_{opt}, ER_2 from (1), (2) and (3).
4. If $ER_1 \le ER_2$, decreasing the rate, the new rate deal $= 0.925 * er$, the expected revenue $= ER_2$,

   else, preserve the rate, the expected revenue = ER_1

Output

decission on the new rate and the new trading scheme parameters

5. Modeling of optimal decision

The decision support system can be used by sales officers to monitor the performance, and if the last agreement is considered by the company or by the business partner to be a non-optimal agreement the decision support system can give some options to be chosen so that the next agreement is hoped to be an optimal one.

We assume that the performance of the last argument can be reviewed if it has been elapsed at least 80% of its duration, The performance indicator we use depends on the trading scheme of the last agreement, which we describe in the following table.

Table 5. The first performance indicators of the trading schemes

Trading scheme	The first performance indicator	The performance classes	The conditions
Flat Rate	the effecive rate	good	if the effective rate is higher than the critical rate
		bad	if the effective rate is less than or equal to the critical rate
Class Tiering	C=(the total first volume deal- the aggregate volume) / the total first volume deal	good	if $C > 0.5$
		bad	if $C \leq 0.5$
Revenue Capped	the effecive rate	good	if the effective rate is higher than the critical rate
		bad	if the effective rate is less than or equal to the critical rate
Volume Commitment	the expected aggregate volume at the end of the agreement period	good	if the expected aggregate volume exceeds the first volume deal
		bad	if the expected aggregate volume does not exceed the first volume deal

We also use another performance indicator, i.e. the trend of the traffic volume in last six months, which we denote by d_2.

The decision for the next agreement is obtained by following the algorithm below.

Algorithm 1

Input

- the last trading scheme and its parameters (rate deal/volume deal/revenue deal)
- the traffic volume in the last six months

Steps

1. Evaluate A, d_2 by using model (1).
2. Evaluate er (the effective rate in the last month).
3. Evaluate the first performance indicator.
4. If the first performance indicator is good, the last agreement can be preserved, go to , otherwise go to 5.
5. We have four cases:

Case 1 (if the last trading scheme is Flat Rate)	If $d_2 > 0$ preserve the last trading scheme parameters, else if $critical_{rate} < 0.925 * er$ then call **Simulation 1,** else convert to Volume Commitment trading scheme with: • the first volume deal $= EV$ • the first rate deal $= er$.
Case 2 (if the last trading scheme is Volume Commitment)	Try to preserve the last trading scheme parameters. If the business partner wants to change the parameters, the minimum first volume deal for a month is EV.
Case 3 (if the last trading scheme is Class Tiering)	If $d_2 > 0$ preserve the last trading scheme parameters, else if $critical_{rate} < 0.925 * er$ then call **Simulation 1,** else change the trading scheme parameters to: • the first volume deal $= EV$ • the first rate deal $= er$.
Case 4 (if the last trading scheme is Revenue Capped)	Convert to Volume Commitment trading scheme with the minimum first volume deal for a month $= EV$ and the first rate deal $= critical_{rate}$.

then go to 6.

6. end

Output

chosen trading scheme and its (new) parameters.

6. Modeling of optimal decision

In this section we give some examples of the implementation of the Algorithm 1. If the last agreement has come to be review, we give the trading scheme parameters for all trading schemes that can be chosen.

The first example is the agreement between the company and business partner P_1 to destination ROC[1], where the last trading scheme is Volume Commitment. From Figure 6 we can see that the first volume deal is too high and can impose a loss to the business partner due to volume shortfall at the end of the agreement period. If the business partner realize it, and want to change the agreement, our decision is to preserve the trading scheme but the trading scheme parameter can be changed as

[1] one of six destinations.

can be seen in Figure 7. If the optimal decision above is followed, then in the next six month the traffic volume will follow the combination of the red trend and the yellow trend shown in figure 8.

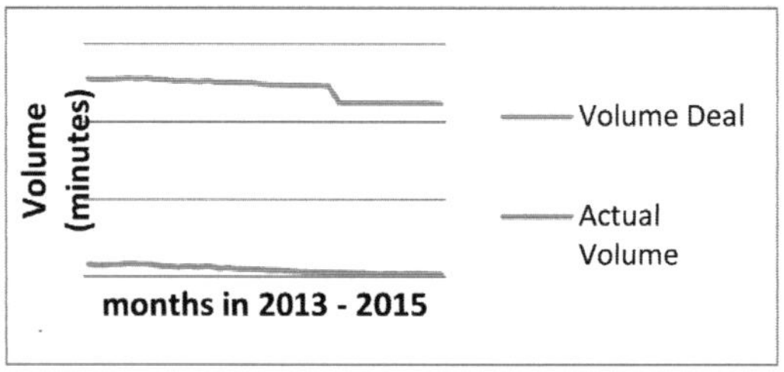

Figure 6. The chart of the first volume deal and the actual volume of the company and business partner P_1

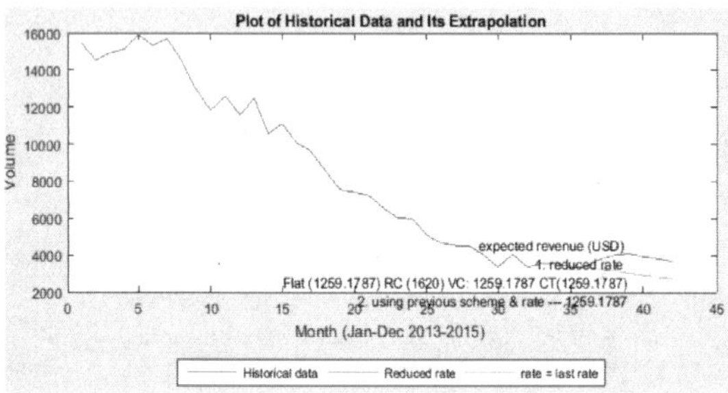

Figure 8. The expected revenue at the next agreement between the company and business partner P_1, if the optimal decision is followed.

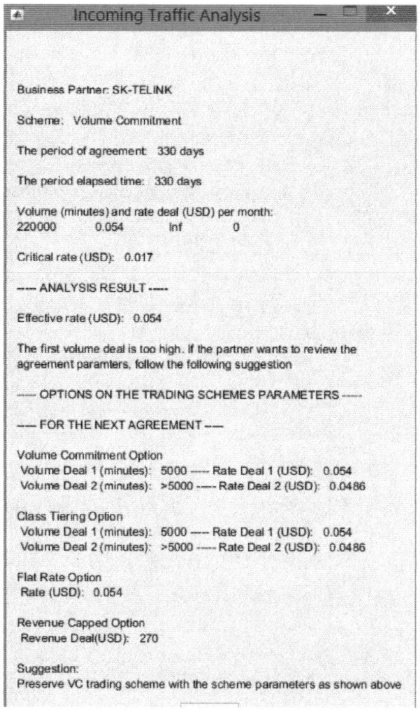

Figure 7. The optimal decision for the next agreement between the company and business partner P_1

The second example is the agreement between the company and business partner P_2 to the same destination and the same trading scheme as the first example. Unlike the first example, the first volume deal is not too high (see Figure 9) so that the business partner get benefit from the second rate deal application due to the actual volume which exceed the first volume deal. This reasonable volume deal in fact yields an increasing trend in the volume, and this condition is what we try to achieve by our optimal decission. The trading scheme parameters for the next agreement can be seen in Figure 10, and the expected volume and the expected revenue due to the optimal decision can be seen in Figure 11.

Actually, we still have many interesting examples of decisions for various trading schemes. But due to space limitation, we just present the two examples here.

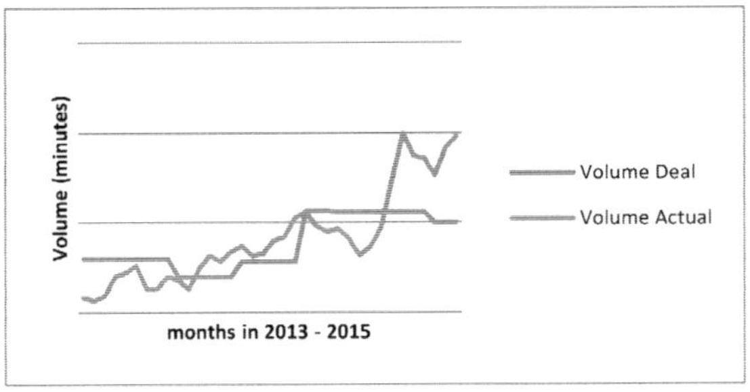

Figure 9. The chart of the first volume deal and the actual volume of the company and business partner

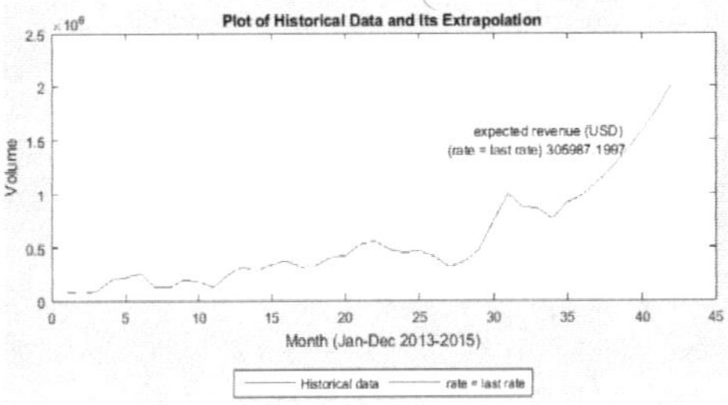

Figure 11. The expected revenue at the next agreement between the company and business partner P_2, if the optimal decision is followed.

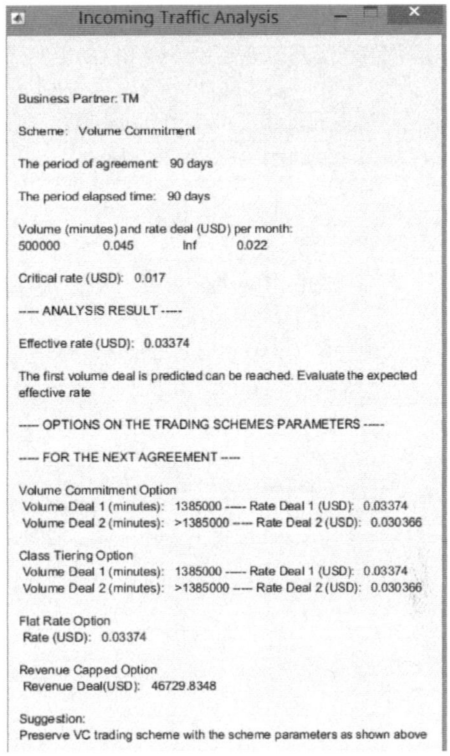

Figure 10. The optimal decision for the next agreement between the company and business partner P_2

7. Conclusion

In this paper we derive an algorithm for obtaining an optimal decision for the next agreement between the company and a business partner. The algorithm is derived through a mathematical modeling process which consists of trading scheme analysis, historical data analysis, modeling the future traffic volume and the expected future revenue. The decision we provide are the trading scheme that can be chosen and its trading scheme parameters. If the optimal decision is followed, we can calculate the expected future revenue. A software tool incorporating our methodology is currently in in the trial stage, where this software also has a feature where any user can simulate the future traffic volume and the expected future revenue if trading scheme and the trading scheme parameters chosen by the user are different from the optimal ones. Another feature that can be found in the software is the optimal decision for outgoing analysis, where the mathematical model for the derivation of this optimal decision can be found in other article we has been composing (see Hadianti and Uttunggadewa [2]).

Acknowledgment

The research is a conjunction with the collaborative research between PT Telkom International and PT LAPI ITB in year 2016. The authors thank to the research assistants of this research for providing data analysis and numerical computations.

References

[1]. Chan-Olmsted, Sylvia, and Mark A. Jamison. "Rivalry Through Alliances: Competitive Strategy In the Global Telecommunication Market." *European Management Journal*, 2001.

[2]. Hadianrti and Uttunggadewa, "The optimal routing in international-voice-services business", in preparation.

[3]. Meisner, Stefan M. "Global Telecommunications Competition a Reality: United States Complies with WTO Pact." *American University International Law Review 13, no.5*, 1998: 1345-1381.

[4]. Metty, Theresa, et al. "Reinventing the Supplier Negotiation Process at Motorola." *Interfaces Vol. 35 No. 1* (INFORMS), 2005: 7-23.

[5]. PT. Telkom International, personal communication.

[6]. Quinn, Phil, Bruce Andrews, and Henry Parsons. "Allocating Telecommunications Resources at L. L. Bean, Inc." *Interfaces Vol. 21* (INFORMS), 1991: 75-91.

[7]. Scott, Sheridan, Alberto Heimler, and Yasmin Carrim. *ICN Working Group on Telecommunications Services.* Cape Town, South Africa: International Competition Network, 2006.

IOP Conf. Series: Materials Science and Engineering **166** (2017) 012007 doi:10.1088/1757-899X/166/1/012007

The development of scientific identification theory to conduct operation research in education management

S. Hardhienata
Department of Computer Science and Doctoral Program of Education Management, Pakuan University, INDONESIA
E-mail: s-hardh@indo.net.id

Abstract. Operations research is a general method used in the study and optimization of a system through modeling of the system. In the field of education, especially in education management, operations research has not been widely used. This paper gives an exposition of ideas about how operations research can be used to conduct research and optimization in the field of education management by developing SITOREM (Scientific Identification Theory for Operation Research in Education Management). To clarify the intent of the idea, an example of applying SITOREM to enhance the professional commitment of lecturers associated with achieving the vision of university will be described.

1. Introduction

Operations research is the application of scientific methods to find optimal solutions and decision-making of a problem by taking into account the resources and the existing restrictions. In the analysis and solution of the problems mentioned above usually done by using modeling and optimization. In the field of education management there are a lot of analysis and problem solving is done by using statistical models. Statistical model is equations formed from a statistical sample to describe the relationship between variables that is affected (the dependent variable) with the variables that influence (independent variables). Most studies that use statistical models in the field of education management, the discussion is stopped on the finding that there is a relationship or a positive influence between independent variable with or towards the dependent variable. This paper provides a description of how the research findings are followed up by finding optimal solutions so that research results can give a more precise answer and more concrete benefits to the issues raised.

2. Scientific Identification Theory

The scientific identification theory is a theory that describes the steps in systems analysis, modeling, and simulation. Starting from the real system which is to be studied, an observation or experimentation is conducted to obtain data from the real system. Based on the data from the real system, an analysis of the system is performed to built an abstract model. The abstract model is a description of the real system in our minds. From the abstract model, we create a mathematical model or a simulation model which is depends on the situation and steps that we choose to resolve the model. If we choose the mathematical model then we resolve the model through a mathematical deduction, while for the simulation model the solution is typically done by using a computer. The data that obtained from the mathematical deduction or from the computer simulation is then compared with the data from real system to see how far the model corresponds to the real system that will be studied (validation). The steps of the scientific identification theory can be visualized using a diagram as below:

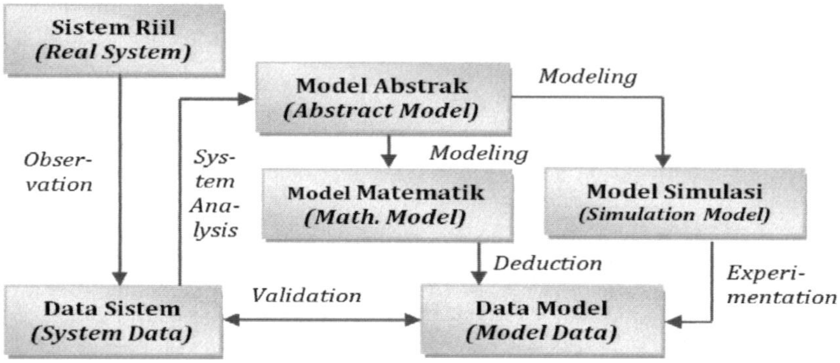

Figure 1. The scientific identification theory

3. The Development of Scientific Identification Theory for Operation Research in Education Management (SITOREM)

For the purpose of operations research in education management, we need to add the scientific identification theory mentioned above with statistical model and steps to obtain an optimal solution. That mean, from the abstract model can be produced mathematical model, simulation model, and/or statistical model, and based on the models we perform an optimization process using appropriate method, namely, mathematical optimization method, optimization through algorithms that implemented on a computer, or other optimization methods. The diagram below depicts the steps of scientific identification theory which has been developed for operation research in education management, which we named as SITOREM.

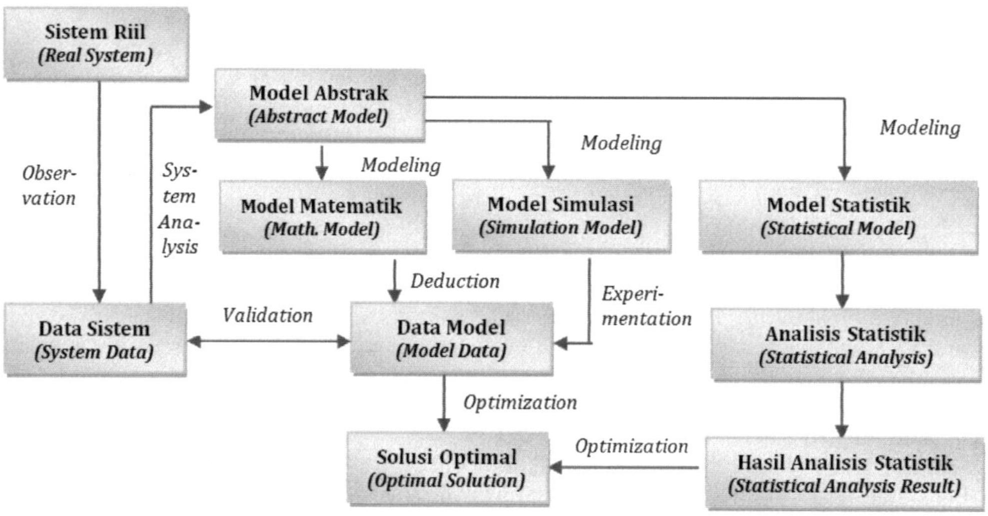

Figure 2. The steps of operation reseach in education management (SITOREM)

4. The Application of SITOREM

In this chapter, a description regarding the idea and the application of SITOREM will be presented. As an illustration, the explanation will be taken as part of the research results from Setyaningsih 2016 [4], namely the research on the influence of empowerment and academic culture towards professional commitment of lecturers.

In the study of Setyaningsih [4] was synthesized that Professional Commitment is loyalty, engagement, and integrity of a lecturer in performing the tasks on the basis of skills, expertise, and responsibility, with indicators: Sooth, Consistency, Improvement Desire, Skills Development, and Dedication and Devotion. Empowerment is the awarding authority and responsibility of the leader to employees to carry out tasks in a creative and innovative ways, in response to various changes dynamically as the capabilities of employees, with indicators: Discretion, Authority, Creativity and Innovation, and Open Communication. While the academic culture is a basic assumption to understand the beliefs, responsibilities, norms, values, artifacts and spirit of academic, embraced by all campus residents and used as a basis in carrying out tasks and solving problems with indicators of Applied Values, Academic Spirit, Responsibility, Tradition Assessment, Support and Cooperation. The constellation of research and statistical models as shown in Figure 3.

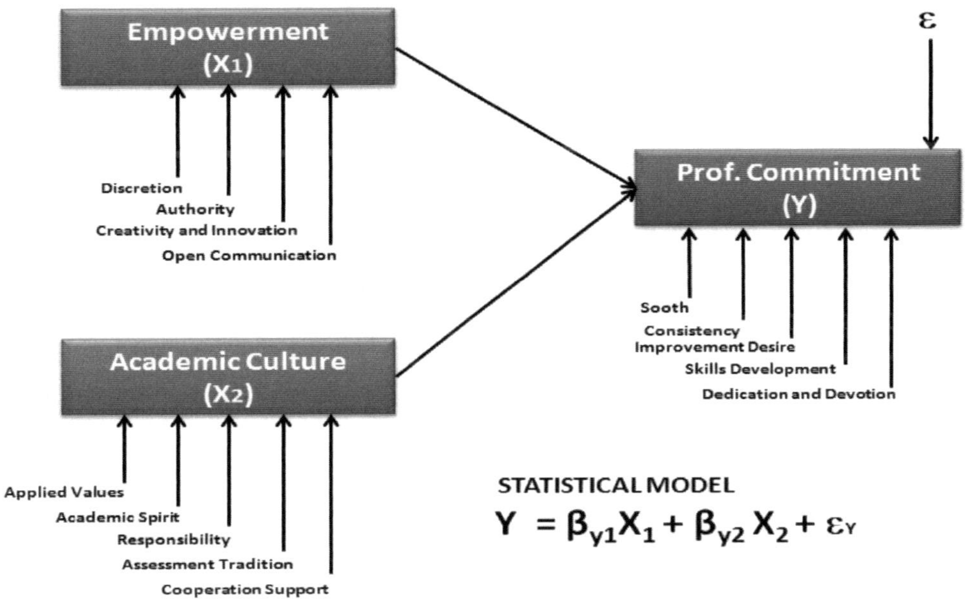

Figure 3. Research Constellation and Statistical Model of [4]

The objective of that research was to investigate whether there exists any direct positive effect of empowerment and academic culture towards lecturer professional commitment at the University of Pakuan, Bogor, Indonesia. A total of 135 from 337 university lecturers were sampled to collect data which was then analyzed using path analysis quantitative method.

The result shows that there exists a positive direct effect of empowerment towards professional commitment and academic culture towards professional commitment with path coeffisient $\rho_{y1} = 0{,}34$ dan $\rho_{y2} = 0{,}17$ respectively. This means that increasing the quality of empowerment can enhance the professional commitment and conducive academic culture can enhance the professional commitment.

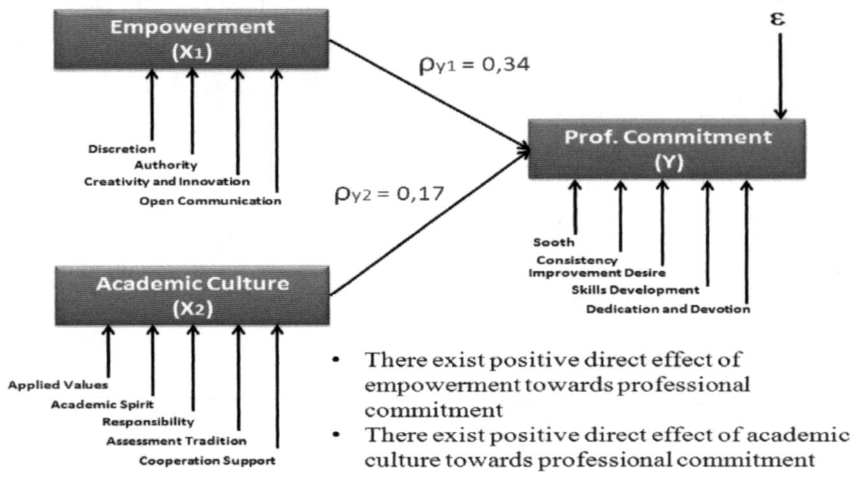

Figure 4. Results of the research of Setyaningsih [4]

The implication is that to improve the professional commitment of lecturers is necessary to increase the quality of empowerment and improvement of academic culture.

Furthermore, we perform optimization by giving weighting to each indicator as shown in Figure 5 below.

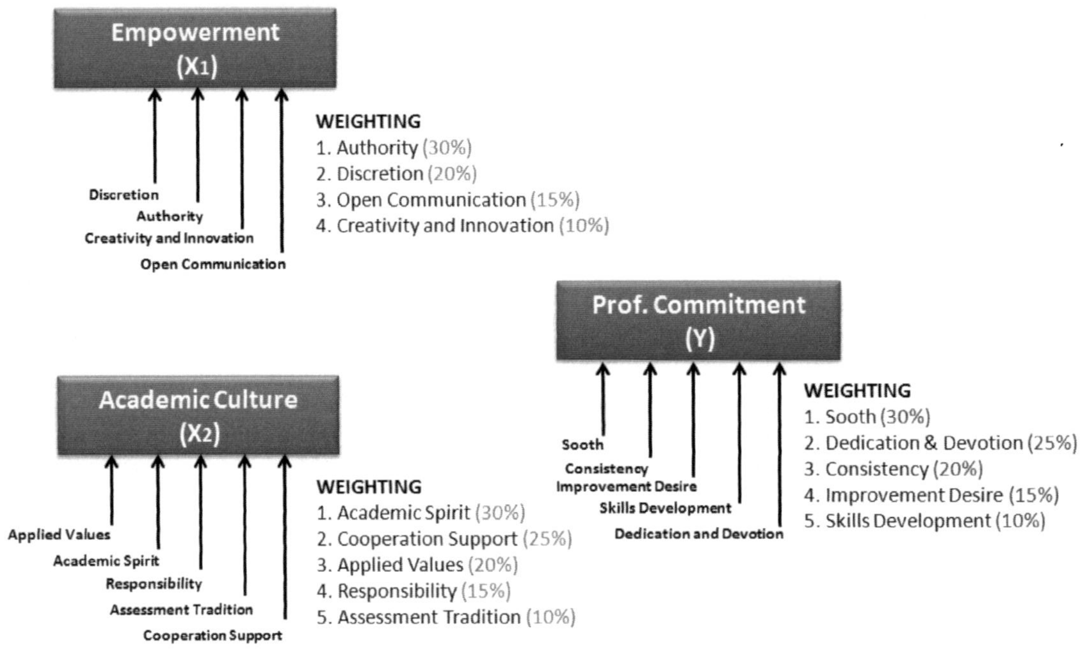

Figure 5. Weighting factors / indicators of each variable

Then we do the combination between the results weighted with the calculation of the indicators obtained from the studies in the field as shown in Figure 6 below.

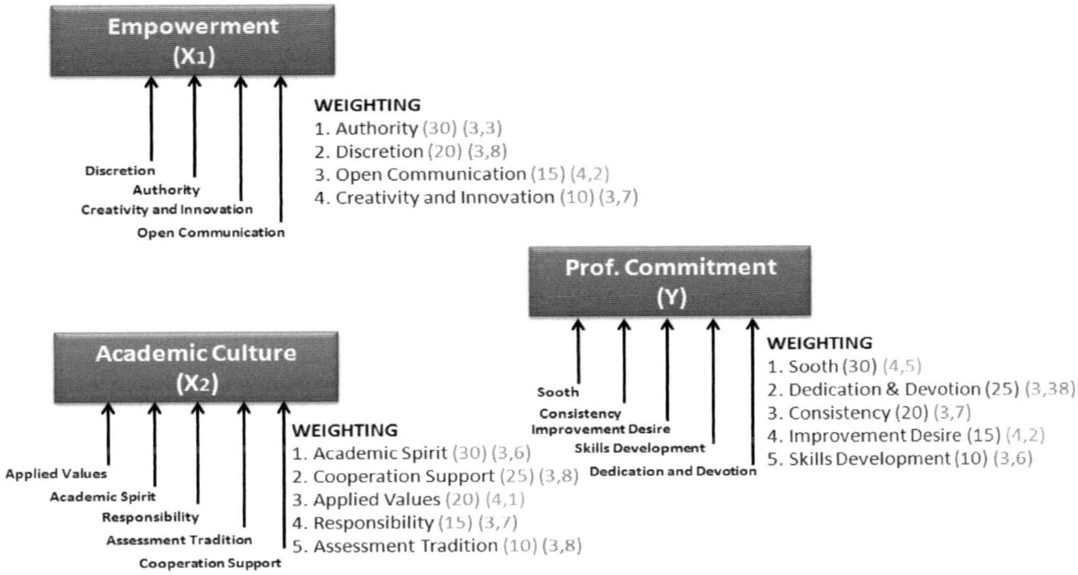

Figure 6. The results of the weighting and the results of the calculation of the factor obtained from the research in the field

From the results shown in the Figure 6 then we make a priority sequence of actions to improve or repair the state of the variables. The results are as shown in Figure 7 below.

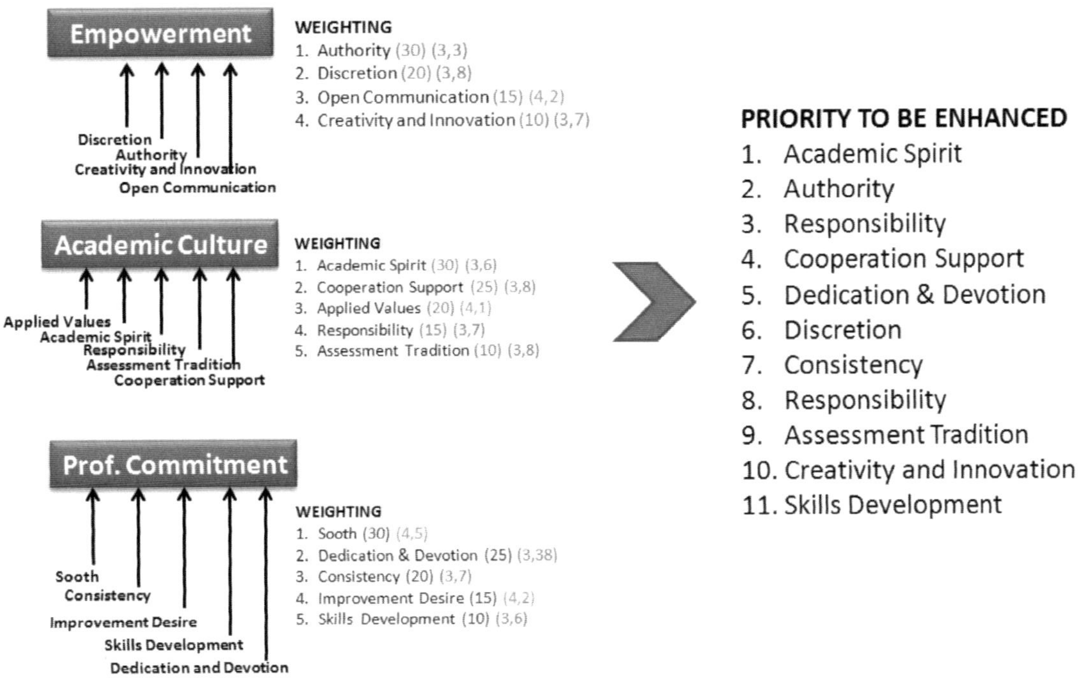

Figure 7. The end result from the optimization towards the factors of the variables

5. Discussion

Based on the description of the steps in Chapter 4, it appears that the results of the study, during which only stops on proof of the positive influence between independent variables and the dependent variable, can be followed up by performing the optimization process.

It gives results that the factors of the studied variables can be sorted out, which is good and which is not a good situation, and what kind of priority improvements can be done so that it can give a positive impact quickly.

With the resulting optimal solution, then the questions raised at the beginning of the study can be solved effectively and efficiently.

6. Conclusions

From the description that has been given in previous chapters can then be concluded as follows:

- SITOREM scientific recognition theory has been developed and can be used to conduct operations research in education management.
- In SITOREM, to obtain the optimal solution of the issues raised, we do a follow-up to give weight to any factors / indicators of the variables. We also consider the value of each of these factors based on the results of the research in the field.
- With the optimal solutions and priorities of the improvements to be made to the factors that the situation is not good, then it can provide concrete and efficient answers to the issues raised at the beginning of the study.

References

[1]. Bell, G., Warwick, J., Galbraith, P., “Higher Education Manajement and Operational Research”, Sense Publischers, Rotterdam/Boston/Taipei, 2012.

[2]. Cheng, T. C. E., “Operation Research and Higher Education Administration”, Journal of Educational Administration, Vol. 31, ISSN: 0957-8234, 1993.

[3]. Hardhienata, S. “Analisis Sistem, Pemodelan, dan Simulasi, e-Book for e-Learning”, Program Studi Ilmu Komputer, Universitas Pakuan, 2015.

[4]. Setyaningsih, S. “Pengaruh Pemberdayaan, Budaya Akademik, Kompetensi Pedagogik, dan Trust terhadap Komitmen Profesi Dosen di Universitas Pakuan ”, Desertasi, Universitas Negeri Jakarta, 2016.

[5]. Smith, M. J. Statistical Analysis Handbook – A comprehensive handbook of statistical concepts, techniques, and software tools, The Winchelsea Press, Winchelsea, UK, 2014.

Optimization of investment portfolio weight of stocks affected by market index

E. Azizah[1], E. Rusyaman[2] and S. Supian[3]

[1,2,3]Department of Mathematics, Padjadjaran Univesity, Indonesia.

E-mail: azizahelfa@gmail.com

Abstract. Stock price assessment, selection of optimum combination, and measure the risk of a portfolio investment is one important issue for investors. In this paper single index model used for the assessment of the stock price, and formulation optimization model developed using Lagrange multiplier technique to determine the proportion of assets to be invested. The level of risk is estimated by using variance. These models are used to analyse the stock price data Lippo Bank and Bumi Putera.

1. Introduction

In the face of risky investment opportunities, investment options cannot rely solely on the expected profit rate. Investors should be willing to bear the high risk if it expects to earn high rates of return [3]. Therefore, the stock price valuation analysis, the selection of a combination of stocks in a portfolio optimum, and measure the risk of investing in the capital market is very necessary. This is where the mathematical model has a very important role to analyze the formation of the portfolio [7].

The mathematical model, particularly single index model seeks to facilitate the analysis of investment. This is because in a single index model assumed that the correlations of returns between stocks occur due to the stock response to changes in the Composite Stock Price Index (general market index).When the market improves (indicated by the market index), the price of individual stocks also increased, and vice versa [4]. Single index model is used for valuation of stocks in an investment.

According to Dowd [2], in conditions of risky investments, a strategy that can be done to reduce the magnitude of the risk of investment is to build a portfolio. To establish the optimum portfolio can be done one of them with the help of optimization using the techniques of Lagrange Multiplier. Meanwhile, to measure the magnitude of risk can be used to estimate the variance (*Var*). Keep in mind, that the *Var* has been used widely and become a common standard for the calculation of risk in financial investments [5].

In this paper aims to analyze the stock price assessment with a single index model, determine the weight of the combined portfolio with Lagrange multiplier technique, and estimate the magnitude of the risk by using the size of the *Var* in portfolios that were formed.

2. Research methodology

In line with the stage of stock price valuation analysis, portfolio weight optimization, and estimation of the amount of investment risk, thus the first to do is to calculate stock returns and market returns as shown below.

2.1. *Calculating the Stock Return and Market Return*

To determine the expected return and variance of stock is required data of stock returns and market returns of each study period. According to Gruber et al. [4], stock returns for each period can be calculated using the following formula:

$$R_{it} = \ln\left(\frac{P_{it+1}}{P_{it}}\right), \tag{1}$$

where R_{it} is the stock returns i in the period t, and P_{it} is the stock price i in the period t. The market index is represented by the composed stock price index (CSPI). In the same way the market return is calculated by using the following formula:

$$R_{mt} = \ln\left(\frac{IHSG_{t+1}}{IHSG_t}\right), \tag{2}$$

where R_{mt} is the market return in the period t, and $IHSG_t$ is $IHSG$ in the period t. Then, the return data will be used for the establishment of a single index model below.

2.2. *Establish a Single Index Model*

Based Gruber et al. [4], the main characteristic of the single index model is the model is acceptable if and only if the underlying assumptions are met. Single index model has some characteristics as follows:

1. the basic equation $R_i = a_i + \beta_i R_m$ and $a_i = \alpha_i + e_i$, so that the single index model form:
   $$R_i = \alpha_i + \beta_i R_m + e_i, \text{ with } i = 1,2,...,k, \tag{3}$$
   where k is the number of shares that were analysed.
2. Based on the formation of the equation, the average e_i is zero, or $E(e_i) = 0$
3. Based on assumptions:
   a. Index unrelated (uncorrelated) with unique return.
      $E[e_i(R_m - \overline{R}_m)] = 0$
   b. Stocks correlated simply because it provides a response to the market.
      $E[e_i.e_j] = 0$, for $i \neq j$
4. Based on the definition
   a. variance e_i is $E[e_i^2] = \sigma_{ei}^2$
   b. variance R_m is $E[(R_m - \overline{R}_m)^2] = \sigma_m^2$

Where a_i is a unique stock return i, α_i is the expectations of unique return, β_i is a measure of sensitivity stock i to market, and e_i *residual error* from *unique return*. The above return data used to estimate the α_i and β_i.

2.3. *Estimating α_i and β_i*

In the assessment of stocks, the market risk that is worth considering is the systematic risk. Therefore, investors need to estimate the amount of beta as a measure of market risk. To estimate the beta can be done with the least squares method. From equation (3) can be obtained by the number of squares residual error [6]:

$$\varepsilon_i = \sum_{t=1}^{n} e_{it}^2 = \sum_{t=1}^{n} (R_i - \alpha_i - \beta_i R_{mt})^2 \tag{4}$$

Derivate equation (4) respects to parameter α_i and β_i will be obtained system of equations as follows:

$$\frac{\partial \varepsilon_i}{\partial \alpha_i} = 0 \text{ and } \frac{\partial \varepsilon_i}{\partial \beta_i} = 0. \tag{5}$$

By completing the system of equations (5) for the parameters β_i will be obtained:

$$\hat{\beta}_i = \frac{n\sum_{t=1}^{n} R_{mt}R_{it} - \sum_{t=1}^{n} R_{mt}\sum_{t=1}^{n} R_{it}}{n\sum_{t=1}^{n} R_{mt}^2 - (\sum_{t=1}^{n} R_{mt})^2} \tag{6}$$

While for parameter α_i can be calculated by using equation:

$$\hat{\alpha}_i = \overline{R}_i - \hat{\beta}_i \overline{R}_m \tag{7}$$

The next step is to determine the expected return and variance stock as follows.

2.4. Estimating Expected Return and Stock Variance

Based on the characteristics of single index model, expectations of individual stock returns are:

$$E[R_i] = E[\alpha_i + \beta_i R_m + e_i]$$
$$= E[\alpha_i] + E[\beta_i R_m] + E[e_i]$$
$$\overline{R}_i = \alpha_i + \beta_i \overline{R}_m \tag{8}$$

Equation (2.8) shows that the expectations of individual stock returns consisting of a unique return (not affected the market) and the level of benefits associated with the market (market related return).

Variance individual stock returns using a single index can be calculated by the equation [6]:

$$E[(R_i - \overline{R}_i)^2] = \beta_i^2 E[(R_m - \overline{R}_m)^2] + 2\beta_i E[e_i(R_m - \overline{R}_m)] + E[e_i^2] \tag{9}$$

Because of the characteristics based on Index unrelated (uncorrelated) with unique return and the definition of variance, then the equation (9) becomes:

$$\sigma_i^2 = \beta_i^2 \sigma_m^2 + \sigma_{ei}^2 \tag{10}$$

Equation (10) shows that the variance of individual stock returns is composed of two parts, namely the unique risk (σ_{ei}^2) and risks associated with the market $(\beta_i^2 \sigma_m^2)$. Furthermore, it also estimated covariance and correlation between stocks [6], [1].

2.5. Estimating Covariance and Correlation Between Stocks

To determine the amount of variance between the stock model single indexes can be done as follows:

$$Cov(R_i, R_j) = E[(R_i - \overline{R}_i)(R_j - \overline{R}_j)]$$

$$= \beta_i \beta_j E[(R_m - \overline{R}_m)] + \beta_j E[e_i (R_m + \overline{R}_m)]$$
$$+ \beta_i E[e_j (R_m - \overline{R}_m)] + E[e_i . e_j] \quad (11)$$

Based on the characteristics (3a) and (3b), equation (11) is written as:

$$Cov(R_i, R_j) = \beta_i \beta_j \sigma_m^2 \quad (12)$$

From equation (12) it appears that the covariance between the stocks is affected by market risk (this is consistent with the assumption). This is what is known as the single index model. Furthermore, based on the equation (12), the correlation coefficient between stocks, can be expressed as [6], [1]:

$$Cov(R_i, R_j) = \rho_{ij} \sigma_i \sigma_j = \beta_i \beta_j \sigma_m^2$$

or

$$\hat{\rho}_{ij} = \frac{\beta_i \beta_j \sigma_m^2}{\sigma_i \sigma_j} \quad (13)$$

After each individually assessed and selected, then determine the model of expectation and variance of portfolio return follows.

2.6. *Forming Expectations Return and Variance Portfolio Model*

Suppose w_i represents the proportion (weight) of assets invested in stocks i in the portfolio formation, thus the portfolio return is:

Based on the equation (13) expected portfolio return is obtained as follows [8]:

$$E[R_p] = E[\sum_{i=1}^{k} w_i R_i] = E[\sum_{i=1}^{k} w_i (\alpha_i + \beta_i R_m)]$$

Or

$$\overline{R}_p = \sum_{i=1}^{k} w_i \alpha_i + \overline{R}_m \sum_{i=1}^{k} w_i \beta_i = \alpha_p + \beta_p \overline{R}_m \quad (14)$$

With $\alpha_p = \sum_{i=1}^{k} w_i \alpha_i$ and $\beta_p = \sum_{i=1}^{k} w_i \beta_i$

While, the variance of portfolio return can be determined as follows [8], [3]:

$$\sigma_p^2 = E[(R_p - \overline{R}_p)^2]$$
$$= \sum_{i=1}^{k} w_i^2 \sigma_i^2 + \sum_{i=1}^{k} \sum_{j=1}^{k} w_i w_j Cov(R_i, R_j); i \neq j$$
$$\sigma_p^2 = \beta_p^2 \sigma_m^2 + \sum_{i=1}^{k} w_i^2 \sigma_{ei}^2 \quad (15)$$

To get the proportions right assets, portfolio optimization carried out as follows.

2.7. *Doing weight optimization of combination portfolio*

Suppose an investor has an aversion factor (risk aversion) θ, it can be estimated by comparison of the two portfolios. If an investor with risk aversion θ, with two portfolios that have average rate of return $\overline{R}_p$ and $\tilde{R}_p$, along with the variance σ_p^2 and $\tilde{\sigma}_p^2$, the comparison between the two is by equating portfolio [7]:

$$\bar{R}_p - \theta\sigma_p^2 = \tilde{R}_p^2 - \theta\tilde{\sigma}_p^2 \quad (16)$$

Resolving equation (16) for θ will be obtained:

$$\hat{\theta} = \frac{\bar{R}_p - \tilde{R}_p}{\sigma_p^2 - \tilde{\sigma}_p^2} \quad (17)$$

Optimization aims to maximize $\bar{R}_p - \theta\sigma_p^2$, so the portfolio optimization problems can be formulated as follows:

The objective function:

$$f(w_1, w_2, ..., w_k) = (\sum_{i=1}^{k} w_i\alpha_i + \bar{R}_m \sum_{i=1}^{k} w_i\beta_i) - \theta(\beta_p^2\sigma_m^2 + \sum_{i=1}^{k} w_i^2\sigma_{ei}^2) \quad (18)$$

Constraint functions:

$$g(w_1, w_2, ..., w_k) = 1 - \sum_{i=1}^{k} w_i = 0 \quad (19)$$

From equation (18) and (19) subsequently formed as a summation function Lagrange, objective function and constraint functions, the Lagrange multiplier factor as follows:

$$L(w_1, w_2, ..., w_k; \lambda) = f(w_1, w_2, ..., w_k) + \lambda g(w_1, w_2, ..., w_k) \quad (20)$$

Derivate partially respect to $w_1, w_2, ..., w_k$ and λ, it will be obtained [7]:

$$\partial L / \partial w_1 = 0 \Rightarrow \lambda = \alpha_1 + (\bar{R}_m - \theta\sigma_m^2)\beta_1 + 2w_1\sigma_{e1}^2 \quad (21)$$

$$\partial L / \partial w_2 = 0 \Rightarrow \lambda = \alpha_2 + (\bar{R}_m - \theta\sigma_m^2)\beta_2 + 2w_2\sigma_{e2}^2 \quad (22)$$

And so on up:

$$\partial L / \partial w_k = 0 \Rightarrow \lambda = \alpha_k + (\bar{R}_m - \theta\sigma_m^2)\beta_k + 2w_k\sigma_{ek}^2 \quad (23)$$

$$\partial L / \partial \lambda = 0 \Rightarrow 1 - \sum_{i=1}^{k} w_i = 0 \quad (24)$$

Resolving the equation system, will be obtained $w_1, w_2, ..., w_k$. Match the equation (23) by (21) and the finish will be obtained:

$$w_k = \frac{(\alpha_k - \alpha_1) + (\beta_k - \beta_1)(\bar{R}_m - \theta\sigma_m^2) + 2w_1\sigma_{e1}^2}{2e_{e1}^2} \quad (25)$$

And so on until the equalization equation (22) by (21) and the finish is obtained:

$$w_2 = \frac{(\alpha_2 - \alpha_1) + (\beta_2 - \beta_1)(\bar{R}_m - \theta\sigma_m^2) + 2w_1\sigma_{e1}^2}{2e_{e1}^2} \quad (26)$$

and

$$w_1 = 1 - \sum_{i=2}^{k} w_i \quad (27)$$

The values of this proportion $w_1, w_2, ..., w_k$ is showing the composition of asset allocation (funding) will be invested in each stock in the portfolio formation, so that the rate of return on the maximum investment portfolio [8]. Finally, the estimated variance will do the following.

3. Case Analysis

The data is analysed daily stock price of Lippo Bank and Bumi Putera and Stock Price Index (CSPI) performance of 129 working days in 2015. Then the data is each determined the value of log returns using equation (2.1) and (2.2). Based on descriptive statistics, each return log data has a mean and variance as given in Table 1 below.

Table 1 Mean and Variance of Log Return Data

Log Return Data	*Mean*	Variance
Log Return Lippo	0,000718	0,00010588
Log Return Bumi Putera	0,001261	0,01575025
Log Return IHSG	0,017440	0,05112120

Regression between log return Lippo stock and ISHG obtained equation LIPPO= 0,000664 + 0.00310 IHSG, while regression between log returns of Bumi Putera obtained equation BUMPU = 0.0019 + 0.0378 IHSG. Squared Residual of regression LIPPO is $\sigma_{eL}^2 = 0{,}0001063$ and squared residuals of the regression Bumi Putera is $\sigma_{eB}^2 = 0{,}0158000$. Often, to determine the weight of the investment portfolio is determined based on the comparison of the value of stock returns last log data, where the value of last log data from Lippo stock return is 0.0057907 and last log data value of stock returns from Bumi Putera is 0.028029; thereby weighting the portfolio's stake in Lippo is 0.17 and the portfolio weights for shares in Bumi Putera is 0.83. Using these values variance portfolio weights are calculated using the formula (2.16), the result is $\sigma_p^2 = 0.104$.

It will be different results if used Lagrange Multiplier optimization approach. If the equation (2.18) was approached by several levels challenger risk (risk taker): $0 < a \leq 1$ and several levels of risk-averse: $-1 \leq a < 0$; then weighting the portfolio for investors for the risk taker and risk aversion investor can be shown in Figure 1 below.

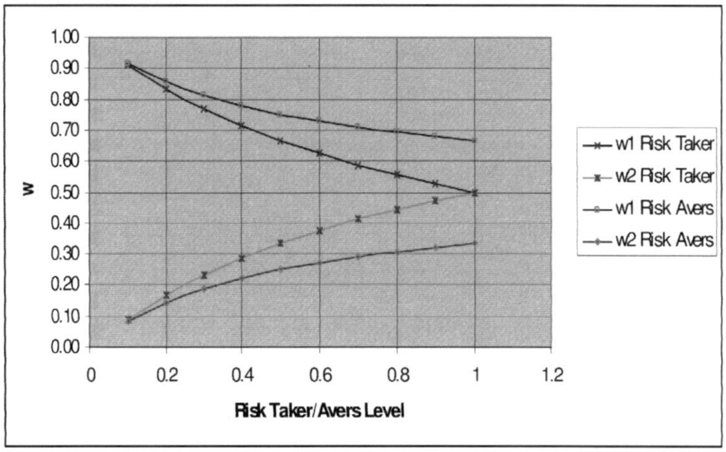

Figure 1. Weight portfolio for investors risk taker and risk averse

It appears that for investors who have a level of risk taker tends to halve its capital allocation with equal weights, which are 0.5 to stock in Lippo and 0.5 for stock in Bumi Putera. As for investors who have a level of risk aversion one would halve its capital allocation with a weight of 0.67 for the Lippo stock and 0.33 for the Bumi Putera stock.

4. Conclusions

Based on results and discussions above can finally be concluded and suggestions for other researchers who are interested in further researching this study. The result of the weight optimization portfolio for investors risk taker with a level of risk taker/avers high tends to divide two equal allocations of investment funds in stocks in its portfolio. As for investors with a level of risk taker / avers higher proportionally while maintaining different weighting the allocation of funds to the investment shares in its portfolio.

Acknowledgment

We thank the program of the Academic Leadership Grant (ALG), Faculty of Mathematics and Natural Sciences, Universitas Padjadjaran (Indonesia), which has provided facilities to conduct a research and publication.

References

[1] Alexader, C. (Editor). (1999). *Risk Management and Analysis*. Volume 1: Measuring and Modelling Financial Risk. New York : John Wiley & Sons Inc.
[2] Dowd, K. (2002). *An Introduction to Market Risk Measurement*. United State American: John Wiley & Sons Inc.
[3] Bodie *et al.* (1999). *Investment*. Fourth Edition. Singapore : Irwin / McGraw-Hill.
[4] Gruber *et al.* (1991). *Modern Portfolio Theory and Investment Analysis*. Singapore: John Wiley & Sons, Inc.
[5] Jorion, P. (2004). Bank Trading Risk and Systemic Risk. Third draft: December 2004.
[6] Ruppert, David. 2004. *Statistics and Finanace an Introduction*. New York: Springer.
[7] Sukono. (2002). Model Matematika Pembentukan Portofolio Investasi Optimum Dengan Menggunakan Indeks Tunggal. Jurnal Matematika Integratif. Vol. 1, No. 1, April 2002.
[8] Tadelilin, E. (2001). *Analisis Investasi*. Yogyakarta: BPFE.

Simulation of product distribution at PT Anugrah Citra Boga by using capacitated vehicle routing problem method

T. Lamdjaya[1] **and E. Jobiliong**[2]

[1,2]Department of Industrial Engineering, Universitas Pelita Harapan, Tangerang, Indonesia.

E-mails: timotius.lamdjaya@gmail.com, eric.jobiliong@uph.edu

Abstract. PT Anugrah Citra Boga is a food processing industry that produces meatballs as their main product. The distribution system of the products must be considered, because it needs to be more efficient in order to reduce the shipment cost. The purpose of this research is to optimize the distribution time by simulating the distribution channels with capacitated vehicle routing problem method. Firstly, the distribution route is observed in order to calculate the average speed, time capacity and shipping costs. Then build the model using AIMMS software. A few things that are required to simulate the model are customer locations, distances, and the process time. Finally, compare the total distribution cost obtained by the simulation and the historical data. It concludes that the company can reduce the shipping cost around 4.1% or Rp 529,800 per month. By using this model, the utilization rate can be more optimal. The current value for the first vehicle is 104.6% and after the simulation it becomes 88.6%. Meanwhile, the utilization rate of the second vehicle is increase from 59.8% to 74.1%. The simulation model is able to produce the optimal shipping route with time restriction, vehicle capacity, and amount of vehicle.

1. Introduction

Transportation of goods is an important task in running a business of today. The amounts of daily expenses on fuel, equipment, maintenance of equipment and wages are noteworthy to be concerned. It is therefore obvious to attempt to reduce the amount of money spent on transportation as even small improvements can lead to huge improvements in absolute terms. Several approaches are taken by the company to minimize the cost, one could improve the infrastructure and other could look at the operation research techniques. Toth and Vigo claim that the use of computerize planning of the distribution process often leads to savings in the area of transportation costs [1]. Operation research has been quite successful in the transportation area [2]. OR techniques are applied within the airline, trucking, shipping industries and other optimization matter. Consequently, the optimization within transportation is a subject that is used and sought-after in the real industry and not just a topic studied in academia. The real industry need solution methods that are fast, easy to apply to variety of problem characteristics, precise, and more robust.

One problem in the field of transportation related OR that has been given a lot of attention in the scientific literature is the method of vehicle routing problem (VRP). The VRP is a widely studied combinatorial optimization problem that was introduced in 1959 by G B Dantzig and J H Ramser [3] The VRP is defined as the problem of serving a number of customers with a fleet of vehicles, minimizing the cost of distributing the goods. The cost of travelling between each pair of customers and the depot is given. The task is to find a route for each vehicle that have the less total cost. Typically, the solution has to obey several other restrictions, such as capacity of the vehicles or desired visit times at customers. In which a quantity d_i of a single commodity is to be delivered to each customer $i \in N = \{1, \cdots, n\}$ from a central depot {0} using k independent delivery vehicles of identical capacity C. Delivery is to be accomplished at minimum cost, with $C_{ij} \geq 0$ denoting the transit cost from i to j, for $0 \leq i, j \leq n$. The matrix cost and distance is assumed symmetric, i.e., $C_{ij} = C_{ji}$ and $C_{ii} = 0$. Therefore, the solution for this problem consists of a partition of N into k routes $\{R_1, ..., R_k\}$, each satisfying $\sum_{j \in R_i} d_j \leq C$, and a corresponding permutation of each route specifying the service ordering. This problem is naturally associated with the complete undirected graph consisting of nodes $N \cup \{0\}$, edges E, and edge-traversal costs $C_{ij}, \{i, j\} \in E$ [4]. By associating a binary variable with each edge in the graph, we obtain the following integer programming formula. In this thesis the term of vehicle routing problem is used for the method of Anugrah Citra Boga company to determine the route for each vehicle.

2. Research Method

The research within an area like vehicle routing problem start with visiting Anugrah Citra Boga Company for observation and interview to supervisor and distribution manager. After doing a quick glance of the company activity, the next step is identifying problem of Anugrah Citra Boga. The main problem of the company is lack of efficiency in distribution routing that cause a high shipping cost. Afterward start with studying some literature that connected to optimization and distribution topic.
The purpose of this thesis is creating a simulation model for the distribution of goods at Anugrah Citra Boga company for the area of Jakarta, Depok, Bekasi, and Cikarang. Collecting data is needed to calculate the distribution cost of each vehicle.

Distribution route per day, service time in each customer, distance and, shipping costs are collected as data. Then build the model using AIMMS software with the capacitated vehicle routing problem method, based on figure 2.1 the transportation model start from a depot which is the factory and then distribute the goods to each customer by using vehicles as their transporter. Each vehicle has their own schedule route according to the optimal time in every customer. After obtaining the simulation result of each day, compare the shipping time of simulation model and historical data. Analyse the result of comparison and create a sample of graph between simulation model and historical data. So it can conclude whether the simulation model work well or not.

In building a model a vehicle scheduling is needed to optimize the distribution and minimize the cost. Anugrah Citra Boga use their own transportation system to distribute their goods to the customer, the company should deliver the goods from their home base to several locations in different direction. This problem could add some problem for the distribution system start with amount of vehicle that need for the shipping, vehicle capacity and the best routing for optimize the shipping cost. Although there are problems that may occur in the distribution system, vehicle scheduling could help on fixing this problem. Some rule that help on making a scheduling is start with appoint the farthest point from the depot. Next step is find a location that have the fastest time service (travel time + service time). Then continue finding the best location for the optimal distribution route. Use this method for several vehicles to find out the best distribution route.

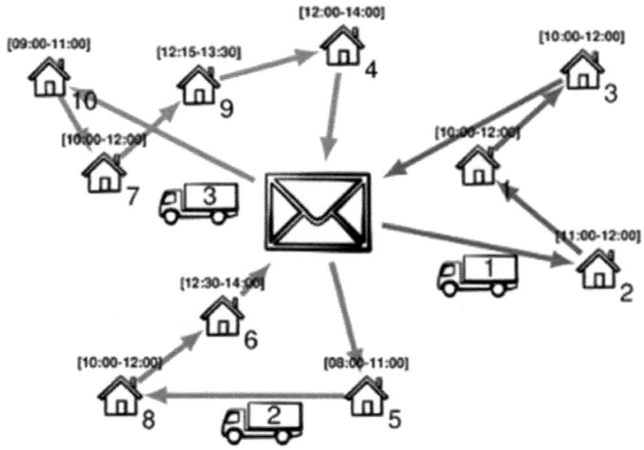

Figure 2.1 VRP scheduling

3. Modelling and Solution Methods

The art and science of modelling can be divided into two disciplines. The first one is concerned with modelling a problem occurring in real life. The point that should be considered are the description of the real life problem given to the modeller may be ambiguous, so a good model should be expressed such that there are no ambiguities (the model should be written in mathematical notation, but textual can be sufficient as well). Next, the model must represent the real life problem reasonably well and last the model should not be unnecessarily complicated.

The second discipline in modelling is transforming a model into an equivalent model that in some way is easier to solve using existing techniques. The new model should have the same solution as the original model given the same input. **[5]**

The set of feasible solutions is so large that require computer to calculate the model simulation and evaluate the cost of a trillion solutions in a blink of an eye. There are several type of solution methods that are typically employed to solve these types of problem heuristics, approximation algorithms, and exact methods. However, in this thesis the model simulation use heuristics method as their type of solution methods.

- **Heuristics**. A solution that quite fast on finding a feasible solution with reasonable quality. The heuristics method is tested empirically and based on the experiment of data history, this are the best solution on solving real life problem because of the speed and the ability to handle large amount of data in instances. There is a special class of heuristics which is metaheuristics that could provide a general framework for heuristics that can be applied to many problems, also a high solution quality is often obtained using this method. **[6]**

  The vehicle routing problem use heuristics method as their base on solving the distribution model. Step that are required on building the model are; 1.) simulate the distribution channel by using the current vehicle from depots to customer that unserved (with time and vehicle restrictions). 2.) Deliver the goods to nearest customers (customer needs $\leq$ vehicle capacity). 3.) Continue delivering item until all customers are well served and at last send the vehicle to depots (notes each route that the vehicle travel during the distribution). 4.) Repeat the simulation model for all vehicle to accomplish the distribution of goods.

The simulation program that used in the thesis is software simulation AIMMS. AIMMS is a software system design to create optimization model who focus on the shortest total time in each vehicle. So it assumes that by optimizing the total time it can create an optimal cost. Total time are the calculation of service time add by travel time for each customer. In capacitated vehicle routing problem there are restriction that require to be added to perform as the real life problem. The restriction that needed are vehicle capacity over 200 box for each vehicle, time capacity for 1 vehicle over 480 minutes / 8 hours.

4. Building The Model

In creating a model distribution an index is needed to set the main data of model such as customer (*i* & *j*) and vehicle (*k*). Model simulation require identifier to proceed the calculation of route those identifiers are index, parameter, variable and constraints. Step of building the model are shown in the flow chart of figure 4.1. To build the model some data are require to extend the model to present as the real life problem. The declaration for the model start with customers with the index code of *i* and *j*, the different between index *i* and *j* are the declaration of from and too nodes. *i* is refer to from customer *i* and next *j* is referring to too customer. Those identifiers for index domain parameter and variables. The parameter that need to be added to fill the model simulation are number of customers, vehicle speed, customer demand, number of vehicles, and customer selection. Some data that need to determine to show the customers location are the X coordinate and Y coordinate for each location. Then creating matrix of distance between each location for the routing model. Parameter service time of each customer receive from one-month observation over distribution of PT Anugrah Citra Boga. Next input require for vehicle are the load capacity and time available in each day. Next for the formula of Travel Time = (Distance(i,j)/VehicleSpeed*60)+ServiceTime(i). Next for the variables there are 6 variable which is variable X(i,j,k) for the model to determine customer need to be access with the vehicle. Next the binary data for second variable with the name Y(i,k) to determine which vehicle will handle the distribution route. Next for the variable total time to calculate each vehicle travel time. For the constraint of the model to create the simulation as the problem real life. First for the constraint customer assignment to create restriction for each vehicle to only deliver one time per customer. Next constraint for time restriction and capacity restriction in each shipping. For calculating the model simulation, a mathematical programming is needed so to optimize the cost, model distribution uses total time as their objective on optimizing the model with the type of calculation is mixed integer programming. For the appearance, creating a new section for the route colour can help people to determine which vehicle deliver the customer goods. At last for the execution in solving the problem, creating main execution is needed to proceed the model simulation to work as the simulation of the real life problem. [7]

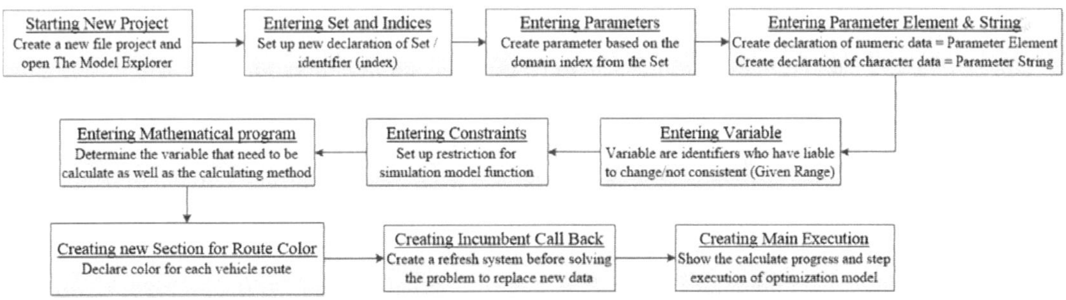

Figure 4.1 Flow chart building simulation model

5. Result and Discussion

The distribution system in PT Anugrah Citra Boga have a number of customers that have a daily routine of shipment in each day. The comparison between the model simulation and real life data is based on the total time for each vehicle. Provided that, in this paper to analyse between real life data and simulation data it takes Monday 2 May 2016 as the sample. On Monday the number of customers need to be deliver are 9 customers, which is bakso dimas, hero distribution centre (DC), hypermart distribution centre (DC), kimbo, Grogol traditional market, Senen traditional market, yusuf catering, carefour ITC Depok and at last hypermart Depok. Based on the real life data, for 9 customers the company use 2 vehicles to accompany all shipment of goods. Two route are made, the first one is south route in delivering 3 customers as Senen market, carefour ITC Depok, and hypermart Depok. Next for the second vehicle they use east route and start from Grogol market to bakso dimas, yusuf catering, kimbo, hypermart DC, and hero DC. The real life data for the distribution are shown as figure 5.1.

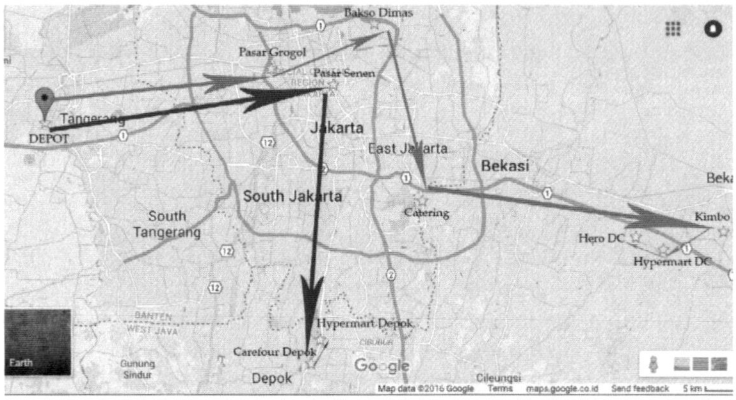

Figure 5.1 Distribution Route Real Data

5.1.Comparison Route

According to the data on Monday 2 May 2016, the total time of delivery is 790 minutes by using 2 vehicles. For each vehicle with the route as shown in figure 5.1 the delivery time for red route is 505 minutes which is more than the worker time (480 minutes). On the other side, the blue route delivery time is 285 minutes which is far from the standard worker time. Imbalance of vehicle distribution may occur an additional cost for red route vehicle and idle time for blue route vehicle. After calculating by using model distribution the result is 791 for total time of 2 vehicles, it seems that the simulation time is higher than the real data. However, the delivery time for each vehicle didn't exceed the capacity of worker time. The amount of delivery time for the first vehicle and second vehicle are 462, 329 respectively. By optimizing the delivery time for each vehicle it prevents from additional cost of worker overtime and minimize idle time. The comparison of simulation model distribution and historical data distribution is shown as table 5.1 and present in figure 5.2.

Table 5.1 Comparison real route and simulation model distribution on Monday 2 May 2016

Data Customer		Real Route		Simulation Route	
Code	Customer Name	Vehicle 1	Vehicle 2	Vehicle 1	Vehicle 2
2	Bakso dimas	1	1	1	1
3	Hero DC	6	8	7	6
4	Hyper DC	2	13	5	2
5	Kimbo	7	15	3	8
6	Grogol	5		4	13
7	Catering	4			15
8	Senen	3			
13	ITC depok				
15	Hyper depok	1	1	1	1
Departure		4:15	4:15	4:15	4:15
Arrival		12:40	9:00	11:57	9:44
Time Durations (minutes)		505	285	462	329
Total Time (minutes)		790		791	

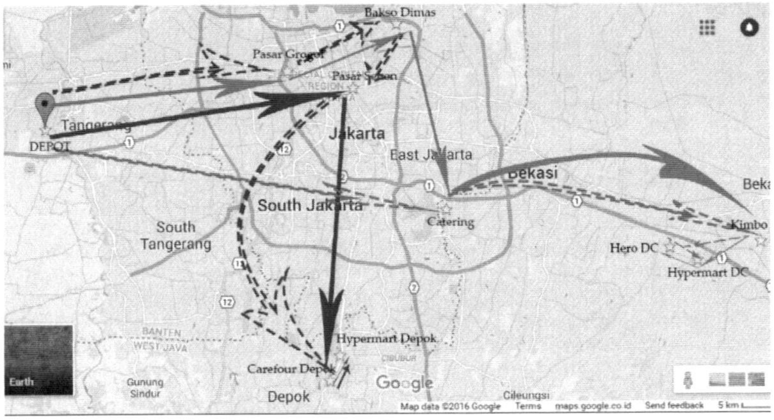

Figure 5.2 Comparison route of simulation and historical data

5.2. Utilization

Further calculation for each day over a week there are variant numbers of time for vehicle 1 and vehicle 2. To compare data simulation and historical data in this thesis it takes 6 days of working day which is Monday 2 May 2016, Tuesday 3 May 2016, Wednesday 11 May 2016, Thursday 5 May 2016, Friday 6 May 2016 and Saturday 7 May 2016. For the comparison time of each vehicle it shown at table 5.2.

Table 5.2 Data comparison delivery time

Day / Date	Delivery Time (minutes)			
	Real Data		Simulation Data	
	Vehicle 1	Vehicle 2	Vehicle 1	Vehicle 2
Monday / 2 May 2016	505	285	462	329
Tuesday / 3 May 2016	465	255	294	413
Wednesday / 11 May 2016	585	345	476	364
Thursday / 5 May 2016	435	175	432	193
Friday / 6 May 2016	545	375	463	479
Saturday / 7 May 2016	195	0	185	0
Total Time in 1 Week	2730	1435	2312	1778

The total time capacity for each vehicle in 1 week is 2610 minutes, with a calculation of half day on Saturday. Based on the real data for vehicle 1 it takes 2730 minutes on the total time delivery which is exceed the capacity of time. The extra time will be paid as the worker overtime cost. The amount of time that vehicle 1 use it exceeds the utilization over 100% with number of 104.6%. Next, for vehicle 2 it has shipment time 1435 minutes with the capacity over 5 days is 2400 minutes. The utilization of vehicle 2 is 59.8% and it quite low for utilization rate of vehicle because around 40% of time are idle for vehicle 2. By optimize the distribution route using the AIMMS model it create a higher efficiency over vehicle 1 and vehicle 2. The calculation in one week it turns out that by using simulation model it optimize the distribution for vehicle 1 to prevent from worker overtime. The rate of vehicle 1 become 88.6% it is high for a utilization rate. On the other hand, the simulation model increases the delivery time of vehicle 2 from 1435 minutes to 1778 minutes. Increasing the amount of time will affect the utilization rate from 59.8% to 74.1%. The graph of delivery time comparison shown as figure 5.3.

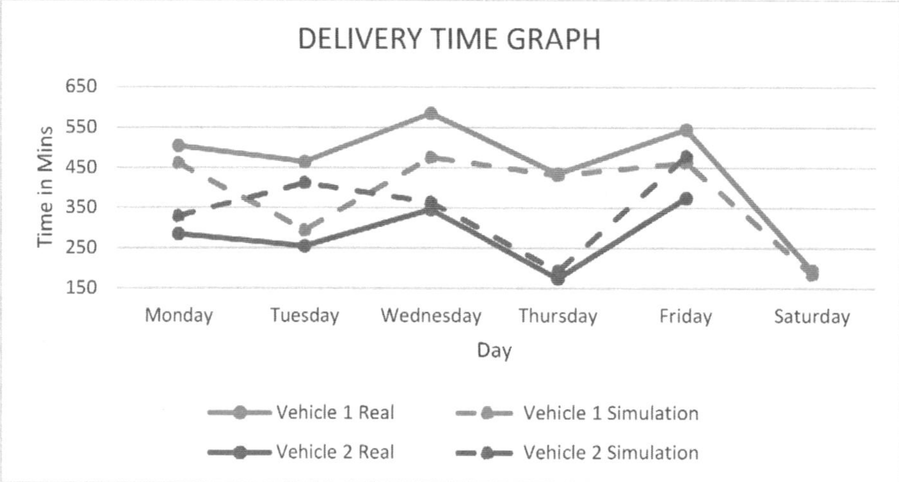

Figure 5.3 Comparison delivery time graph

5.3. Cost comparison

Next step is cost comparison between historical data and simulation data. The cost of each day over a week will be shown as table 5.3. Total cost obtained from fuel which is Indonesia Rupiah /minutes add by the worker overtime and wage.

Table 5.3 Comparison shipping cost data

	Shipping cost data	
Day / Date	Real Data	Simulation Model
Monday / 2 May 2016	Rp608,100	Rp558,490
Tuesday / 3 May 2016	Rp530,800	Rp525,730
Wednesday / 11 May 2016	Rp615,900	Rp577,600
Thursday / 5 May 2016	Rp487,900	Rp493,750
Friday / 6 May 2016	Rp658,800	Rp617,380
Saturday / 7 May 2016	Rp326,050	Rp322,150
Total Cost in 1 Week	Rp3,227,550	Rp3,095,100

Based on the comparison data over shipment cost in one week. The total cost of real data is Rp3,227,550 for both vehicles. However, the total cost that obtained from the model simulation is Rp3,095,100. The difference in cost between real data and simulation data is Rp132,450 for 1 week or Rp529,800 per month. The company could make a saving around 4.1% by using the model simulation. The graph of shipment cost comparison is shown as figure 5.4.

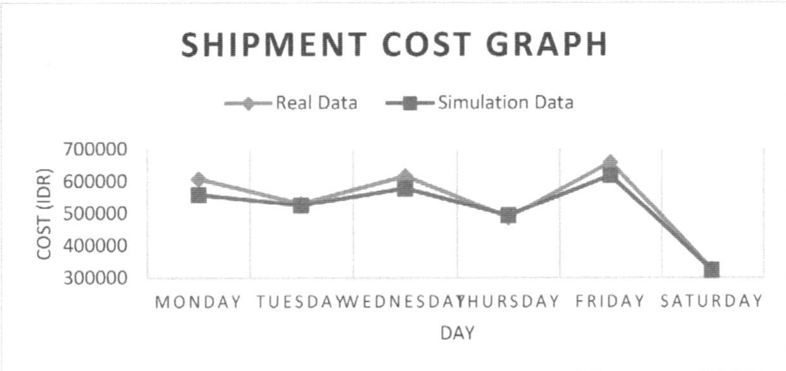

Figure 5.4 Comparison shipment cost graph

6. Conclusion

Transportation of goods is an important issue that need to be concern to minimize the cost. The thesis explains about observation and calculation model of distribution in PT Anugrah Citra Boga. The distribution software model use to optimize the routing is AIMMS software, by comparing the total cost between historical data and model simulation it concludes that this software could minimize the distribution cost around 4.1% or Rp529,800 in one month. Furthermore, the calculation of model using capacitated vehicle routing problem method could optimize the utilization of each vehicle. For the first vehicle, according to the historical data they use to have 104.6% utilization rate which is affect the worker overtime. On the other hand, for vehicle 2 the utilization rate quite low with a number around 59.8%. Using the simulation model it can increase the efficiency of the two vehicle to be more optimal in delivering goods. The utilization rate by the simulation model become 88.6% and 74.1% respectively for both vehicles 1 and 2.

References

[1] P. Toth and D. Vigo, The Vehicle Routing Problem, Philadelphia: Society for Industrial and Applied Mathematics, 2002.

[2] R. L. Rardin, Optimization In Operations Research, International Edition ed., United State of America: Prentice-Hall, Inc, 1998.

[3] R. T.K, K. L, P. W.R and T. L.E, "VRP," *On the Capacitated Vehicle Routing Problem,* pp. 1-19, 2001.

[4] R. S. Russell and B. W. T. , Operation Management, New Jersey: Prentice Hall, 2003.

[5] B. Kallehauge and M. O.B.G., "Lagrangean Duality Applied on Vehicle Routing with Time Windows," *Technical Report,* 2001.

[6] R. Stefan, "Heuristics vehicle routing problems," *Heuristic and exact algorithms for vehicle routing problems,* pp. 24-35, 2005.

[7] K. Heerink, "AIMMS tutorial for professionals," Paragon Decision Technology, The Netherlands, 2012.

[8] R. H. Ballou, Basic Business Logistics Transportation, Material Management, Physical Distribution, Prentice-Hall International ed ed., vol. 2, New Jersey: Prentice-Hall, 1987.

[9] R. H. Ballou, Business Logistics / Supply Chain Management, 5th Edition ed., New Jersey: Pearson Education, 2004.

[10] A. Wahyujati, "Operation Research 2," Jakarta, 2009.

Assessment of IT solutions used in the Hungarian income tax microsimulation system

I. Molnar[1] and S. Hardhienata[2]

[1]Retired Faculty, 4990 Sentinel Dr., Bethesda, MD, USA.
[2]Department of Computer Science and Doctoral Program of Pakuan University, INDONESIA.

E-mails: molnar01@aol.com, s-hardh@indo.net.id

Abstract. This paper focuses on the use of information technology (IT) in diverse microsimulation studies and presents state-of-the-art solutions in the traditional application field of personal income tax simulation. The aim of the paper is to promote solutions, which can improve the efficiency and quality of microsimulation model implementation, assess their applicability and help to shift attention from microsimulation model implementation and data analysis towards experiment design and model use. First, the authors shortly discuss the relevant characteristics of the microsimulation application field and the managerial decision-making problem. After examination of the salient problems, advanced IT solutions, such as meta-database and service-oriented architecture are presented. The authors show how selected technologies can be applied to support both data- and behavior-driven and even agent-based personal income tax microsimulation model development. Finally, examples are presented and references made to the Hungarian Income Tax Simulator (HITS) models and their results. The paper concludes with a summary of the IT assessment and application-related author remarks dedicated to an Indonesian Income Tax Microsimulation Model.

1. Introduction

Microsimulation is a method that is able to handle the high complexity of the socioeconomic problems by creating and studying a model that makes intensive use of the statistical data of the observed objects, also called *micro-units* of the socioeconomic system, e.g., person, family or household. The microsimulation models use simulation techniques in order to study the behavior of micro level units. To receive the best possible set of performance indicators of the socioeconomic system, the microsimulation model can be optimized using multi-objective nonlinear optimization algorithms. Microsimulation models can be integrated in e-government environments and provide support for decision making for government administrators.

Microsimulation models have the following main elements: simulation data, simulation model, simulation parameters and simulation results. The initial model data, intermediate and/or final simulation result data, are stored for further analysis. The simulation model consists of algorithms, which describe the behavior of micro-units and represent their environment. Simulating the micro-units' state and behavior (the changes of state over time) and the change of micro-units as a result of policy changes is often called "aging the data". Static aging refers to the adjustment by re-weighting the used sample based on some aggregate control variable (e.g., the composition of the sample by gender), while dynamic aging refers to adjusting each micro-unit attribute (e.g., income) by recalculating them at each time period, one period at a time.

Special care is taken to do the data analysis and the estimation of simulation model parameters, the latter of which are stored together with the simulation model. Given the model responses of micro-units at unit level, microsimulation models can estimate aggregate effects and aggregate changes by grouping, creating distributions, tabulating or summing up unit-level individual model results in order to make statements about the various characteristics of the population as a whole. The microsimulation model is working in an experimental framework in order to study the effects of policy changes on all micro-units, and on the microsimulation model behavior itself. Model calculations are executed by a digital computer.

Microsimulation models are generally accepted and frequently used to prepare political decisions in Australia (e.g., [7], [15], [25]), Europe (e.g., [3], [4], [10], [11], [12], [14], [28]), Canada and the USA (e.g., [2], [5], [9]. Economies in transition also use microsimulation models to study problems in demography, pension systems, healthcare and taxation.

In this paper the authors pursue an IT-oriented approach and assess the impact of new technology in enabling a wider spectrum of microsimulation model applications worldwide. The technology was developed during the past decade along with the applications (e.g., HITS) and it is essential to summarize all the lessons learned during the process. As will be clearly demonstrated using the example of the HITS, the presented technologies are able to satisfy all relevant user requirements. Moreover, by using this state-of-the-art IT, government policy decisions are transparently connected with the use of underlying microsimulation models and their implementation. With this, the foundation for the microsimulation model integration into an e-government application framework is established. The general applicability of the technology makes it easy to implement national income tax microsimulation models; a project plan for creating an Indonesian Income Tax Microsimulation System is shortly presented.

2. Personal Income Tax Microsimulation

2.1 Macro- and Micro-economic Considerations

Tax is a financial charge or levy imposed by a state or administrative authority on a taxpayer to finance various public financial obligations. Taxes are enforced by the tax authorities and tax evasion is punished by law. When state expenditures (e.g., expenditures on economic infrastructure, education, health care system, military, scientific research, culture, public works and government operations) exceed tax revenues, the state accumulates debt.

The tax system is also used by the government to redistribute resources among individuals or groups of individuals. According to the definitions created by the Organization for Economic Co-operation and Development (OECD), income tax is imposed on income of individuals or business entities. Individual or personal income tax is determined on the basis of the taxable income according to the tax law. Tax rates may vary (e.g., progressive or regressive) or are constant (flat).

Progressive tax is considered to be reducing economic inequality. The re-distributive property of the tax system, especially that of the personal income tax (along with the different type of property taxes), plays an important role in controlling long term inequality. Recent results of Th. Piketty [23] attracted additional research efforts and as a consequence, extended personal income tax models (e.g., tax-benefit models) are being introduced, with which an "optimal" set of policy control parameters (e.g., a construct of tax rates and tax benefits) should be determined (see [10] and [28]).

The Indonesian economy uses a progressive income tax system (see [29] and [30]) and has a vital interest to establish a tax system, which satisfies related economic and social requirements, while supporting and motivating the economic growth and social transition of the country. The provided IT solutions enable the creation and study of complex personal income tax microsimulation models and the quick adaptation of the technology itself. The present paper does not discuss in further details the macro- and micro-economic issues related to the income tax modeling.

2.2 Data Collection, Storage, Retrieval and Analysis

One of the major components of microsimulation models (and therefore of the whole microsimulation modelling environment) are the micro units' data: these are initial simulation model data, as well as intermediate and/or final simulation data. Initial data of microsimulation models are usually collected from cross-sectional and/or longitudinal surveys. Cross-sectional surveys collect data about a sample population for a single period of time (e.g., a survey of tourist expenditures), while longitudinal surveys collect data about the same sample population (also called panel) for several periods of time (e.g., a household statistical survey). Microsimulation models could also use a time-series of cross-sectional surveys, which collect data at periodic intervals (e.g., micro-census). Special techniques (e.g., imputing, merging, synthetic data) have been developed to improve data quality and use additional data sources available (see e.g., [16], [17], [24], [26]). In addition, the microsimulation model consists of a series of economic indicators and model parameters, which are also stored for simulation and used for further analysis and computations.

The model behavior in microsimulation models is defined by algorithms, which, among others, reflect the micro units' behavioral rules, describe economic processes and represent their impact. By using this methodology, special care is taken to perform data analysis and estimate the microsimulation model parameters.

The microsimulation model is working in an experimental framework in order to study the effects of policy changes on the microsimulation model. Given the model responses of micro-units at unit level, the microsimulation models can estimate aggregate effects and aggregate changes by grouping, creating distributions, tabulating or summing up unit-level model results in order to make statements about the various characteristics of the population as a whole and helping to determine "winners" and "losers" (see e.g., [14]).

The modelling process and numerical computation of simulation results can be riddled with different kind of errors; e.g., the model is never a perfect representation of the system, parameter-estimations are rarely perfect, while the numerical computation might also contain rounding and/or method-related errors. Validation checks the modelling process and its final result, the conceptual model, and whether the model was able to represent the studied system in a satisfactory way. Verification checks the computer model whether it is/was executed properly and whether the results were calculated correctly. It is the responsibility of the modeler to make a final decision about the model and the acceptance/rejection of the model results. Therefore, model verification and validation also use various sophisticated statistical methods and techniques (see [16]).

Taking the development and the application environment of microsimulation models into account, one can say that data are collected, stored, retrieved and processed in a distributed way in different databases at different locations, moreover, they are maintained by various, mainly government authorities. Most of the data are available in the form of different time series, in such a way that data content is hard to define, it might change with time and data integrity and accuracy are difficult to maintain. One of the biggest practical problems could be the management of the same data content under different names and different data content under the same name. Some of the simulation modelling problems are traditionally solved by using synthetic data sets (e.g., merging, imputing), which means that artificial data sources (and methods) are applied instead of traditional system data sampling. In addition to the "real system's" data, different types of simulation data are regularly stored and retrieved. These characteristics of microsimulation models' "messy" data sets are reflections of the characteristics of very large scale social science systems, which are also very "messy" (see [8]).

Despite the data-related problems, to the best of the author's knowledge, as of now there are no significant efforts to use standardized database content, data structure and access and data retrieval for microsimulation models. It is clear that most of the recently published studies focus on platform dependent, non-portable microsimulation applications (e.g., PC-oriented or supercomputer-based) and provide related data management solutions (see e.g., [1] and [28]). Little attention has been paid to the management of large data sets for distributed microsimulation databases and distributed applications. Portable and architecture neutral network-oriented technologies have been largely ignored. The authors are convinced that the development of networked multi-platform microsimulation applications is nowadays not just necessary but also technically possible (see [18], [19] and [27]).

2.3 Model classes and general software requirements

The authors call data systems "model-oriented", when those were designed and used to support applications based on a class of mathematical models. In most cases presented in the literature, data systems have been designed and used for a specific microsimulation model. Because data represented in the microsimulation modelling environment originate in different organizations (Central Statistical Office, Central Bank, Ministry of Finance, etc.) and are generated by different sources (i.e., measured, synthetic, simulated data), the simultaneous use of data by simulation models can be considered a special type of data-driven application (see [27]).

Behavior-driven models, as opposed to data-driven models, reflect the behavioral patterns of micro-units and the impact of policy change on them. Behavior driven models are frequently realized using agent-based modelling (e.g., [20]). In agent-based models, an initial configuration of multiple interacting agents is provided and then the model is driven by agent interactions, which are observed over time without further intervention. Similar to ASPEN (see [1]), several model objects can act as an agent (e.g., individuals, governments, markets, firms, banks, etc.). Based on the agents' characteristics, agents can be grouped, thus creating agent classes.

Despite the different modelling approaches, both model classes, data-driven and behavior-driven models, must handle model data and methods in the same way. In both cases large amount of data must be analyzed and processed in a way that model independence and data independence are provided. Therefore, general IT and a series of particular microsimulation software development environment requirements are applied. These requirements help to establish a broad framework for different microsimulation applications:

- Platform-independent hardware and software solutions based on open standards.
- Data and network security.
- User friendliness, standardized user interfaces.
- Network-oriented data and model access using a database management system.
- Distributed model development, model execution and data analysis.
- Efficiency of software development during the whole software life cycle.

2.4 Applied IT Solutions

Some of the listed requirements can be realized by using Service Oriented Architectures (SOA) (e.g., web-service technologies), as suggested in [19]. The large and complex data systems can be best managed using a database management system, which consists of crucial information about all individual microsimulation data available in the data system; i.e., a meta-database, which consists of data about microsimulation data. Based on the applied microsimulation models, the meta-database of microsimulation data must reflect the supported microsimulation model classes. The major advantage of using a meta-database for microsimulation data consist of, among others, increased data quality and overall cost-efficiency. The other listed criteria, like network-oriented data access and data analysis, as well as the efficiency of software development, require different technologies, which are sometimes integrated with the core database management system (e.g., Oracle's Business Intelligence solutions support data warehousing, data analysis, data mining and report generation).

When using a meta-database for microsimulation data, all data can be best referred to by using the appropriate data references of the meta-database. Because all methods which can be executed on the microsimulation database data can also be considered as data, it is evident that the most efficient solution to store the methods themselves is to store them in a database(s), i.e., by creating appropriate method-database(s). Having method-databases requires the same data management practice as using a meta-database of data; i.e., a meta- database for microsimulation methods must be created.

The model-oriented, data-driven microsimulation architecture, presented in [27], uses the following IT solutions:

- *Meta-database,*
- *Data warehouse,*
- *Object-oriented data modelling (OODM)* see [13],
- *Object oriented programming (OOP)* see [6],
- *Service Oriented Architecture (SOA)* see [19] and [20].

The authors consider the SOA and the meta-database as *key technologies* for microsimulation modelling; therefore these technologies will be discussed subsequently. Meta-database and the related object-oriented database technologies help to manage the complex data sets related to microsimulation applications. Service-oriented architectures are able to satisfy several application requirements, such as distributed data processing and model computations, security, platform independence and overall software development efficiency through the whole software life cycle.

2.5 Use of Microsimulation Meta-Database

The author considers a fundamental principle of the microsimulation application and development environment that two types of meta-databases are to be developed, one for the microsimulation data and another for the microsimulation model methods (see also in [27]). The advantage of this solution is that large amount of "messy" data can be managed comfortably along with the microsimulation model development results accumulated over a long period of time and substantiated in different model versions.

2.5.1 Meta-database for microsimulation data

The meta-database stores data about each microsimulation data into four groups:

- *Data content:* used for identification and usability decisions. Among others, the following data are stored: unique id, mnemonic, description, validation, version, usability role, and accessibility.
- *Data access:* access-related data are stored in specific order, depending on the storage medium, the location, the type of access (e.g., local or networked), the applied protocol (e.g., http, ftp, soap), etc.
- *Data processing:* most of the data are used for further calculations; therefore, algorithms and/or methods used for processing these data must be identified. The full specification of each algorithm and/or method is given as data content in two different ways: using direct statements (all processing steps of the algorithm are described directly, there are no loosely coupled method calls) and using indirect statements (the algorithm consists of loosely coupled method calls; e.g., Java RMI, Microsoft COM).
- *Other data:* technical data might also be stored e.g., action log, access log, error log.

The architecture with extended meta-database for microsimulation data is presented in Figure 1.

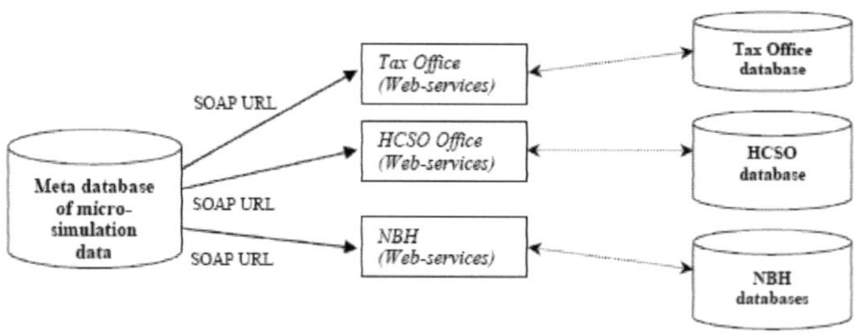

Figure 1. The architecture with extended meta-database.

2.5.2 Meta-database for microsimulation methods

A meta-database for microsimulation methods can be considered as a separate meta-database. In this case all meta-database data groups, except the data processing, can be used for the method definition. (In special cases, the meta-database for microsimulation methods can be directly integrated into the data processing data group of the meta-database for microsimulation data). The architecture with extended meta-database for microsimulation methods is presented in Figure 2.

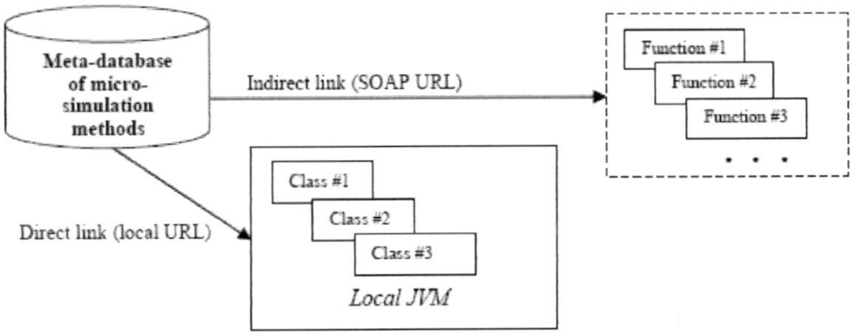

Figure 2. The architecture with extended meta-database for microsimulation methods.

2.6 Other software components

As presented above, the meta-databases for microsimulation are characterized by networked remote access and therefore both methods and data are accessed using information stored in meta-databases. Methods can produce both data-driven and behavior-driven models and therefore, agent-based modelling can be realized by not requiring separate software components. If special agent-based modelling tools or software components must be used, they can be integrated into this environment and invoked upon request. Microsimulation systems described here can be realized using the following major software components (see Figure 3).

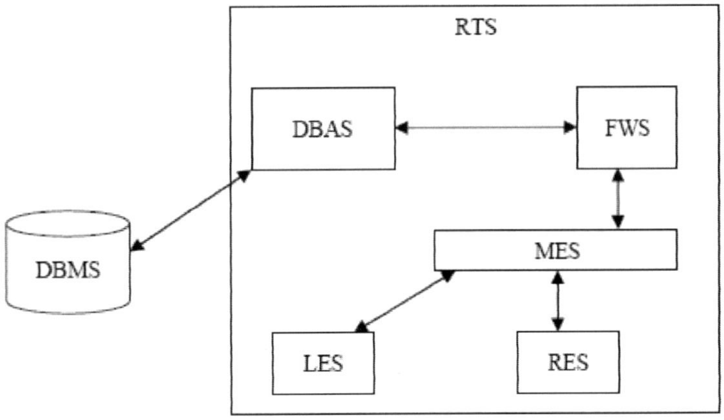

Figure 3. Major Software Components

- *RunTime System (RTS):* runs all software components. The JBoss Application Server has been used, which is an Open Source J2EE implementation.
- *FrameWork System (FWS):* synchronizes the components' run, working as the centre of the architecture communicating with the components. Implemented in J2EE.
- *Remote Execution System (RES):* runs all loosely coupled methods (for both remote data access and remote method execution). Implemented in J2EE, uses PHP running in Apache WebServer.

- *Local Execution System (LES):* runs local methods. Implemented in J2EE.
- *Messaging System (MES):* manages communication between/among the components. The JBossAS JMS has been used.
- *DataBase Access System (DBAS):* provides access to all local databases (incl. meta-databases). EIS Connection Pool, which is implemented in the Application Server.
- *DataBase Management System (DBMS):* manages all local databases (incl. meta-databases). MySQL Server 5 open source RDBMS has been used.

Using the carefully selected software components, the implementation of RTS, DBAS, DBMS and MES is a simple task of parameter setting. The implementation of FWS, LES, and RES, furthermore the population of the microsimulation meta-database, including the programming of agent-based microsimulation methods, require significant efforts.

All components of the HITS (incl. the operating system) were implemented as open source software. The extended LAMP ((Linux, Apache, MySQL, PHP), and JBoss provide robustness, scalability, reliability and data security. The applied web solutions and technologies also provide great organizational flexibility by separating developer competencies, which allows much higher life-cycle efficiency.

2.7 Data- and behavior-driven microsimulation models

The development of both data-driven and behavior-driven microsimulation models consists of two phases, which can be realized to a great extent independently from each other. In phase one the software system is implemented, while in phase two, the meta-database(s) and database(s) are populated. Both phases require different team compositions and professional background.

The HITS model implementation uses standard technologies, which are easy to adapt and modify, especially taking in consideration that the source code is fully available under GNU license. The Indonesian Income Tax Microsimulation System can build around these solid foundations; extensions and modifications are easy to implement. The HITS model scenarios were implemented using the multi-tier architecture:

- *Client tier* contains the microsimulation model and the client stub, which invokes the tax income forecast web service using SOAP on HTTP.
- *Application tier* contains the Oracle HTTP Server (actually an Apache), which receives request and sends response, furthermore the MySQL Application Server with MySQL Container for J2EE to store the web service, which performs the model calculations for income tax forecasting.
- *Information tier* runs MySQL database, which includes the microsimulation model data.

The software elements of the presented architecture were selected based on cost efficiency: Open Source, Free or Public Domain software have been used. The elements of the software architecture can be replaced by any other software-component of the same functionality, but additional resources are needed to ensure a smooth transition.

2.8 Assessment of the microsimulation model development environment

The authors consider the SOA and the meta-database as key technologies for the microsimulation modelling; therefore these technologies will be discussed below. Meta-database and the related object-oriented database technologies help to manage the complex data sets related to microsimulation applications. Service-oriented architectures are able to satisfy several application requirements, such as distributed data processing and model computations, security, platform independence and overall software development efficiency through the whole software life cycle (see [19] and [20]).

One of the most important features of the architecture and especially the meta-database concept is that there is a great flexibility in using various data-sets for the microsimulation models provided. Any databases (e.g., longitudinal, cross-sectional, survey-related, semi-synthetic, or synthetic) can be selected and used along with its interface to the simulation model. The choice assumes that the appropriate microsimulation model will also be developed (or has already been developed). In order to show this flexibility, instead of household statistical survey data, HITS scenarios use anonymous national tax office data (about 4.5 million personal income tax records were processed). Because all data (not just sample data of a given sample size) are related to the tax paying population, a higher statistical accuracy is anticipated. A consequence of this is that a better quality and more accurate aging process is anticipated. These factors increase the quality of the modelling process (model parameters and global variables are estimated using the original [full] population) and the data quality, as well. The authors think that these quality improvements are significant. Later developments, especially some contributions of the IMA 2015 provide a positive feedback showing an increased use of administrative data-sets and the acceptance of publicly-accessible microsimulation models.

The interfaces developed for the databases make it possible to use a series of data analysis tools, like SAP, SPSS, R, or other individual tools. Separate databases can be defined also for macroeconomic data or for data analysis results (using the data warehouse concept). When taxpayers' data are analyzed, individual and full population-related data are collected and stored. The "aging" process uses these data using the 'component-based income prediction' method. Predicting future personal income, the main components of the personal income is to be determined by using previously defined (calculating the parameters and variables) individual and group-population data. The personal tax is calculated and is always based on the current tax regulations applied to the individual and the individual's calculated (predicted) income. Statistical methods are applied to accumulate statistics based e.g., on income group (location, sex, or any other individual-related data property), in order to create useful indicators for political decision makers. The applied methodology separates data sets and simulation models efficiently, while providing sufficient information for decision makers, including also the national income tax budget component.

In order to illustrate the basic principles, the HITS scenarios base-line model and flat income tax microsimulation will be presented based on [21]. Other scenarios are discussed in detail in [20] and [27]. The tax contribution of income deciles to the national budget (in case of different flat tax rate scenarios) are presented in Figure 4.

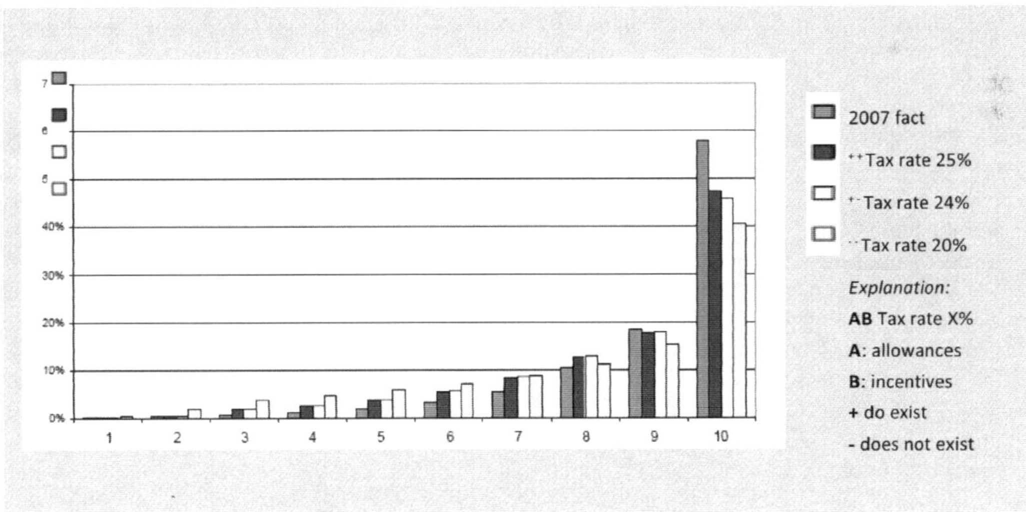

Figure 4. The tax contribution of income

3. Conclusion

The paper and the references present a generally applicable methodology and a microsimulation model development environment to create and use microsimulation models within a platform-independent, distributed and networked software environment. The basic idea of the solution is that the data and model complexity is managed by a multidimensional distributed database, which makes extensive use of meta-database data, which are also included in the very same distributed environment using Web Services. In order to increase the flexibility of data storage and data processing, the relationships stored in databases are described using object-oriented data modelling and programming concepts; the computation engine uses the stored data and methods from the same database. The resulting microsimulation model development environment is able to support the implementation of data-driven and behavior-driven static and dynamic microsimulation models. The resulting architecture does not require, but allows and supports the use of specialized agent-based modelling software.

Using the presented technologies, all relevant user requirements, listed as objectives, can be satisfied. The experiences with models of different sizes show results, on which one can build microsimulation applications used for administrative purposes. However, in case of large and complex microsimulation models, users need to handle a long sequence of data definitions and method calls, which might be a source of different errors. As a solution, the development of a modelling language or role-based further software developments might be appropriate.

With the increasing use of model-based policy decisions and the increased professional interest in microsimulation, the authors sincerely believe that the current practice of political decision-making will be radically changed by including a phase of impact analysis of policy changes before the decision is made. The importance of new technologies, especially the use of IT therefore cannot be overestimated. It is suggested to employ technological solutions which provide long-term, high quality and flexible solutions and use the available resources in the most efficient way. Advantages of scientific research and development can only be fully utilized if the political decision-makers and practical economists are aware of the new research results on the one hand, and the researchers are aware of the needs of practical decision makers, on the other hand. The technological solutions provided will redirect the attention to the simulation modelling of socio-economic systems and supporting the use of model results; the knowledge produced during the modelling process is substantiated in the final product, which could form the basis of a knowledge-based policy decision.

The conversion of a traditional microsimulation model into a web-based model application makes it possible to integrate legacy systems into new integrated information systems. The new technology based on SOA provides real solutions for distributed and secure data access, which is a major concern in most government applications, and also ensures distributed model development and analysis. A rational final stage of the development is the integration of the microsimulation model-development environment into an e-government environment.

The HITS model applications underline the importance of a real world environment and can be included into microsimulation model bases, forming the first elements of a microsimulation-based Decision Support System. These systems could be extremely useful for scientific research and state administration alike. With the separate management of data and methods, independent alternative model solutions can be developed parallel and compared in an easy way.

Unlike in the case of engineering models, socioeconomic models are not developed and used within the framework of a rational, often standardized process, but rather according to the policy makers' decisions, which again are often missing the deep scientific arguments. In order to extend the range of applications and provide wide availability of the microsimulation models, a change of policy-makers' world view, as well as changes in administrative regulations of data access must be implemented. In order to support this process, applied IT solutions will have to be made available in the near future.

The first considerations and results in the adaptation of the HITS to the Indonesian Income Tax System is an ongoing effort, which will be converted into a project in the near future. Beyond the assessment of ICT solutions used in the HITS, the research team will investigate the requirements of the Indonesian Income Tax Microsimulation System, the applicability of the use of a new non-linear optimization algorithm to determine optimal tax strategies, furthermore, determine the prerequisites of the practical use of the Indonesian Income Tax Microsimulation System and the international dissemination of its results.

References

[1] N. Basu, R.J. Pryor, T. Quint T. Arnold, ASPEN: A Microsimulation Model of the Economy (Sandia Report, Sandia National Laboratories, SAND96-2459 UC905, Albaquerque, New Mexico, 1996).
[2] W.W. Beach, R.A. Rector, R.S. Hederman, A.B. Goyburu, T. Kane The Candidates' Tax Plans: Comparing the Economic and Fiscal Effects of the Bush and Kerry Tax Proposals. (The Heritage Center for Data Analysis, Washington, D.C., 2004).
[3] P. Belyó, I. Molnar, Introduction of Capital Income Taxation: a Microsimulation Study, Simulationstechnique 18th Symposium in Erlangen (Ed.: F. Huelsemann, M. Kowarschik, and U. Ruede) Forschritte in der Simulationstechnik FS15, ISBN 3-936150-41-9, SCS European Publishing House e. V., Erlangen, 2005, 688-693.
[4] P. Belyó, I. Molnar, Use of Microsimulation Models in Political Decision Making, Public Finance Quarterly, LI/3, HUISSN 0031-496-X, Budapest, Hungary, 353-365.
[5] L.E. Burman, The Individual Alternative Minimum Tax, President's Advisory Panel on Federal Tax Reform, (The Urban Institute, The Tax Policy Center, Washington D.C., 2005)
[6] P. Coad, J. Nicola, Object –oriented Programming (New Jersey, Prentice Hall, second edition, 1993).
[7] J. Creedy, G. Kalb. Behavioural Microsimulation Modelling for Tax Policy, Analysis in Australia: Experience and Prospects (The University of Melbourne, Melbourne Institute Working Paper No. 2/05, ISSN 1328-4991 (Print),ISSN 1447-5863 (Online), 2005).
[8] P. Davidsson, Agent Based Social Simulation: A Computer Science View, Journal of Artificial Societies and Social Simulation 5(1), 2002, (http://jasss.soc.surrey.ac.uk /5/1/7.html), last visited: 4/30/2016).
[9] Ch. Edwards, A Proposal for a "Dual-Rate Income Tax" (Cato Institute, Tax & budget bulletin, No. 22, Washington D.C., 2005).
[10] O. Lelkes, Tax-Benefit Microsimulation Models In Eastern Europe, International Journal of Microsimulation, 1(1), 2007, 54-56.
[11] J. Flory, S. Stöwhase, MIKMOD-ESt: A Static Microsimulation Model of Personal Income Taxation in Germany, International Journal of Microsimulation, 5(2), 2012, 66-73.
[12] W.G. Gale, J. Rohaly, Effects of Tax Simplification Options on Equity, Efficiency, and Simplicity: A Quantitative Analysis (Urban Institute, Brookings Institution, Tax Policy Center, Washington, D.C., 2002).
[13] P. Gray, K.G. Kulkarni. N.W. Paton, Object-Oriented Databases A Semantic Data Model Approach (New York, Prentice-Hall, 1992).
[14] P. Haan, V. Steiner, Distributional and Fiscal Effects of the German Tax Reform 2000. A Behavioral Microsimulation Analysis (German Institute for Economic Research, Discussion Papers #419, Berlin, 2004)
[15] A. Harding, N. Warren, An introduction to microsimulation models of tax reform (NATSEM. National Centre for Social and Economic Modelling, University of Canberra, 1998).
[16] J. Li, C. O'Donoghue, A survey of dynamic microsimulation models: uses, model structure and methodology. (International Journal of Microsimulation, 6(2), 2013, 3-55).
[17] R.J.A. Little, D.B. Rubin, Statistical analysis with missing data (New York, Wiley-Interscience, 2002)
[18] I. Molnar, Gy. Lipovszki, Data Stream Management in Income Tax Microsimulation Models, Proceedings of The 23rd European Modeling & Simulation Symposium (EMSS2011), Rome, Italy, 2011, 394-399.
[19] I. Molnar, Microsimulation Model Development Environments International Journal of Simulation Systems, Science and Technology, 6(5), ISSN:1473-804x [Online], ISSN: 1473-8031 [Print], 2005, 33-41.
[20] I. Molnar, I. Sinka, Toward Agent-Based Microsimulation -- Another Approach Proceedings of the First Asia international Conference on Modelling & Simulation. IEEE Computer Society, Washington, DC, 2007, 403-408.

[21] I. Molnar, I. Sinka, P. Belyo, Cs.G. Toth, D. Bathori, Hungarian individual income tax microsimulation. Proceedings of the seventh International Conference on Simulation in Industry and Services, Brussels, Belgium, 2009, 187-211.
[22] G. Orcutt, M. Greenberger, J. Korbel, and A. Rivlin, Microanalysis of socioeconomic systems: a simulation study (Harper & Brothers. New York, 1961).
[23] Th. Piketty, Capital in the twenty first century (The Belknap Press of Harvard University Press, Cambridge, Massachusetts, 2014).
[24] D.B. Rubin, Multiple Imputation for Nonresponse in Surveys (New York, John Wiley & Sons, 2004).
[25] R. Scutella, M. Harris, J. Creedy, A. Duncan, Microsimulation modelling of taxation and the labour market: the Melbourne Institute tax and transfer simulator (Cheltenham, United Kingdom, Edward Elgar Publishing, 2002).
[26] D. Schofield, J. Polette, A comparison of data merging methodologies for extending a microsimulation model (NATSEM STINMOD Technical Paper No. 11. National Centre for Social and Economic Modelling, University of Canberra, 1998)
[27] I. Sinka, I. Molnar 2006 "Modell-oriented Data-driven Architectures for Microsimulation" (Simulationstechnique 19th Symposium in Hannover. In: ASIM Forschritte in der Simulationstechnik (Ed.: M. Becker, H. Szczerbicka), SCS Publishing House e. V., ISBN 3-936150-49-4, Erlangen, 2006. 35-41).
[28] H. Sutherland, Final report: EOROMOD: an integrated European benefit-tax model (EUROMOD Working Paper No. EM9/01, 2001)
[29] - : Taxation and Investment in Indonesia 2014: Reach, Relevance, Reliabitiy, Indonesian Highlight 2016, A Publication of Deloitte Touche Tohmatsu Limited, 2016.
[30] - : Indonesian Pocket Tax Book 2015, PWC Indonesia, www.pwc.com/id, 2015.

IOP Conf. Series: Materials Science and Engineering **166** (2017) 012011 doi:10.1088/1757-899X/166/1/012011

A combination method for solving nonlinear equations

B.P. Silalahi[1], R. Laila[2] and I.S. Sitanggang[3]

[1,2]Department of Mathematics, Bogor Agriculture University, Bogor, INDONESIA
[3]Department of Computer Science, Bogor Agriculture University, Bogor, INDONESIA

Emails: bibparuhum@gmail.com, lailarahmah.ella@gmail.com

Abstract. This paper discusses methods for finding solutions of nonlinear equations: the Newton method, the Halley method and the combination of the Newton method, the Newton inverse method and the Halley method. Computational results in terms of the accuracy, the number of iterations and the running time for solving some given problems are presented.

1. Introduction

In this paper we will discuss methods for finding solutions of nonlinear equations. The Newton method is one of the best methods to determine the root solution of nonlinear equations (Sánchez 2009). In its development the Newton method is also used to find the optimum point of an optimization problems (Silalahi 2014). The Newton method is one of the best techniques to solve nonlinear equations and optimization problems. This method is very easy to implement and often converges quickly, when the iteration begins quite close to the desired root (Kumar 2012).

Some researchers have been looking for the most effective and efficient methods for solving nonlinear equation problems. Some modified Newton methods have been developed by some researchers including Werakoon and Fernando (2000). They modified the Newton method using the trapezoidal rule to produce Newton trapezoidal method. This method has cubic convergence which is better than the Newton method. The results of Werakoon and Fernando have sparked a lot of researches on the Newton method. The study was carried out to obtain a better method for searching root values of nonlinear equation problems. Subsequent researchers are Frontini (2003) and Ozban (2004). They use arithmetic Newton method and harmonics Newton method, both of these methods also produce cubic convergence. Subsequent research conducted by Homeier (2005) that modifying Newton method by using the inverse function and also produces cubic convergence. Then Noor (2006) modified the Halley method with the basic concepts of the Newton method and produce a new development of the Halley method with better performance.

Based on the researches that has been done, this research will use the combination of the the Newton method, the Newton inverse method combined and the Halley method to obtain better iteration, running time and convergent to the exact value.

2. Combination and Formulation of the Newton Method, the Newton Inverse Method and the Halley Method

The Newton method, the Newton inverse method and the Halley method have the same characteristics, they are the iterative methods. For finding a root of a function, the Newton method requires an initial guess point x_0. Newton method does not require a large cost for searching the root of a function, so that in the combination: the formula of the Newton method is placed in the first step. In general the formula of the Newton method for finding the value of the root of a function is shown in (1), (Luenberger and Ye 2008),

$$x_n^* = x - \frac{f(x)}{f'(x)} \quad (1)$$

with $x = x_0$ is initial guess point, $f(x)$ is a function value by substituting the value of x, and $f'(x)$ is the first derivative function value at x.

After determined the initial method, the next step is to add the Newton inverse method formula in the second step of finding the root of a function. According to Homeier (2005), the Newton inverse method produces fewer number of iterations in the search of the root of a function. The result point x_n^* in (1) is used as a value that is going to be substituted to find the root value in the second step using the Newton inverse formula as in (2), (Homeier 2005),

$$\bar{x}_n = x - \frac{f(x)}{2}\left(\frac{1}{f'(x)} + \frac{1}{f'(x_n^*)}\right) \tag{2}$$

with $f'(x_n^*)$ states the first derivative function value at x_n^* which is obtained from the calculation of (1).

The Halley method contains the second derivative and thus requires much time in the calculation process. Therefore, this method is placed in third step for finding a solution approach of a given function. The general form of the formula Halley method is as follows (Noor et al. 2006),

$$x_{n+1} = \bar{x}_n - \frac{2f(\bar{x}_n)f'(\bar{x}_n)}{2(f'(\bar{x}_n))^2 - f(\bar{x}_n)f''(\bar{x}_n)} \tag{3}$$

with $f(\bar{x}_n)$ states a function value at $\bar{x}_n$, $f'(\bar{x}_n)$ states the first derivative function value at $\bar{x}_n$, and $f''(\bar{x}_n)$ states the second derivative function value at $\bar{x}_n$.

These three combination methods are applied in each iteration. The result of the combination and formulation methods then used to create an algorithm: combination of the Newton Method, the Newton Inverse Method and the Halley Method (NIH algorithm) as follows:

NIH algorithm (Newton Invers Halley algorithm)

Step 1. Given the initial function $f(x)$

Step 2. Given the initial guess point $x_0 \epsilon \mathbb{R}^n$ and the limits of tolerance $0 < \varepsilon < 1$

Step 3. Given a maximum point of iteration

Step 4. Calculate $x_n^* = x_0 - \frac{f(x_0)}{f'(x_0)}$

Calculate $\bar{x}_n = x_0 - \frac{f(x_0)}{2}\left(\frac{1}{f'(x_0)} + \frac{1}{f'(x_n^*)}\right)$

Calculate $x_{n+1} = \bar{x}_n - \frac{2f(\bar{x}_n)f'(\bar{x}_n)}{2(f'(\bar{x}_n))^2 - f(\bar{x}_n)f''(\bar{x}_n)}$

Step 5. Calculate $b = x_{n+1} - x_n$

Step 6. If $b < \varepsilon$ or iteration reaches the maximum point, go to step 4.

3. The functions for Testing

The functions that will be used for data test of computing, with α is the root of each function described below (Werakoon and Fernando 2000):

$f_1(x) = x^3 + 4x^2 - 10$ $\alpha = 1.36523001341448$

$f_2(x) = sin^2(x) - x^2 + 1$ $\alpha = 1.40449164821621$

$f_3(x) = x^2 - e^x - 3x + 2$ $\alpha = 0.257530285439771$

$f_4(x) = \cos(x) - x$ $\alpha = 0.739085133214758$

$f_5(x) = (x-1)^3 - 1$ $\alpha = 2.0$

$f_6(x) = x^3 - 10$ $\alpha = 2.15443469003367$

$f_7(x) = x\exp(x^2) - sin^2(x) + 3\cos(x) + 5$ $\alpha = -1.20764782713013$

$f_8(x) = x^2 sin^2(x) + \exp[x^2\cos(x)\sin(x)] - 28$ $\alpha = 4.82458931731526$

$f_9(x) = \exp(x^2 + 7x - 30) - 1$ $\alpha = 3.0$

4. Results

This research was carried out by using hardware and software as follows : personal computer with the following specifications: Processor Intel (R) Pentium (R) CPU B970@2.30Ghz, RAM 2 GB, mouse and keyboard. The software used is as follows: Operating system Windows 7 Ultimate 32-bit, Microsoft Excel 2013 and a mathematical software.

Computational test is conducted by comparing the numerical results of only the Newton method, only the Halley method and the combination method NIH. The functions $f_n(x)$ is used for testing by different initial values. In this test the tolerance ε is 10^{-4}.

4.1 The number of iterations
The computational results for the number of iterations of each method with tolerance $\varepsilon = 10^{-4}$ are shown in Table 1. Based on the computational results, the proposed NIH method is better than the Newton method and the Newton inverse method. The total number of iterations for all experiments is 69 which is less than two other method greatly. The starting point x_0 and the tolerance also affect the number of iterations. If we use a starting point or initial point that is close enough to the root value, then the iteration process becomes smaller.

4.2 The running time
The running time of each method with tolerance $\varepsilon = 10^{-4}$ are shown in Table 2. The computational results show that the NIH method has the smallest total number of the running time. But we still need further exploration, because for some functions the Newton method is the best.

Table 1 Comparison of the number of iterations of each method with tolerance $\varepsilon = 10^{-4}$

$f(x)$	x_0	The number of iterations of each methods		
		Newton	**Halley**	**NIH**
f_1	-0.5	130	74	7
	1	4	4	3
	2	4	4	3
	-0.3	53	53	11
f_2	1	5	5	3
	3	5	5	3
f_3	2	2	2	2
	3	5	4	3
f_4	1	3	44	3
	1.7	4	4	3
	-0.3	5	5	3
f_5	3.5	6	4	3
	2.5	5	4	3
f_6	1.5	5	4	3
f_7	-2	7	5	4
f_8	5	8	7	4
f_9	3.5	11	5	4
	3.25	7	4	4
Total		269	237	69

4.3 The accuration
The computational results for accuracy (the root value) of each method with tolerance $\varepsilon = 10^{-4}$ are shown in Table 3. Table 3 shows that almost all of the three methods find the expected value of the root of all the given functions. The root values obtained from computational test using a tolerance $\varepsilon = 10^{-4}$ are not different significantly when compared with the root value obtained from Weerakoon and Fernando. But for f_8 there are differences between root value of Werakoon and Fernando (4.8246) with the Newton method (3.4374), the Halley Method (3.4374), and the NIH method (4.6221). From table, it can be seen that the root value of the NIH method is closer to the real root value compared to the Newton method and the Halley Method. The NIH Method seem is better than Newton method and the Halley Method for finding a solution of "difficult functions" such as f_8.

IOP Conf. Series: Materials Science and Engineering **166** (2017) 012011 doi:10.1088/1757-899X/166/1/012011

Table 2. Comparison of the running time of each method with tolerance $\varepsilon = 10^{-4}$

$f(x)$	x_0	Running time (ms)		
		Newton	Halley	NIH
f_1	-0.5	0.1204	0.0822	0.0280
	1	0.0049	0.0072	0.0096
	2	0.0049	0.0059	0.0100
	-0.3	0.0438	0.0705	0.0193
f_2	1	0.0063	0.0107	0.0107
	3	0.0064	0.0108	0.0160
f_3	2	0.0031	0.0064	0.0117
	3	0.0070	0.0122	0.0172
f_4	1	0.0013	0.0067	0.0088
	1.7	0.0044	0.0087	0.0094
	-0.3	0.0053	0.0087	0.0128
f_5	3.5	0.0066	0.0089	0.0131
	2.5	0.0055	0.0088	0.0090
f_6	1.5	0.0054	0.0074	0.0088
f_7	-2	0.0108	0.0143	0.0208
f_8	5	0.0149	0.0246	0.0242
f_9	3.5	0.0146	0.0182	0.0207
	3.25	0.0092	0.0151	0.0170
Total		0.2748	0.3273	0.2671

Table 3. Comparison of the root value of each method for tolerance $\varepsilon = 10^{-4}$

$f(x)$	x_0	root value		
		Newton	Halley	NIH
f_1 1.3652	-0.5	1.3652	1.3652	1.3652
	1	1.3652	1.3652	1.3652
	2	1.3652	1.3652	1.3652
	-0.3	1.3652	1.3652	1.3652
f_2 1.4045	1	1.4044	1.4044	1.4044
	3	1.4044	1.4044	1.4044
f_3 0.2575	2	0.2575	0.2575	0.2575
	3	0.2575	0.2575	0.2575
f_4 0.7391	1	0.7391	0.7391	0.7391
	1.7	0.7391	0.7391	0.7391
	-0.3	0.7391	0.7391	0.7391
f_5 2.0000	3.5	2.0000	2.0000	2.0000
	2.5	2.0000	2.0000	2.0000
f_6 2.1544	1.5	2.1544	2.1544	2.1544
f_7 −1.2076	-2	-1.2076	-1.2076	-1.2076
f_8 4.8246	5	3.4374	3.4374	4.6221
f_9 3.0000	3.5	3.0000	3.0000	3.0000
	3.25	3.0000	3.0000	3.0000

4. Conclusions

The results showed that a combination of the Newton method, the Newton inverse method and the Halley method (NIH) can be used for finding the root solutions of the given nonlinear functions. Based on the computational results, the combination method in general has better performance in the terms of the number of iterations compared to the Newton method and the Halley method. The computational results show that the combination method has the smallest total number of the running time, but we still need further exploration, because for some functions the Newton method is the best. In terms of accuracy, the NIH method obtains results closer to the roots than other methods especially for the functions that are quite difficult.

Acknowledgment
This work was supported by the directorate general of research strengthening and the development, the ministries of research, technology, and higher education republic of Indonesia (No. 079/SP2H/LT/DRPM/II/2016).

References

[1] Frontini M, Sormain E. 2003. Some Variant of Newton's Method with Third-Order Convergence. *Applied Mathematics and Computation* 140: 419-426.
[2] Homeier HHH. 2005. On Newton-type Methods with Cubic Convergence. *Journal of Computational and Applied Mathematic* 176: 425-432.
[3] Kumar M, Singh AK, Srivastava A. 2012. Various Newton-type Iterative Methods for Solving Nonlinear Equations. *Journal of the Egyptian Mathematic Society* 21,334-339.
[4] Luenberger DG, Ye Y. 2008. *Linear and Nonlinear Programing Third Edition*. Stanford, California: Springer.
[5] Noor MA, Khan WA, Hussain A. 2006. A new modified Halley method without second derivatives for nonlinear equation. *Applied Mathematics and Computation*. 189: 1268–1273
[6] Ozban AY. 2004. Some New Variants of Newton's Methods. *Applied Mathematic Letters* 17: 677-682.
[7] Sánchez MG. 2009. Improving Order and Efficiency: Composition with a Modified Newton's Method. *Journal of Computational and Applied Mathematics* 231, 592-597.
[8] Silalahi BP. 2014. Sharper analysis of upper bound for the iteration complexity of an interior-point method using primal-dual full-Newton step algorithm. *Far East Journal of Mathematical Sciences* 95(1): 69-80.
[9] Weerakoon S, Fernando T.G.I. 2000. A Variant of Newton's Method with Accelerated Third-Order Convergence. *Applied Mathematic Letters* 13: 87-93.

IOP Conf. Series: Materials Science and Engineering **166** (2017) 012012 doi:10.1088/1757-899X/166/1/012012

Modeling Fourth Harmonic Generation in a Non-Vicinal (111) Diamond Lattice

Hendradi Hardhienata

Theoretical Physics Division, Department of Physics, Bogor Agricultural University, Jl. Meranti, Gedung Wing S, Kampus IPB Darmaga, Bogor 16680, Jawa Barat, INDONESIA

E-mail: `hendradi@ipb.ac.id`

Abstract. In this work it is shown how Rotational Anisotropy Fourth Harmonic Generation (RAFHG) data from nonvicinal (111) diamond lattice obtained by M. Downer Group can be very well modelled using the modified nonlinear bond model. Extending the bond model previously developed for second harmonic generation to higher harmonics is straightforwardly achieved by modification of the nonlinier susceptibility in the form of adding more direct product of bond vectors. Because a diamond latttice is centrosymmetric, FHG radiation measured in reflection should mainly come from the interface because it is bulk forbidden due to parity simmetry. Previous work showed that the nonlinear bond model can very well describe Rotational Anisotropy Second Harmonic Generation (RASHG) and Rotational Anisotropy Third Harmonic Generation (RATHG) data and here I present similar agreement for FHG.

1. Introduction

In a series of papers [1-3] Dave Aspnes group at North Carolina University have investigated higher harmonic generation in low symmetry crystal using a phenomenological model by assuming that the radiating sources originated from anharmonic charge oscillation along the atomic bonds. They convincingly showed through a pioneering breakthrough work which also marked the birth of the so-called simplified bond hyperpolarizabilit model (SBHM) [1] that the required independent parameters to describe Rotational Anisotropy Second Harmonic Generation (RASHG) intensity from diamond lattice surfaces were less than people have previously thought using an earlier phenomenological model where the nonlinier susceptibility tensor were obtained from Group Theory (GT) [4].

The important link between SBHM and GT was shown by our work in Ref. [5] which revealed that the bond model is in agreement but also more restrictive than GT, requiring less independent parameters. We also proved that Kleinman Symmetry is an inherent property of SBHM [6] and that the model can be extended to reproduce RASHG experiment result for noncentrosymmetric lattices such as zincblende [7], predicting bulk quadrupolar contribution when the sample is probed using arbitrary polarization input angles [8], and Electric Field Induced Second Harmonic (EFISH) where in the latter, the fourth rank tensor can be simplified into a third rank tensor by matrix contraction due to a direct current electric field [9]. In this work I show, for the first time, that SBHM can also be applied successfully to reproduce experimental data from fourth harmonic generation (FHG) RA experiments on a non-vicinal diamond, e.g. Si(111) lattice that were performed by Aktsipetrov and M. Downer Group [10].

IOP Conf. Series: Materials Science and Engineering **166** (2017) 012012 doi:10.1088/1757-899X/166/1/012012

Due to parity symmetry dipolar fourth harmonic generation sources are forbidden within the bulk but are broken at the SiO-Si surface.

2. Results and Discussion

The SBHM for fourth harmonic generation follows straight forwardly from second harmonic generation, but introducing a fifth rank tensor instead in the nonlinear dipole polarization $\mathbf{P}_{\mathbf{D}}^{(\mathbf{4}),\mathrm{int}}$:

$$\begin{aligned}\mathbf{P}_{\mathbf{D}}^{(\mathbf{4}),\mathrm{int}} &= \chi_D^{(4)} \cdots \mathbf{E}_{\mathrm{int}}\mathbf{E}_{\mathrm{int}}\mathbf{E}_{\mathrm{int}}\mathbf{E}_{\mathrm{int}} \\ &= \frac{N}{V}\sum_{j=1}^{n}\left(\alpha_{4j}\hat{\mathbf{b}}_j(\phi)\hat{\mathbf{b}}_j(\phi)\hat{\mathbf{b}}_j(\phi)\hat{\mathbf{b}}_j(\phi)\hat{\mathbf{b}}_j(\phi)\right)\cdots\mathbf{E}_{\mathrm{int}}\mathbf{E}_{\mathrm{int}}\mathbf{E}_{\mathrm{int}}\mathbf{E}_{\mathrm{int}}\end{aligned} \quad (1)$$

where N/V is the total number of anharmonic dipoles per unit volume that radiates FHG which is just a fitting scaling factor in SBHM, α_{4j} is the fourth harmonic nonlinier hyperpolarizability, $\hat{\mathbf{b}}_j(\phi)$ is the freely rotating z-axis and $\mathbf{E}_{\mathrm{int}}$ is the driving field at the interface with frequency ω.

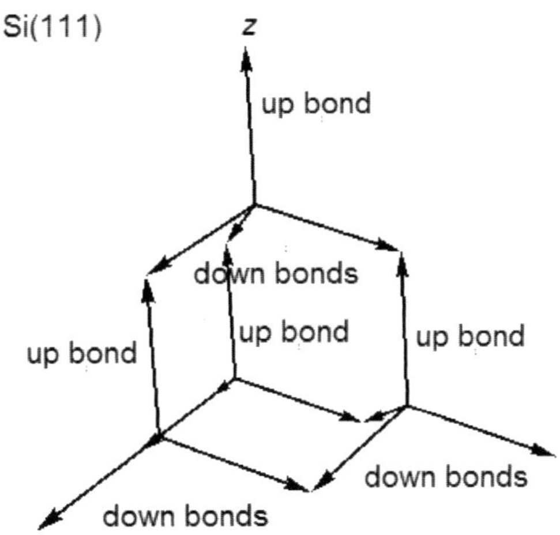

Figure 1. Direction of the up and down bonds, the z-axis is directed paralel to the up bonds

The j-th bond vector is iterated for a total of 4 bonds due to the Si(111) interface/surface sp3 hybridization molecule symmetry which belong to the C_{3v} molecular point group. Because in this case they are bond vector in the (111) orientation they take the form:

$$\hat{\mathbf{b_1}} = \begin{pmatrix} 0 \\ 0 \\ 1 \end{pmatrix} \quad \hat{\mathbf{b_2}} = \begin{pmatrix} \sin\beta \\ 0 \\ \cos\beta \end{pmatrix}$$

$$\hat{\mathbf{b_3}} = \begin{pmatrix} -\frac{1}{2}\sin\beta \\ \frac{\sqrt{3}}{2}\sin\beta \\ \cos\beta \end{pmatrix} \quad \hat{\mathbf{b_4}} = \begin{pmatrix} -\frac{1}{2}\sin\beta \\ -\frac{\sqrt{3}}{2}\sin\beta \\ \cos\beta \end{pmatrix} \quad (2)$$

here the angle between two tetragonal bonds is always $\beta = 109.47°$. Because the sample is rotated about the z-axis each of the bond vector above can be rotated arbitrarily around the azimuthal angle ϕ:

$$\hat{\mathbf{b}}_{\mathbf{j}}(\phi) = R_z(\phi)\hat{\mathbf{b}}_{\mathbf{j}} \tag{3}$$

where $R_z(\phi)$ is the rotation matrix around the z-axis:

$$R_z(\phi) = \begin{pmatrix} \cos(\phi - \Delta\phi) & -\sin(\phi - \Delta\phi) & 0 \\ \sin(\phi - \Delta\phi) & \cos(\phi - \Delta\phi) & 0 \\ 0 & 0 & 1 \end{pmatrix} \tag{4}$$

here the angular offset $\Delta\phi$ is adjusted because the bond azimuthal orientation in the experiment might be unknown. A graphical description of the Si(111) bond is presented in Fig. 1.

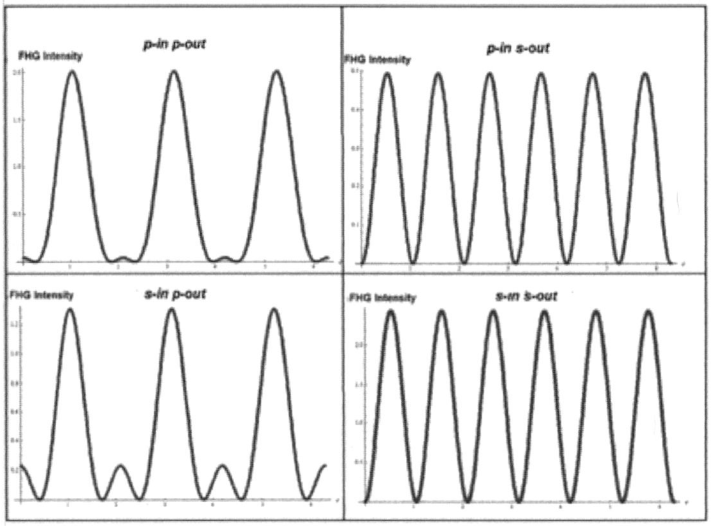

Figure 2. FHG intensity for a complete azhimutal rotation for the 4 polarizations showing very good intensity profile with Fig. 1 in Ref. [10]

Because the material is inversion symmetric the dipolar contribution to FHG inside the bulk $\mathbf{P}_{\mathbf{D}}^{(4),\text{bulk}}$ is zero. When the FHG far field from the interface is calculated using:

$$\mathbf{E}_{ff,\text{surf}}^{FHG}(\mathbf{r}) \propto \left[(\bar{\mathbf{I}} - \hat{\mathbf{k}}_{\mathbf{o4}\omega,\text{int}}\hat{\mathbf{k}}_{\mathbf{o4}\omega,\text{int}}) \cdot \mathbf{P}_{\mathbf{D}}^{(4),\text{int}}\right] \tag{5}$$

where $\hat{\mathbf{k}}_{\mathbf{o4}\omega,\text{int}}$ is the outgoing FHG wavevector for the four types of polarization: *pp* (here meaning *p*-fundamental input *p*-FHG output), *ps*, *sp* and *ss*. Fig. 2 depicts the four polarization FHG intensity which is in agreement with theoretical prediction of interface dipolar FHG sources as the sole dominant contributor of FHG because it can fit the RA spectroscopy nonlinear intensity in Fig. 1 of M. Downers Group in Ref. [10] with very good agreement by squaring Eq. (5) alone, taking just two different hyperpolarizability e.g. $(\alpha_{41}) = (\alpha_u)$ and $(\alpha_{42}) = (\alpha_{43}) = (\alpha_{44}) = (\alpha_d)$ where (α_{4u}) is FHG hyperpolarizability of the up bonds and (α_{4d}) is FHG hyperpolarizability of the three down bonds. In the simulation, the hyperpolarizability values are set to 3+0.5I but should be fitted to experimental data. The incomming and outgoing angles were obtained by Snell's law as 30^0 because the sources comes from the Silicon Oxide-Silicon interface.

3. Conclusion

It has been demonstrated that the FHG intensities for the four polarizations obtained by the bond hyperpolarizability model are in agreement with experimental results showing all the nonlinear intensity peak phases and profiles correctly. The model also confirms theoretical prediction that for centrosymmetric lattices such as Si, the sole main contributor to FHG is the same as in SHG namely coming from dipolar interface sources.

4. References

[1] Powell, G.D., Wang, F. F., and Aspnes D. E. (2002), Simplified bond- hyperpolarizability model of second harmonic generation, Phys. Rev. B 35 1129
[2] E. J. Adles and D. E. Aspnes, Application of the anisotropic bond model to second-harmonic generation from amorphous media, Phys. Rev. B 77, 165102
[3] E. J. Adles and D. E. Aspnes, The anisotropic bond model of nonlinear optics, Phys. Stat. Sol. A 205, 728731 (2008).
[4] Sipe, J. E., Moss, D.J., and van Driel, H. M., Phenomenological theory of optical second- and third-harmonik generation from cubic centrosymmetric crystals, (1987), Phys. Rev. B 35 1129
[5] Alejo-Molina, A, Hardhienata, H. and Hingerl, K., Simplified bond-hyperpolarizability model of second harmonic generation, group theory, and Neumann's principle, J. Opt. Soc. Am. B. 31, 3 (2014).
[6] Alejo-Molina, A, Hardhienata., The role of Kleinman Symmetry in the Simplified Bond Hyperpolarizability Model, IOP Conf. Ser.: Earth Environ. Sci. 31 012020 (2016)
[7] H. Hardhienata, A. Alejo-Molina, C. Reitbck, A. Prylepa, D. Stifter, and K. Hingerl, Bulk dipolar contribution to second-harmonik generation in zincblende, J. Opt. Soc. Am. B., Vol. 33, No. 2, 195 - 201 (2016).
[8] Cornelia Reitbck, David Stifter, Adalberto Alejo-Molina, Kurt Hingerl, Hendradi Hardhienata, Bulk quadrupole and interface dipole contribution for second harmonic generation in Si(111), Journal of optics 02/2016; 18(3):035501. (2016)
[9] Adalberto Alejo-Molina, Kurt Hingerl, Hendradi Hardhienata, Model of third harmonic generation and electric field induced optical second harmonic using simplified bond-hyperpolarizability model, Journal of the Optical Society of America B 04/2015; 32(4) (2015).
[10] R.W. Kempf, P.T. Wilson, J.D. Canterbury, E.D. Mishina, O.A. Aktsipetrov, M.C. Downer, Third and fourth harmonic generation at Si-SiO2 interfaces and in Si-SiO2-Cr MOS structures, Appl. Phys. B 68, 325332 (1999)

A collaborative vendor-buyer production-inventory systems with imperfect quality items, inspection errors, and stochastic demand under budget capacity constraint: a Karush-Kuhn-Tucker conditions approach

N.A. Kurdhi[1], R.A. Nurhayati[2], S.B. Wiyono[3], S.S. Handajani[4] and T.S. Martini[5]

[1,2,3,5]Department of Mathematics, Sebelas Maret University, INDONESIA
[4]Department of Statistics, Sebelas Maret University, INDONESIA

E-mail: arfa@mipa.uns.ac.id[1], rizki.aprilia71@gmail.com[2], sabuwhy@gmail.com[3], ssulistijowati@yahoo.com[4] and titinsmartini@yahoo.com[5]

Abstract. In this paper, we develop an integrated inventory model considering the imperfect quality items, inspection error, controllable lead time, and budget capacity constraint. The imperfect items were uniformly distributed and detected on the screening process. However there are two types of possibilities. The first is type I of inspection error (when a non-defective item classified as defective) and the second is type II of inspection error (when a defective item classified as non-defective). The demand during the lead time is unknown, and it follows the normal distribution. The lead time can be controlled by adding the crashing cost. Furthermore, the existence of the budget capacity constraint is caused by the limited purchasing cost. The purposes of this research are: to modify the integrated vendor and buyer inventory model, to establish the optimal solution using Kuhn-Tucker's conditions, and to apply the models. Based on the result of application and the sensitivity analysis, it can be obtained minimum integrated inventory total cost rather than separated inventory.

1. Introduction

Nowadays, integrated inventory management has enjoyed a great deal of attention. The integration model exists because the inventory replenishment decisions are treated separately from the viewpoints of buyer and vendor based on economic order quantity (EOQ) and economic production quantity (EPQ). The results shows that the optimal EPQ solution for the vendor is not acceptable to the buyer, and vice versa. To identify the efficient solutions, both buyer and vendor should coordinate and collaborate to achieve good inventory management. The supply chain collaborative advantages from a firm's perspective have been showed by Chao and Zhang [5], such as process efficiency, offering flexibility, business synergy, quality and innovation. Goyal [7] was a pioneer in the study of the integrated inventory model. He developed a joint economic lot-sized model for single buyer and single vendor with infinite production rate. Banerjee [3] developed the model by incorporating a finite production rate and following a lot-for-lot policy for the vendor. After that, more researchers considered different cases in integrated inventory model about permissible delay in payments [6], inflation [4], service level constraint [18], price-dependent demand [22], and fuzzy random framework [23]. However, a common unrealistic assumption of the above integrated inventory models is that all items received by a buyer are of good quality.

In a real system, due to the imperfect production process of the vendor, damage in transit, or other unforeseeable circumstance, goods that were received by the buyer often contain some defective items. Further, Goyal et al. [8] and Huang [12] developed integrated vendor-buyer inventory models for items with imperfect quality. They assumed that items of poor quality detected in the screening process of a lot are sold at a discounted price. Several researchers, including Hsu and Hsu [10], Lin [20], Bag and Chakraborty [2], Kurdhi et al. [16], Jindal and Solanki [13] proposed integrated vendor-buyer production-inventory model under imperfect quality items in various conditions. However, the above mentioned papers assumed that there is no error in the inspection process. Khan [14] stated that there are still wastes generated by errors in screening, although there have been advancements in technology. This means that the screening errors may occur with imperfect quality in practice. Moreover, Khan [14] asserted that the inspection accuracy is influenced by some factors. More specifically, these factors can be categorized into three groups: inspector related factors, task related factors and environmental factors. Lin [19] proposed an integrated vendor-buyer model for items with imperfect quality and inspection errors. He suggested two types errors usually could be found in inspection process. These are Type I error, in which the inspector may incorrectly classify a non-defective item as defective; and Type II error, in which a defective item is classified as non-defective. The Lin's [19] model assumed that both the Type I and Type II inspection errors are known constants. Hsu and Hsu [11] then developed a production-inventory model with defective items and inspection errors, where the both type of errors are viewed as random variable. There are more papers related to inspection errors on integrated model such as Widianto et al. [25], Khan et al. [15], Priyan and Uthayakumar [21].

In the integrated inventory management system with both imperfect quality items and inspection errors, the integrated inventory model under deterministic demand to be widely discussed. However, it is noticed that the inventory literature on the integrated inventory model under stochastic demand discussing screening errors is quite sparse. Nearly all integrated models that consider the inspection errors assume no uncertainty in demand and shortages are not allowed. On the other hand, Al-Salamah [1] proposed two EOQ models with stochastic demand, imperfect quality, and inspections errors. In the models, shortages are backordered and demand during lead time has a certain probability density function. However, the Al-Salamah's [1] models focused on determining optimal policy for the buyer only (one-echelon model). These models neglect the opportunity that buyer and vendor can cooperate and negotiate with each other to obtain a better integrated policy. We also notice that in the inventory models under inspection errors, shortages, reorder point, lead time reduction, and budget constraint were not considered. Separately, Priyan and Uthayakumar [5] considered budget capacity constraint on an integrated inventory model. On the model, the buyer has limited capacity to purchase products. On the other hand, there is an upper bound on the purchase of products. According to Kurdhi et al. [16, 17], in many practical situations, lead time can be reduced, by an additional crashing cost, customer service level improved, inventory in safety stocks reduced, and the competitive edge in business increased; in other words, it is controllable.

Summarizing the above description, we seek to investigate a two-echelon production-inventory model by considering shortage backlogging, imperfect quality items, inspection errors, lead time, budget capacity constraint, and stochastic demand. In this study, the percentage of defective item and lead time demand are assumed to follow a uniform distribution and a normal distribution, respectively. It is also assumed that the partial backorder situation is considered. This means that the shortages are partial backordered and partial lost sale with a certain backorder rate. The aim is to minimize the integrated inventory system total cost by optimizing the order quantity, reorder point, lead time, and the number of deliveries per batch production run. This paper continuous with notations and assumptions in Section 2. Section 3 and 4 present model formulation and solution methodology, respectively. A numerical example and sensitivity analysis are discussed in Section 5. Finally, conclusions are given in Section 6.

IOP Conf. Series: Materials Science and Engineering **166** (2017) 012013 doi:10.1088/1757-899X/166/1/012013

2. Notations and Assumptions

To develop the proposed model, the following notations and assumptions are introduced.

2.1. Notations

Q_p : size of production batch of items at the vendor
Q : size of the deliveries from the vendor to the buyer (decision variable)
r : reorder point
k : safety factor (decision variable)
n : number of deliveries per batch production run (decision variable)
D : annual demand of the buyer
P : annual production rate at the vendor
y : inspection rate
S_v : setup cost per production run for the vendor
S_B : ordering cost per order for the buyer
γ : probability that an item produced is defective
$f(\gamma)$: probability density function of γ
e_1 : probability of a Type I error (classifying a non-defective item as defective)
e_2 : probability of a Type II error (classifying a defective item as non-defective)
c_s : buyer's inspection cost per unit
c_w : vendor's unit cost for producing a defective item (warranty)
c_{aB} : buyer's cost of a post-sale defective item
c_{av} : vendor's cost of a post-sale defective item
c_a : cost of accepting a defective item ($c_a = c_{aB} + c_{av}$)
c_r : cost of rejecting a non-defective item
h_v : holding cost per unit per year for the vendor
h_B : holding cost per unit per year for the buyer
F : transportation cost per delivery
B_1 : number of items that are classified as defective in each delivery of Q units
B_2 : number of items that are returned from the market in each delivery of Q units
T : time interval between successive deliveries of Q units
T_1 : period during which the vendor produces
T_2 : period during which the vendor supplies from inventory
T_c : cycle time
L : length of lead time for the buyer (decision variable)
$C(L)$: lead time crashing cost
α : buyer's shortage cost (penalty cost)
α_0 : buyer's marginal profit
β : fraction of the demand during the shortage period that will be backordered, $\beta \epsilon [0,1]$
p : buyer's purchasing price per unit item
B : buyer's maximum available budget to purchase products
X : lead time demand with finite mean DL and standart deviation $\sigma\sqrt{L} \geq 0$
$E(\cdot)$: mathematical expectation
x^+ : maximum value of x and 0, i.e., $x^+ = \max\{x, 0\}$
$*$: superscript representing optimal value.

2.2. Assumptions

1. There are a single vendor and a single buyer with a single product.
2. The buyer's inventory is continuously reviewed and replenishments are made whenever the inventory level falls to the reorder point r. The reorder point r = expected demand during lead

time + safety stock (SS), and $SS = k \times$ (standard deviation of lead time demand), i.e., $r = DL + k\sigma\sqrt{L}$, where k is the safety factor.

3. The production processes are imperfect and may produces defective items. The defective percentage (γ) is variable random uniformly distributed with probability density function $f(\gamma)$. The inspection process is also imperfect. There are two types of possibilities. The first is type I of inspection error (when a non-defective item classified as defective) and the second is type II of inspection error (when a defective item classified as non-defective).
4. The lead time L has m mutually independent components. The ith component has a normal duration b_i, a minimum duration a_i, and a crashing cost per unit time c_i. Furthermore, for convenience, we rearrange c_i such that $c_1 \leq c_2 \leq \cdots \leq c_m$. Then, it is clear that the reduction of lead time should first occur on component 1 (because it has the minimum unit crashing cost), and then component 2, etc. If we let $L_0 = \sum_{j=1}^{m} b_j$ and L_i be the length of components $1, 2, \ldots, i$ crashed to their minimum duration, then L_i can be expressed as
$$L_i = \sum_{j=1}^{m} b_j - \sum_{j=1}^{i}(b_j - a_j), i = 1,2,\ldots,n;$$
and the lead time crashing cost $C(L)$ per cycle for a given $L \in [L_i, L_{i-1}]$, is given by
$$C(L) = c_i(L_{i-1} - L) + \sum_{j=1}^{i-1} c_j(b_j - a_j).$$
5. Shortage is allowed for the buyer and partially backordered.
6. The purchasing cost for all products is limited, mathematically,
$$pQ \leq B.$$

3. Model Formulation

In this paper, an integrated single-vendor and single-buyer inventory model involving defective items, inspection errors, stochastic demand, controllable lead time, and partial backorder under budget capacity constraint. The coordination mechanism is such that the vendor receives the buyer's demand and produces the single product at a finite production rate P. The vendor replenishes the order in a number of equal-sized shipments. It is assumed that the vendor's production processes are imperfect and devective items may be produced. Thus, once the buyer receives the lot-size Q, a 100% screening process is conducted. The length of the screening process is $t_1 = Q/y$ year. The screening process is also imperfect in that an inspector may incorrectly classify a non-defective product as defective (Type I error), or a defective product as non-defective (Type II error). It is assumed that the customers who buy the defective items will detect the quality problem and return them to the buyer and receive a good item from the buyer. Later, the buyer returns all items classified as defective and those returned from the customers to the vendor, and receives a full price refund from the vendor. Both the vendor and the buyer incur a post-sale failure cost (e.g., loss of good will) for the items returned from the market. The goal of this study is to simultaneously optimize the order quantity, safety factor, lead time and the number of lots delivered from vendor to buyer with the objective of minimizing the total supply chain integrated cost and the constraint is satisfied.

3.1. The buyer's cost formulation

The buyer's total inventory cost consists of ordering cost, transportation cost, screening cost, post-sale failure cost for the buyer, crashing lead time cost, holding cost, and shortage cost. The behaviour of the inventory level over time for the buyer is described in Fig.1. The buyer's inventory system with random lead time demands, the inventory level is continuous reviewed and new order is triggered until the inventory level declines to r units. If the lead time demand exceeds the reorder point, the inventory system will experience shortages. The shortages are partial lost sale and partial backordered with backorder rate β. Let $(X - r)^+ = \max\{X - r, 0\}$. Then the number of demand backordered is $\beta E[(X - r)^+]$, and the number of demand lost is $(1 - \beta)E[(X - r)^+]$, where $E[(X - r)^+]$ is the expected inventory shortage at the end of a cycle. We note that the lead time demand, X, follows a normal distribution with *p.d.f.* $f(\cdot)$, mean DL and standard deviation $\sigma\sqrt{L}$; and the reorder point $r = DL + k\sigma\sqrt{L}$ (assumption 2). Then we have

$$E[(X-r)^+] = \int_r^\infty (x-r)f(x)dx = \sigma\sqrt{L}\Psi(k),$$

where $\Psi(k) = \phi(k) - k[1-\Phi(k)]$, and ϕ and Φ are standard normal distribution and cumulative distribution function, respectively. Hence, the shortage cost for the buyer per production cycle will be

$$n[\alpha + \alpha_0(1-\beta)]\sigma\sqrt{L}\Psi(k),$$

where α and α_0 represent the buyer's penalty cost and marginal profit.

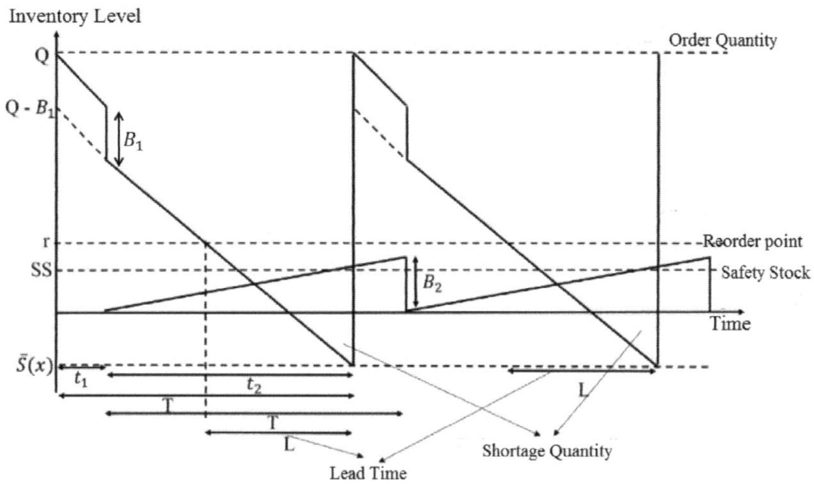

Figure 1. Behavior of the buyer's inventory level over time

By definition, the number of items that are classified as defective include those that are non-defective, $Q(1-\gamma)$, and incorrectly classified as defective (with probability e_1), and those that are defective, $Q\gamma$, and classified as defective (with probability $1-e_2$); thus, we get

$$B_1 = Q(1-\gamma)e_1 + Q\gamma(1-e_2).$$

The number of defective items returned from the market are those that are defective, $Q\gamma$, and incorrectly classified as non-defective (with probability e_2); thus, we have

$$B_2 = Q\gamma e_2.$$

Since we assume that the defective items returned from the market are replaced with good items, the inventory will be depleted at a rate of $D' = D + \frac{B_2}{T}$. By definition, the cycle length

$$T = \frac{Q-B_1}{D'} = \frac{Q(1-\gamma)(1-e_1)}{D}.$$

Hence, the holding cost per production cycle is

$$HC_B = nh_B\left(B_1t_1 + \frac{B_2T}{2}\right) + nh_BT\left(\frac{Q-B_1}{2} + r - DL + (1-\beta)\sigma\sqrt{L}\Psi(k)\right)$$

$$= nh_B\left[\frac{Q^2[(1-\gamma)e_1+\gamma(1-e_2)]}{y} + \frac{Q(1-\gamma)(1-e_1)}{D}\left(\frac{Q}{2}[1-(e_1+\gamma)+\gamma(e_1+2e_1)] + k\sigma\sqrt{L} + (1-\beta)\sigma\sqrt{L}\varphi(k)\right)\right].$$

The ordering cost, transportation cost, screening cost, and crashing lead time cost per delivery cycle are S_B, F, c_iQ, and $C(L)$, respectively. There are n deliveries in one production cycle, then the ordering cost, transportation cost, screening cost, and crashing cost per production cycle are nS_B, nF, nc_sQ, and $nC(L)$, respectively. Further, the buyer incur a post-sale failure cost c_{aB} for each item being returned from the market. Then, the post-sale failure cost for the buyer per production cycle is $nc_{aB}Q\gamma e_2$.

Then, the buyer's total cost in a production cycle is the sum of ordering, transportation, screening, post-sale failure, crashing, holding, and the shortage costs

$$ETC_B(Q,k,L) = nS_B + nF + nC(L) + nc_sQ + nc_{aB}Q\gamma e_2 + n[\alpha + \alpha_0(1-\beta)]\sigma\sqrt{L}\Psi(k)$$
$$+ nh_B\left[\frac{Q^2[(1-\gamma)e_1+\gamma(1-e_2)]}{y} + \frac{Q(1-\gamma)(1-e_1)}{D}\left(\frac{Q}{2}[1-(e_1+\gamma)+\gamma(e_1+2e_1)] + k\sigma\sqrt{L} + (1-\beta)\sigma\sqrt{L}\varphi(k)\right)\right].$$

3.2. The vendor's cost formulation

The vendor prepares for the repeating flow of orders size $Q_p = nQ$ from the buyer by producing items in in batches of size Q_p and by planning to have each batch delivered to the buyers in n deliveries, each with a lot of Q units. Fig. 2 shows the vendor's holding cost per cycle. After adding setup, warranty, type I and type II errors, and holding costs, the expected cost per year for the vendor is

$$ETC_B(n,Q) = \frac{DS_v}{nQ(1-E[\gamma])(1-e_1)} + \frac{D[c_we[\gamma]+c_r(1-E[\gamma])e_1+c_{av}E[\gamma]e_2]}{(1-E[\gamma])(1-e_1)}$$
$$+h_v\left[\frac{QD}{P(1-E[\gamma])(1-e_1)} - \frac{nQD}{2P(1-E[\gamma])(1-e_1)} + \frac{(n-1)Q}{2}\right].$$

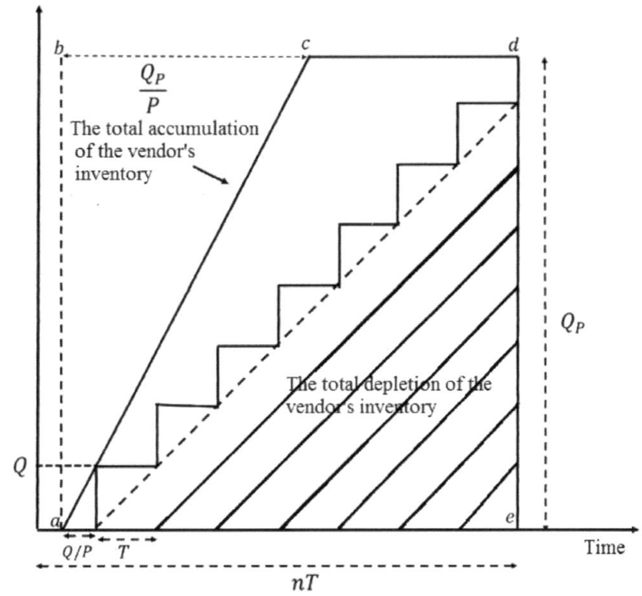

Figure 2. Time-weighted inventory for the vendor

3.3. An integrated vendor-buyer production-inventory model

The expected total annual cost of the vendor and buyer is

$$ETC(n,Q,k,L) = \frac{D\left(S_B+F+C(L)+[\alpha+\alpha_0(1-\beta)]\sigma\sqrt{L}\varphi(k)\right)}{Q(1-E[\gamma])(1-e_1)} + \frac{D[c_s+c_{aB}E[\gamma]e_2]}{(1-E[\gamma])(1-e_1)}$$
$$+h_B\left[\frac{QD\left((1-E[\gamma])e_1+E[\gamma](1-e_2)\right)}{y(1-E[\gamma])(1-e_1)} + \frac{QE[A]}{2(1-E[\gamma])(1-e_1)} + k\sigma\sqrt{L} + (1-\beta)\sigma\sqrt{L}\varphi(k)\right]$$
$$+\frac{DS_v}{nQ(1-E[\gamma])(1-e_1)} + \frac{D[c_we[\gamma]+c_r(1-E[\gamma])e_1+c_{av}E[\gamma]e_2]}{(1-E[\gamma])(1-e_1)}$$
$$+h_v\left[\frac{QD}{P(1-E[\gamma])(1-e_1)} - \frac{nQD}{2P(1-E[\gamma])(1-e_1)} + \frac{(n-1)Q}{2}\right].$$

The purchasing cost for all products is limited, mathematically, $pQ \leq B$. The integrated vendor- buyer inventory model with budget capacity constraint is

Minimize $ETC(n,Q,k,L) =$

$$\frac{D(S_B+F+C(L)+[\alpha+\alpha_0(1-\beta)]\sigma\sqrt{L}\varphi(k))}{Q(1-E[\gamma])(1-e_1)}+\frac{D[c_s+c_{aB}E[\gamma]e_2]}{(1-E[\gamma])(1-e_1)}+h_B\left[\frac{QD((1-E[\gamma])e_1+E[\gamma](1-e_2))}{y(1-E[\gamma])(1-e_1)}+\frac{QE[A]}{2(1-E[\gamma])(1-e_1)}+k\sigma\sqrt{L}+(1-\beta)\sigma\sqrt{L}\varphi(k)\right]+\frac{DS_v}{nQ(1-E[\gamma])(1-e_1)}+\frac{D[c_w e[\gamma]+c_r(1-E[\gamma])e_1+c_{av}E[\gamma]e_2]}{(1-E[\gamma])(1-e_1)}+h_v\left[\frac{QD}{P(1-E[\gamma])(1-e_1)}-\frac{nQD}{2P(1-E[\gamma])(1-e_1)}+\frac{(n-1)Q}{2}\right], \quad (1)$$

subject to $pQ \leq B$.

4. Solution Technique

Taha [24] discussed how to solve the optimum solution of nonlinear programming problem subject to inequality constraints by using the Kuhn-Tucker conditions. The development of the Kuhn-Tucker conditions is based on the Lagrangian method. Suppose that the problem (1) can be written as follow.

Minimize $f(z) = ETC(n, Q, k, L)$, (2)

subject to $g(z) = pQ - B \leq 0$.

A new function, i.e the Lagrangian function $ETC(n, Q, k, L, \lambda)$ is formed by introducing Lagrangian multiplier λ, then we have

$$ETC(z,\lambda) = f(z) - \lambda g(z)$$
$$= \frac{D\left(S_B+F+C(L)+[\alpha+\alpha_0(1-\beta)]\sigma\sqrt{L}\varphi(k)\right)}{Q(1-E[\gamma])(1-e_1)}+\frac{D[c_s+c_{aB}E[\gamma]e_2]}{(1-E[\gamma])(1-e_1)}+\frac{DS_v}{nQ(1-E[\gamma])(1-e_1)}+\frac{D[c_w e[\gamma]+c_r(1-E[\gamma])e_1+c_{av}E[\gamma]e_2]}{(1-E[\gamma])(1-e_1)}+h_B\left[\frac{QD\left((1-E[\gamma])e_1+E[\gamma](1-e_2)\right)}{y(1-E[\gamma])(1-e_1)}+\frac{QE[A]}{2(1-E[\gamma])(1-e_1)}+k\sigma\sqrt{L}+(1-\beta)\sigma\sqrt{L}\varphi(k)\right]+h_v\left[\frac{QD}{P(1-E[\gamma])(1-e_1)}-\frac{nQD}{2P(1-E[\gamma])(1-e_1)}+\frac{(n-1)Q}{2}\right]-\lambda(pQ-B). \quad (3)$$

The Kuhn-Tucker conditions need z and λ to be a stationary point of minimization problem which can be summarized as following:

$$\begin{cases}\nabla f(z)-\lambda\nabla g(z)=0,\\ \lambda g(z)=0,\\ g(z)\leq 0,\\ \lambda\geq 0.\end{cases} \quad (4)$$

By the method of Kuhn-Tucker conditions, consider the two cases $\lambda = 0$ and $\lambda \neq 0$.

i. For $\lambda = 0$, from (3) and (4), we have

$$Q^* = \sqrt{\frac{D\left(nS_B+S_v+nF+nC(L)\right)+nD(\alpha+\alpha_0(1-\beta))\sigma\sqrt{L}\varphi(k)}{M}}, \quad (5)$$

and

$$\Phi(k^*) = 1-\frac{h_B Q(1-E[\gamma])(1-e_1)}{h_B Q(1-E[\gamma])(1-\beta)+D(\alpha+\alpha_0(1-\beta))}, \quad (6)$$

with

$$M = nh_B\left\{\frac{D}{y}[(1-E[\gamma])e_1+E[\gamma](1-e_2)]+\frac{E[A]}{2}\right\}nh_v\left\{\frac{D}{P}-\frac{nD}{2P}+\frac{(n-1)}{2}(1-E[\gamma])(1-e_1)\right\}.$$

The explicit general solution for Q and k cannot be obtained by solving equations (5) and (6) because the evaluation each of the expressions requires knowledge of the value of the other. The value of (Q, k) can be obtained by adopting a similar graphical technique used in Hadley and Within [9]. The same numerical search technique also has been used in Lin [19, 20] and Kurdhi et al. [16, 17]. Therefore, the following iterative algorithm is established to determine the solution of (n, Q, k, L) for $\lambda = 0$.

Algorithm 1

a. Set $n = 1$ and $\lambda = 0$.
b. For each L_i, $i = 0,1,2,\ldots,m$, perform (i) to (vii).
   (i) Start with $k_{i1} = 0$, $\phi(k_{i1}) = 0.39894$, and $\Phi(k_{i1}) = 0.5$.
   (ii) Compute $\varphi(k_{i1})$.
   (iii) Substituting $\varphi(k_{i1})$ into Equation (5) to evaluate Q_{i1}.

(iv) Utilizing Q_{i1} to determine $\Phi(k_{i2})$.
(v) Check $\Phi(k_{i2})$ on standard normal table to determine k_{i2} and then $\phi(k_{i2})$.
(vi) Repeat (i) to (v) until no change occurs in the values of Q_i and k_i. These can be notated by Q_i^* and k_i^*.
(vii) Compute $g(z)$ using Q_i^*.

c. Compute $ETC(n, Q_i^*, k_i^*, L_i)$, $i = 0,1,2, \ldots, m$.
d. Set $ETC(n, Q_n^*, k_n^*, L_n^*) = min_{i=0,1,2,\ldots,m}\{ETC(n, Q_i^*, k_i^*, L_i)\}$, then (n, Q_n^*, k_n^*, L_n^*) is the solution for fixed n.
e. Set $n = n + 1$ and repeat step (b)-(d) to get $ETC(n, Q_n^*, k_n^*, L_n^*)$.
f. If $ETC(n, Q_n^*, k_n^*, L_n^*) \le ETC(n-1, Q_{n-1}^*, k_{n-1}^*, L_{n-1}^*)$, then go to step (e), otherwise go to step (g).
g. Set $ETC(n^*, Q^*, k^*, L^*)$= $ETC(n-1, Q_{n-1}^*, k_{n-1}^*, L_{n-1}^*)$, then $ETC(n^*, Q^*, k^*, L^*)$ is the solution.

ii. For $\lambda \neq 0$, from (3) and (4), we have

$$\lambda^* = -\frac{D(nS_B+nF+C(L)+S_V)}{pnQ^2(1-E[\gamma])(1-e_1)} - \frac{D[\alpha+\alpha_0(1-\beta)]\sigma\sqrt{L}\varphi(k)}{pQ^2(1-E[\gamma])(1-e_1)} + \frac{h_B}{p}\left\{\frac{D\big((1-E[\gamma])e_1+E[\gamma](1-e_2)\big)}{y(1-E[\gamma])(1-e_1)} + \frac{E[A]}{2(1-E[\gamma])(1-e_1)}\right\} + \frac{h_V}{p}\left\{\frac{D}{P(1-E[\gamma])(1-e_1)} - \frac{nD}{2P(1-E[\gamma])(1-e_1)} + \frac{(n-1)}{2}\right\}, \quad (7)$$

$$Q^* = \sqrt{\frac{D\big(nS_B+S_v+nF+nC(L)\big)+nD(\alpha+\alpha_0(1-\beta))\sigma\sqrt{L}\varphi(k)}{W}}, \quad (8)$$

and

$$\Phi(k^*) = 1 - \frac{h_BQ(1-E[\gamma])(1-e_1)}{h_BQ(1-E[\gamma])(1-\beta)+D\big(\alpha+\alpha_0(1-\beta)\big)}, \quad (9)$$

with

$$W = nh_B\left\{\frac{D}{y}[(1-E[\gamma])e_1 + E[\gamma](1-e_2)] + \frac{E[A]}{2}\right\} + nh_v\left\{\frac{D}{P} - \frac{nD}{2P} + \frac{(n-1)}{2}(1-E[\gamma])(1-e_1)\right\} - n\lambda p(1-E[\gamma])(1-e_1).$$

With the same argument as case $\lambda = 0$, the following iterative algorithm is established to determine the solution of (n, Q, k, L) for $\lambda \neq 0$.

Algorithm 2

a. Set $n = 1$.
b. For each L_i, $i = 0,1,2, \ldots, m$, perform (i) to (ix).
   (i) Start with $\lambda_{i1} = 0.1$, $k_{i1} = 0$, $\phi(k_{i1}) = 0.39894$, and $\Phi(k_{i1}) = 0.5$.
   (ii) Compute $\varphi(k_{i1})$.
   (iii) Substituting $\varphi(k_{i1})$ into (8) to evaluate Q_{i1}.
   (iv) Utilizing Q_{i1} to determine $\Phi(k_{i2})$.
   (v) Substituting Q_{i1} and $\varphi(k_{i1})$ into (7) to evaluate λ_{i2}.
   (vi) Check $\Phi(k_{i2})$ on standard normal table to determine k_{i2}, then $\phi(k_{i2})$ and $\varphi(k_{i2})$.
   (vii) Substituting λ_{i2} and $\varphi(k_{i2})$ into (8) to evaluate Q_{i2}.
   (viii) Repeat (iv) to (vii) until no change occurs in the values of Q_i, k_i, and λ_i. These can be notated by $Q_i{}^*$, $k_i{}^*$, and $\lambda_i{}^*$.
   (ix) Compute $g(z)$ using $Q_i{}^*$.
c. Compute $ETC(n, Q_i^*, k_i^*, L_i, \lambda_i^*)$, $i = 0,1,2, \ldots, m$.
d. Set $ETC(n, Q_n^*, k_n^*, L_n^*, \lambda_n^*) = min_{i=0,1,2,\ldots,m}\{ETC(n, Q_i^*, k_i^*, L_i, \lambda_i^*)\}$, then $(n, Q_n^*, k_n^*, L_n^*, \lambda_n^*)$ is the solution for fixed n.
e. Set $n = n + 1$ and repeat step (b)-(d) to get $(n, Q_n^*, k_n^*, L_n^*, \lambda_n^*)$.

f. If $ETC(n,\ Q_n^*,\ k_n^*, L_n^*, \lambda_n^*) \leq ETC(n-1,\ Q_{n-1}^*,\ k_{n-1}^*, L_{n-1}^*, \lambda_{n-1}^*)$, then go to step (e), otherwise go to step (g).

g. Set $ETC(n^*, Q^*, k^*, L^*, \lambda^*) = ETC(n-1, Q_{n-1}^*, k_{n-1}^*, L_{n-1}^*, \lambda_{n-1}^*)$, then $ETC(n^*, Q^*, k^*, L^*, \lambda^*)$ is the solution.

5. A numerical example

In this section, an example is solved using the proposed solution in the previous section. The purpose is to illustrate the solution procedure and conduct some sensitivity analysis for important model parameters. Consider an integrated vendor and buyer inventory model with the following parameters:

Production rate (P) = 160,000 units/year
Demand rate (D) = 50,000 units/year
Inspection rate (y) = 155,200 units/year
Setup cost for vendor (S_V) = \$300/ production run
Ordering cost for buyer (S_B) = \$100/order
Holding cost for vendor (h_v) = \$2/unit/year
Holding cost for buyer (h_B) = \$5/unit/year
Transportation cost (F) = \$25/delivery
Inspection cost (c_s) = \$0.5/unit
The cost of producing a defective item (c_w) = \$50/unit
The cost of rejecting a non-defective item (c_r) = \$100/unit
The buyer's post sale failure cost (c_{aB}) = \$200/unit
The vendor's post sale failure cost (c_{aV}) = \$300/unit
Buyer's shortage cost (α) = \$25/unit
Buyer's marginal profit(α_0) = \$75/unit
Buyer's purchasing price (p) = \$10/unit
Buyer's maximum available budget to purchase products (B) = \$30,000/year.

The lead time has three components with data shown in Table 1. Moreover, we assume that the defective percentage is uniformly distributed with *p.d.f.* as

$$f(\gamma) = \begin{cases} \dfrac{1}{\mu}, 0 \leq \gamma \leq \mu; \\ 0, \text{otherwise.} \end{cases}$$

Table 1. Lead time data.

Lead time component, i	Normal duration, b_i (days)	Minimum duration, a_i (days)	Unit crashing cost, c_i (days)
1	20	6	0.4
2	20	6	1.2
3	16	9	5.0

By the method of Kuhn-Tucker conditions, consider the two cases 1) $\lambda = 0$ and 2) $\lambda \neq 0$. Table 2 shows the solutions of n, Q, k, L, r, and the minimum integrated inventory total cost ETC for $\lambda = 0$.

Table 2. The solutions for $\lambda = 0$.

λ	n	L (weeks)	Q (unit)	$k(r)$ (unit)	$pQ - B$	$ETC(.)$ (\$)
0	1	8	2860.37	2.47194(7702.00)	-1396.26	325244.74
0	2	8	2066.32	2.58600(7702.45)	-4336.79	323597.14
0	3	8	1710.55	2.65040(7702.70)	-12894.50	323432.15
0	3	8	1495.70	2.69540(7702.88)	-10043.00	323663.70

From Table 2, we obtain $n^* = 3$ times, $L^* = 8$ weeks, $Q^* = 1710.55$ units, $k^* = 2.65040$, $r = 7702.70$ units, and the minimum total expected annual cost is \$323432.15. We also have $pQ - B = -12894.50 < 0$, then the lot size satisfy the budget constraint. The Kuhn-Tucker conditions are satisfied, so the solution is optimal and feasible. Table 3 shows the solutions of n, Q, k, L, r, and the minimum integrated inventory total cost ETC for $\lambda \neq 0$. From the table, we get $\lambda^* = 0.190555$, $n^* = 1$ times, $L^* = 8$ weeks, $Q^* = 5119.92$ units, $k^* = 2.25670$, $r = 7701.16$ units, and the minimum total expected annual cost is \$323992.96. However, the lot size does not satisfy the budget constraint, that is $pQ - B = 21199.22 > 0$. Hence, the solution is not feasible.

Table 3. The solutions for $\lambda \neq 0$.

λ	n	L (weeks)	Q (unit)	$k(r)$ (unit)	$pQ-B$	$ETC(.)$ (\$)
0.190555	1	8	5119.92	2.25670(7701.16)	21199.22	323992.96
0.250710	2	8	3973.81	2.35230(7701.53)	9738.11	324329.39
0.306326	3	8	3398.39	2.40981(7701.76)	3983.90	325679.13

In order to study how the parameters affect the integrated optimal solution, the sensitivity analyses for transportation cost (F), vendor's holding cost (h_v), buyer's holding cost (h_B), upper limit of defective percentage (μ), probability of Type I error (e_1), and probability of Type II error (e_2) are performed.

Table 4 shows the optimal solutions for different transportation costs. When the transportation cost increases, the joint total cost also increase. The percentage increase in joint total cost is 0.05%. We may also observe that increasing in transportation cost, will increase the order quantity and decrease the reorder point. The smaller the transportation cost up to \$10, the larger the cost reduction of the integrated model in comparison to independent decision is.

Table 4. Optimal solution for different transportation cost.

F (\$)	L (weeks)	n	Q (unit)	$k\ (r)$ (unit)	ETC Independent (\$)	ETC Integrated (\$)	Cost reduction (\$)
5	8	3	1632.97	2.6660 (7702.76)	322826.05	322794.44	31.61
10	8	3	1652.71	2.6620 (7702.75)	322979.18	322936.44	42.74
15	8	3	1672.21	2.6580 (7702.73)	323132.31	323117.01	15.30
20	8	3	1691.49	2.6542 (7702.72)	323285.25	323275.47	9.78
25	8	3	1710.55	2.6504 (7702.70)	323437.83	323432.15	5.68

Table 5 shows the optimal solutions for different vendor's holding cost per unit. From the table, one can see that the higher the vendor's holding cost is, the greater the joint total cost is. The percentage increase in joint total cost is 0.40%. One can also see that when the vendor's holding cost decreases, it is more profitable to decrease the order quantity and to deliver the items more frequently from the vendor to the buyer. The greater the vendor's holding cost up to \$8, the larger the cost reduction of the integrated model in comparison to independent decision is.

Table 5. Optimal solution for different vendor's holding cost per unit.

h_v (\$)	L (weeks)	n	Q (unit)	$k\ (r)$ (unit)	ETC Independent (\$)	ETC Integrated (\$)	Cost reduction (\$)
2	8	3	1710.55	2.65040 (7702.70)	323437.83	323432.15	5.68
4	8	2	1818.33	2.62970 (7702.62)	325591.62	325526.24	65.38
6	8	1	2568.64	2.51013 (7702.15)	328187.55	327038.22	1149.33
8	8	1	2452.67	2.52640 (7702.22)	329288.64	327869.74	1418.90
10	8	1	2351.09	2.54120 (7702.74)	329839.19	328665.40	1173.79

IOP Conf. Series: Materials Science and Engineering **166** (2017) 012013 doi:10.1088/1757-899X/166/1/012013

Table 6 shows the optimal solutions for different buyer's holding cost per unit. One can see that the smaller the buyer's holding cost is, the greater the advantages to deliver the items more rarely from the vendor to the buyer. Moreover, when the buyer's holding cost increases, the joint total cost also increase. The percentage increase in joint total cost is 0.82%.

Table 6. Optimal solution for different buyer's holding cost per unit.

h_B ($)	L (weeks)	n	Q (unit)	k (r) (unit)	ETC Independent ($)	ETC Integrated ($)	Cost reduction ($)
5	8	3	1710.55	2.65040 (7702.70)	323437.83	323432.15	5.68
10	8	4	1219.21	2.52840 (7702.22)	327036.45	327027.61	8.84
15	8	5	981.97	2.46144 (7701.96)	329769.97	329757.70	12.27
20	8	6	838.54	2.41455 (7701.84)	332056.19	332056.83	0.64
25	8	7	741.07	2.37813 (7701.69)	334094.30	334091.79	2.51

Table 7 shows the optimal solutions for different probability of Type I error. One can see that when the probability of Type I error increases, both the buyer and the vendor incur a higher joint total annual cost. The percentage increase in joint total cost is 82.16%. The smaller the probability of Type I error up to 0.08, the smaller the cost reduction of the integrated model in comparison to independent decision is.

Table 7. Optimal solution for different probability type I error.

e_1	L (weeks)	n	Q (unit)	k (r) (unit)	ETC Independent ($)	$ETC(.)$ Integrated ($)	Cost reduction ($)
0.04	8	3	1710.55	2.65040 (7702.70)	323437.83	323432.15	5.68
0.08	8	3	1757.22	2.65560 (7702.72)	554497.57	554492.15	5.42
0.12	8	3	1805.93	2.66140 (7702.75)	806567.65	806562.20	5.45
0.16	8	3	1856.77	2.66772 (7702.77)	1082649.96	1082644.15	5.81
0.20	8	3	1909.78	2.67470 (7702.79)	1386346.88	1386340.27	6.62

Table 8 shows the optimal solutions for different probability of Type II error. From the table, we may see that the probability of Type II error has a similar impact as the probability of Type I error. When the probability of Type II error increases, the joint total cost also increases. The percentage increase in joint total cost is 6.57%. The smaller the probability of Type II error up to 0.12, the smaller the cost reduction of the integrated model in comparison to independent decision is.

Table 8. Optimal solution for different probability type II error.

e_2	L (weeks)	n	Q (unit)	$k(r)$ (unit)	ETC Independent ($)	ETC Integrated ($)	Cost reduction ($)
0.04	8	3	1710.55	2.6504(7702.7)	323437.83	323432.15	5.68
0.08	8	3	1709.97	2.6505(7702.7)	344701.18	344695.41	5.77
0.12	8	3	1709.39	2.6506(7702.7)	365964.52	365958.66	5.58
0.16	8	3	1708.81	2.6507(7702.7)	387227.87	387221.92	5.95
0.20	8	3	1708.23	2.6508(7702.7)	408491.22	408485.17	6.05

Table 9 shows the optimal solutions for different upper limit of defective percentage. From the table, we may see that when the upper limit of defective percentage increases, the joint total cost and the order quantity tend to increase. The percentage increase in joint total cost is 25.78%. The smaller the upper limit of defective percentage up to 0.08, the smaller the cost reduction of the integrated model in comparison to independent decision is.

Table 9. Optimal solution for different upper limit of defective percentage.

μ	L (weeks)	n	Q (unit)	k (r) (unit)	ETC Independent ($)	ETC Integrated ($)	Cost reduction ($)
0.04	8	3	1710.55	2.65040 (7702.70)	323437.83	323432.15	5.68
0.08	8	3	1732.57	2.65300 (7702.71)	401606.19	401600.57	5.62
0.12	8	3	1755.06	2.65580 (7702.72)	483101.76	483096.14	5.62
0.16	8	3	1778.04	2.65860 (7702.74)	568141.59	568135.91	5.68
0.20	8	3	1801.50	2.66162 (7702.75)	656961.98	656956.16	6.82

6. Conclusions

In this paper, we propose the two-echelon supply chain inventory model by considering shortage backlogging, imperfect quality items, inspection errors, lead time, budget capacity constraint, and stochastic demand. An analytic solution procedure is developed to determine the optimal number of shipments, the size of each shipment, the reorder point, and the lead time. An investigation of the effects of six important parameters (transportation cost, vendor's holding cost, buyer's holding cost, upper limit of defective percentage, Type I error, and Type II error) on the optimal solution is also made. Numerical results show that (1) the joint total annual cost is more sensitive to the variation of the upper limit of percentage defective, Type I error, and Type II error; (2) the cost reduction of the integrated model in comparison to independent decision can be increased by decreasing the transportation cost; or by increasing the upper limit of defective percentage, the probability of Type I and II errors, and the vendor's holding cost.

Possible extensions for future research could be by considering learning in production. In this case, the vendor experiences learning in the production process while some of the units are defective. The effect of learning in buyer's inspection errors on supply chain cost also can be investigated. Furthermore, the other possibility is considering the alternative inventory models such as periodic review.

Acknowledgment

This research is supported by PNBP Sebelas Maret University, Surakarta, Indonesia, year 2016.

References

[1] Al-Salamah M 2012 Economic order quantity with stochastic demand, imperfect quality, and inspection errors *Journal of Engineering and Computer Sciences* **5** 2 pp 131-145.
[2] Bag S and Chakraborty D 2014 An integrated inventory model of imperfect quality products with fuzzy chance constraints *Journal of Uncertainty in Mathematics Science* **2014** jums-00013 19 Pages.
[3] Banerjee A 1986 A joint economic-lot-size model for purchaser and vendor *Decision Sciences* **17** 3 pp 292–311.
[4] Bansal K K and Ahalawat N 2012 Integrated inventory models for decaying items with exponential demand under inflation *International Journal of Soft Computing and Engineering* **2** 3 pp 578-587.
[5] Cao M and Zhang Q 2010 Supply chain collaborative advantage: A firm's perspective *International Journal of Production Economics* **128** 1 pp 358-367.
[6] Chen L-H and Kang F-S 2007 Integrated vendor–buyer cooperative inventory models with variant permissible delay in payments *European Journal of Operational Research* **182** 2 pp 658–673.
[7] Goyal S K 1976 An integrated inventory model for a single supplier-single customer problem *International Journal of Production Research* **15** 1 pp 107–111.
[8] Goyal S K, Huang C K, and Chen K C 2003 A simple integrated production policy of an imperfect item for vendor and buyer *Production Planning & Control* **14** pp 596- 602.
[9] Hadley G and Whitin T M 1963 *Analysis of Inventory Systems* Prentice-Hall, Englewood Cliffs, New Jersey.

[10] Hsu J-T and Hsu L F 2012 An integrated vendor-buyer cooperative inventory model for items with imperfect quality and shortage backordering *Advances in Decision Sciences* **2012** 679083 19 pages.

[11] Hsu J-T and Hsu L F 2012 An integrated single-vendor single-buyer production-inventory model for items with imperfect quality and inspection errors *International Journal of Industrial Engineering Computations* **3** pp 703-720.

[12] Huang C K 2004 An optimal policy for a single-vendor single-buyer integrated production-inventory problem with process unreliability consideration *International Journal of Production Economics* **91** pp 91-98.

[13] Jindal P and Solanki A 2016 Integrated vendor-buyer inventory models with inflation and time value of money in controllable lead time *Decision Science Letters* **5**, pp 81–94.

[14] Khan M 2010 Inventory control in a two-level supply chain with learning, quality and inspection errors *Theses and dissertations* Paper 740.

[15] Khan M, Ahmad A R, and Jaber M Y 2014 An integrated supply chain model with errors in quality inspection and learning in production *Omega* **42** 1 pp 16-24.

[16] Kurdhi N A, Lestari S M P, and Susanti Y 2015 A fuzzy collaborative supply chain inventory model with controllable setup cost and service level constraint for imperfect items *International Journal of Applied Management Science* **7** 2 pp 93-122.

[17] Kurdhi N A, Prasetyo J, and Handajani S S 2015 An inventory model involving back-order price discount when the amount received is uncertain *International Journal of Systems Science* **47** 3 pp 662-671.

[18] Li G, Guan X, Liu M-Q, and Shan M-Y. 2013 Two-echelon inventory model with service level constraint and controllable lead time sensitive to order quantity *Journal of Central South University* **20** 11 pp 3324-3333.

[19] Lin T-Y 2009 Optimal policy for a simple supply chain system with defective items and returned cost under screening errors *Journal of the Operations Research Society of Japan* **52** 3 pp 307-320.

[20] Lin H-J 2013 An integrated supply chain inventory model with imperfect-quality items, controllable lead time and distribution-free demand *Yugoslav Journal of Operations Research* **23** 1 pp 87-109.

[21] Priyan S and Uthayakumar R 2015 An integrated production–distribution inventory system involving probabilistic defective and errors in quality inspection under variable setup cost *International Transactions in Operational Research* doi:10.1111/itor.12202 pp 1-38.

[22] Rad M A, Khoshalhan F and Setak M 2014 Supply chain single vendor – single buyer inventory model with price - dependent demand *Journal of Industrial Engineering and Management* **7** 4 pp 733-748.

[23] Soni H N and Patel K A 2015 Optimal policies for integrated inventory system under fuzzy random framework *The International Journal of Advanced Manufacturing Technology* **78** 5 pp 947-959.

[24] Taha H A 1997 *Operations Research* Prentice Hall New Jersey USA.

[25] Widianto I P, Jauhari W A, and Rosyidi C N 2014 Cooperative Vendor-Buyer Inventory Model with Imperfect Quality and Inspection Errors *Proceedings of the International MultiConference of Engineers and Computer Scientists* **II** pp 990-994.

Analysis of production-inventory decisions in a decentralized supply chain with price-dependent demand

N.A. Kurdhi[1], **S.T. Irsanianto**[2], **Sutanto**[3]

[1,2,3]Department of Mathematics, Sebelas Maret University, INDONESIA

E-mails: arfa@mipa.uns.ac.id, shidiqt@gmail.com, sutantomipa@gmail.com

Abstract. In this paper, we consider a production-inventory supply chain system with single-manufacturer and single-retailer. There are many types of contract that guarantee the supply chain. However, the administrative costs of the contract are usually neglected in real situation. The additional gain from integration may not cover the extra administrative costs may not addressed to supply chain. Therefore, a Stackelberg game and RFM policy are examined in order to investigate its performance on supply chain. The RFM policy is applied because its administrative costs are lower than othe policies. Although RFM policy is not capable of coordinating the channel, it leads to considerable improvements over the channel. The purpose of this research is to present a model of integrated policy, in which the goal is to maximize the whole system profit, and to evaluate decentralized-Stackelberg and RFM policies, in which individual firms in the supply chain have their own objectives and decisions to optimize.

1. Introduction

Generally, a supply chain is composed of independent members, each with its own individual costs and objectives. It is important how the members behave to manage their inventory. The integrated policy should be chosen when the take care about overall system performance. However, each member may be interested in maximing its own profit independently. In this case, the whole channel's performance could not necessarily be optimzed, *i.e.* the decentralized policy. Therefore, the overall system performance may be improved by using a collaboration mechanism. Note that the channel will be collaborated under one contract if i) channel's profit reaches its maximum as in integrated policy; and, ii) none of the members' profit is worse compared with the decentralized policy. There is extensive research on the channel cooperation problem by means of designing efficient contracts, such as revenue sharing contracts, buyback contract, and cost sharing. The supply chain incurs some extra administrative costs by applying those contracts. The additional gain from coordination mat not cover the costs. Alaei et al. [1] considered the production-inventory problem in a two level supply chain which is formulated as a Stackelberg game. They examined the retail fixed mark-up (RFM) in order to investigated its performance on supply chain. In the Stackelberg approach, two firms play a game to obtain Stackelberg equilibrium. The equilibrium is a pair of policies in which each firm maximizes its own profit assuming the other player sets his equilibrium policy. Furthermore, in RFM policy, the manufacturer determines the wholesale price first, that is equivalent to setting the retail price. In this policy, only the retail price must be monitored. Thus, the administrative costs related to this type of contract are lower than those of other contracts. In many industries fixed mark-ups are used such as electronics industry, gasoline dealers, and grocers. Moreover, RFM is also investigated in marketing environent. Li and Atkins [8] consider RFM policy for operation sections and marketing in a single firm. Liou et al. [10] proposed multy-period inventory models. They studied the problem under the Stackelberg approach to achieve the optimal policies. Furthermore, several mult-echelon inventory systems are analyzed by using Stackelberg game.

Many integrated manufacturer-retailer inventory models assumed that the demand rate is constant and is not affected by the retail price of the product. It means that the demand remains the same over the changing of the retail price. Rad and Khoshalhan [11] proposed the integrated inventory model with backorder and assumed that the demand rate is constant. Shah et al. [13] developed the integrated inventory model with the influence of availability of stock goods under the constant demand rate. However, according to Within [14], retail price is one of the important decision variable which influences. Within [14] developed the economic order quantity (EOQ) model with pricing for a buyer that has a price dependent demand with a linear function. Many researchers are encouraged by his work to investigate joint ordering and pricing problems. The focus of these models has been on demand functions (e.g., Lau and Lau [7]), on quantity discount (e.g., Lin and Ho [9]), or on perishable inventories (e.g., Khanra et al. [5]). Chung and Wee [3] proposed joint ordering and pricing problemsin which multiples companies in a supply chain coordinate with each other. Kim et al [6] developed a supply chain consisting of a single manufacturer and a single retailer under joint ordering and pricing policies for price-dependent demand. Chung and Liao [2] also introduced the integrated inventory model that involve price-sensitive demands. Recently, Rad et al. [12] discussed the integrated inventory model that considers operations and pricing decisions, where the demand rate has an iso-elastic function of the selling price.

To the best of knowledge, none of the above-mentioned decentralized models focused on the effects of decentalized policies on the performance of the supply chain comparing to integrated production-inventory models, especially when the demand rate has an iso-elastic function of the selling price. Therefore, in this paper, we consider a supply chain with a single manufacturer and a single retailer in a production-inventory system. An outside supplier supplies raw material to the vendor with zero lead time, and the vendor produces a product, and supplies it to a buyer who in turn supplies it to the consumers. Furthermore, it is assumed that the buyer faces price-dependent demand. End customer demand is assumed to be an iso-elastic function of the selling price to account for the impact of price changes on customer demand. The buyer uses EOQ inventory policy for controlling his costs. The vendor operates on a make-to-order basis and uses a lot-for-lot policy. The research conducted in this paper presents a model of (1) integrated policy, in which the goal is to maximize the whole system profit, and (2) to evaluate decentralized-Stackelberg and decentralized-RFM (Retail Fixed Mark-up) policy, in which individual firms in the supply chain have their own objectives and decisions to optimize.

2. Notations and Assumptions

To develop the proposed model, the following notations and assumptions are introduced.

2.1. Notations

A : Retailer fixed ordering cost
S : Manufacturer fixed setup cost
h : Retailer unit holding cost *per* unit time
H : Manufacturer unit holding cost *per* unit time
T : Cycle time
D : Price-dependent annual market demand
p : Retail price *per* unit (decision variable), $p > 0$
w : Wholesale price *per* unit (decision variable), $c < w < p$
c : Procurement cost *per* unit, $0 < c < w$
k_1 : Time-dependent production cost *per* unit time
k_2 : Technology development cost, *per* one unit increasing on the production rate
l : Lead time

μ : Manufacturer production rate
Q : Order Quantity (decision variable)
TP_r : Retailer profit
TP_m : Manufacturer profit
JTP : The joint total profit

2.2. Assumptions

1. There are single-manufacturer and single-retailer for a single product in this model.
2. Shortage is not allowed.
3. For each unit of product, the manufacturer spends c in production and receives w from retailer. After that, the retailer sell it by p to its customer. The relationship between them is $p > w > c$.
4. The demand rate is a decreasing function of the retail price, $D(p) = \gamma p^{-\beta}$, where $\gamma > 0$ is a scaling factor and $\beta > 1$ is the index of price elasticity. This type of demand function, which has been used by researchers such as Hays and DeLurgio (2009), Lin and Ho [4] and Rad *et al.* [6], for example, is commonly referred to as iso-elastic demand function. For notational simplicity, $D(p)$ and D will be used interchangeably in this article.

3. Model Formulation

In this section, the retailer's and the manufacturer's inventory model are derived.

3.1. Retailer's total profit

We assume that the retailer uses the EOQ inventory policy as shown in Figure 1 for controlling his costs. The retailer's total profit includes the profit of the products, average ordering cost, and average holding cost. Thus the total annual profit of the retailer is given by

$$TP_r(p,Q) = \text{sales revenue of retailer} - \text{ordering cost} - \text{purchasing cost} - \text{holding cost}$$
$$= \left(\gamma p^{-\beta}\right)\left(p - \frac{A}{Q} - w\right) - \frac{hQ}{2}. \quad (1)$$

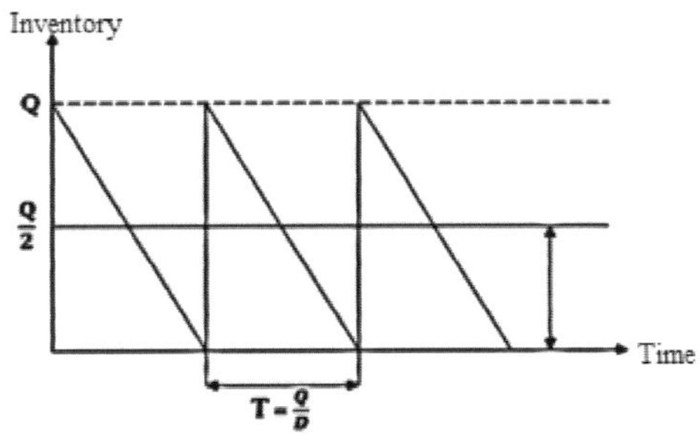

Figure 1. Retailer's inventory level

3.2. Manufacturer's Total Profit.

The manufacturer's inventory level is shown in Figure 2. The manufacturer's total profit consists of sales revenue, holding cost, setup cost, and time-dependent production cost. The setup cost is divided into two parts: one part is fixed for every production period, and another one is an increasing function of the production rate. For instance, assume an assembly line that has the technology for assembling a set of parts that are supplied by a supplier. In this assembly line, time-dependent production cost coincides with the daily production cost. It is necessary to enhance the technology when the production rate exceeds a specific limit. Therefore, it is assumed that the manufacturer incurred a cost called technology development cost for every increasing unit on the production rate. For example, if the manufacturer incurred \$400 for increasing 200 units on the production rate, then the unit technology development cost will be \$2.

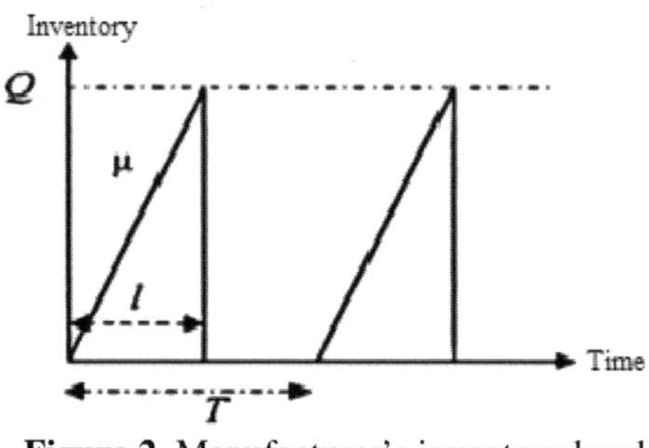

Figure 2. Manufacturer's inventory level

The manufacturer operates on a make-to-order basis using a lot for-lot policy, hence, the manufacturer begins to produce a batch of Q at the rate of μ, as soon as he receives an order and delivers it to the retailer after the lead time. It is assumed that the manufacturer has to produce the product with minimum possible production rate during the lead time, so we have $\mu = Q/l$. Thus, the total annual profit of the manufacturer can be expressed by

$$TP_m(p,Q) = \text{sales revenue of manufacturer} - \text{setup cost} - \text{time-dependent production cost} - \text{holding cost}$$
$$= (\gamma p^{-\beta})\left(w - c - \frac{S}{Q} - \frac{k_1 l}{Q} - \frac{k_2}{l} - \frac{Hl}{2}\right). \quad (2)$$

4. Policies

4.1. Integrated Policy

In this policy, the goal is maximizing the joint total profit (JTP) which is the sum of the total annual profit for retailer (TP_r) and manufacturer (TP_m). Then the problem to be solved is to maximize

$$JTP(p,Q) = (\gamma p^{-\beta})\left(p - c - \frac{(S+A+k_1\, l)}{Q} - \frac{Hl}{2} - \frac{k_2}{l}\right) - \frac{h\,Q}{2}. \quad (3)$$

Where retail price (p) and order quantity (Q) are decision variables.

Proposition 1. *The integrated order quantity Q^* is one of the positive roots of the following equation:*

$$h(Q)^2 - 2^{1+\beta}\gamma(S + A + k_1 l)\left(\frac{\beta(2l(S+A+k_1 l)+2k_2 Q+lQ(2c+Hl))}{lQ(\beta-1)}\right)^{-\beta} = 0. \quad (4)$$

And, the integrated retail price is

$$p^* = \frac{\beta\left(c+\frac{(S+A+k_1 l)}{Q^*}+\frac{H\,l}{2}+\frac{k_2}{l}\right)}{\beta-1}.$$

Proof. The optimal retail price, p^*, is obtained from $\frac{\partial JTP(p,Q)}{\partial p} = 0$. Similarly, from $\frac{\partial JTP(p,Q)}{\partial Q} = 0$, the optimal order quantity is $Q^* = \frac{\sqrt{2\gamma p^{-\beta}(S+A+k_1\, l)}}{\sqrt{h}}$, and Equation (4) can be achieved by simultaneously considering p^* and Q^*. It can be proved that this form of equations either has two positive roots or have no one. Furthermore, the Hessian matrix of $JTP(p,Q)$ is

$$\mathbf{H} = \begin{pmatrix} \frac{\partial^2 JTP(p,Q)}{\partial p^2} & \frac{\partial^2 JTP(p,Q)}{\partial p \partial Q} \\ \frac{\partial^2 JTP(p,Q)}{\partial Q \partial p} & \frac{\partial^2 JTP(p,Q)}{\partial Q^2} \end{pmatrix},$$

where

$$\frac{\partial^2 JTP(p,Q)}{\partial p^2} = -\frac{1}{2lQ}p^{-2-\beta}\gamma\beta(2l(S+A+k_1l)+2k_2Q+l(2c+Hl+2p)Q+(2l(S+A+k_1l)+2k_2Q+l(2c+Hl+2p)Q)\beta),$$
$$\frac{\partial^2 JTP(p,Q)}{\partial p\partial Q} = \frac{\partial^2 JTP(p,Q)}{\partial Q\partial p} = -\frac{\gamma\beta p^{-\beta-1}(S+A+k_1l)}{Q^2},$$
$$\frac{\partial^2 JTP(p,Q)}{\partial Q^2} = -\frac{2\gamma p^{-\beta}(S+A+k_1l)}{Q^3}.$$

Checking the sign of the first and second principal minor determinant of **H**, we get

$$|H_{11}| = \frac{\partial^2 JTP(p,Q)}{\partial p^2} < 0$$
$$|H_{22}| = \frac{1}{lQ^4}(S+A+k_1l)p^{-2(1+\beta)}\beta\gamma^2\big(2l(S+A+k_1l)+2k_2Q+l(2c+Hl+2p)Q+\big(2k_2Q+l(S+A+k_1l+2cQ+HlQ-2pQ)\big)\beta\big) > 0.$$

Therefore, the Hessian matrix is a negative definite matrix, so $JTP(p,Q)$ is a concave function in p and Q. Therefore, (p^*,Q^*) maximizes the integrated inventory model.

4.2. Decentralized (Stackelberg policy)

In Stackelberg policy, manufacturer and retailer are classified as leader and follower, respectively. The manufacturer chooses a strategy first, and then the retailer observes this decision and makes his own strategy. It is necessary to assume that each enterprise is not willing to deviate from maximizing his profit. In other words, each player chooses his best strategy. The manufacturer determines his wholesale price, and acts as a leader by announcing it to the retailer in advance, and the retailer acts as a follower by choosing his retail price and order quantity based on the manufacturer's strategy.

Proposition 2. *The Stackelberg's order quantity and wholesale price are achieved by solving the following system of equations:*

$$(i)\quad h\,(Q^*)^2 - 2\,A\,\gamma\left(\frac{(A+Q^*\,w)\beta}{Q^*(\beta-1)}\right)^{-\beta} = 0,\ and \tag{5}$$

$$(ii)\quad \frac{\left(\frac{(A+Q^*\,w^*)p}{Q^*(\beta-1)}\right)^{-\beta}\left(2l\,(A+Q^*\,w^*)\gamma+\left(2k_2Q^*+l\left(2S+2k_1l+Q^*(2c+Hl-2w^*)\right)\right)\gamma\beta\right)}{2l\,(A+Q^*\,w^*)} = 0, \tag{6}$$

and the Stackelberg's retail price is

$$p^* = \frac{A\beta+Q^*\,w\beta}{Q^*(\beta-1)}.$$

Proof. We need the retailer's reaction function (p^*,Q^*) for given w. It can be proved that the Hessian matrix of TP_r is a negative definite matrix, so TP_r is concave in p and Q. Therefore, the optimal retail price and the optimal order quantity are obtained from $\frac{\partial TP_r(p,Q)}{\partial p} = 0$ and $\frac{\partial TP_r(p,Q)}{\partial Q} = 0$. We get

$$p = \frac{A\beta+Qw\beta}{Q(\beta-1)} \tag{7}$$

and

$$Q = \sqrt{\frac{2A\gamma p^{-\beta}}{h}}. \tag{8}$$

By simultaneously considering equations (7) and (8), we obtain Equation (5). Further, by substituting p to the manufacture's total profit, we have

$$TP_m = \left(\gamma\left(\frac{A\beta+Qw\beta}{Q(\beta-1)}\right)^{-\beta}\right)\left(w-c-\frac{S}{Q}-\frac{k_1l}{Q}-\frac{k_2}{l}-\frac{Hl}{2}\right). \tag{9}$$

The optimal wholesale price is achieved by maximizing Equation (9) with respect to w or equivalently $\frac{\partial TP_m}{\partial w} = 0$. Then we get Equation (6).

4.3. *Decentralized (RFM policy)*

In RFM policy, the wholesale price is determined first by the manufacturer. Next the retailer sets his order quantity. The retailer receives a fixed mark-up ($\alpha = 1 - w/p$). Hence, setting the wholesale price by manufacturer is equivalent to choosing the retail price. Then, the retailer only chooses on value of order quantity and the manufacturer decides the retail price. Note that, the value of α is assumed to be exogenously given, and, not to be endogenously determined. It is important to specify that for which values of α, the RFM policy will be desirable for both members. By substituting $w = (1 - \alpha)p$ to equations (1) and (2), we obtain the total profits of two firms under RFM as below:

$$TP_r(p,Q) = (\gamma\, p^{-\beta})\left(\alpha p - \frac{A}{Q}\right) - \frac{hQ}{2} \tag{10}$$

and

$$TP_m(p,Q) = (\gamma\, p^{-\beta})\left(p - \alpha p - c - \frac{S}{Q} - \frac{Hl}{2} - \frac{k_1 l}{Q} - \frac{k_2}{l}\right). \tag{11}$$

Proposition 3. *RFM's order quantity and retail price are achieved by solving the following system of equations:*

$$(i)\ \frac{h\,(Q)^2}{2\,A} - \gamma\,(p)^{-\beta} = 0, and \tag{12}$$

$$(ii)\ p^{-1-\beta}\gamma\left(2k_2Q\beta + l^2(2k_1 + HQ)\beta + 2l\left((S + cQ)\beta + pQ(\beta - 1)(\alpha - 1)\right)\right) = 0. \tag{13}$$

Proof. Similar to previous section, the retailer's reaction, Q, is obtained from $\frac{\partial TP_r(p,Q)}{\partial Q} = 0$. We have

$$Q = \sqrt{\frac{2A\gamma p^{-\beta}}{h}}.$$

The manufacturer maximizes his total profit by taking the retailer's reaction into account. Furthermore, simplifying $\frac{\partial TP_m(p,Q)}{\partial p} = 0$ results the Equation (13). Moreover, it can be proved that the profit function is concave.

5. Numerical example

Numerical examples given below are for illustrating the feasibility of the above policies. Moreover, the Pareto-improvement region through a numerical study is illustrated. Next, a sensitivity analysis is performed by changing the values of major parameters. We consider the retailer's and the manufacturer's inventory systems with the following data: $D(p) = 300000\, p^{-1.25}$ units per year, $A =$ \$80 per order, $c =$ \$13 per unit, $S =$ \$300, $k_1 =$ \$1000, $k_2 =$ \$0.0002, $l = 0.02$, $h =$ \$1.2, $H =$ \$1. In RFM, we assume that α is equal to 0.32.

Table 1. Solutions of integrated policy, Stackelberg policy and RFM policy.

	Q^*	w^*	p^*	TP_r	TP_m	$JTP(p,Q)$
Integrated policy	1021	-	67.059	-	-	83256
Stackelberg policy	152	77.768	391.479	53865	10792	64657
RFM policy	351	69.667	102.452	29754	51297	81051

The solutions of integrated policy, Stackelberg policy and RFM policy are summarized in Table 1. The table shows that the manufacturer's profit is higher in RFM policy comparing to Stackelberg policy. The opposite results occurs in retailer's profit. However, we can see that RFM profit is higher than Stackelberg profit. Furthermore, we define the competition penalty (ρ) as the difference between the integrated profit and Stackelberg/RFM profit measured as a fraction of the integrated profit. For RFM policy, this value is equal to 3% but increased to 23% in Stackelberg policy. It is important to determine an interval ($\alpha_{min}, \alpha_{max}$) in which both retailer and manufacturer can benefit from RFM policy.

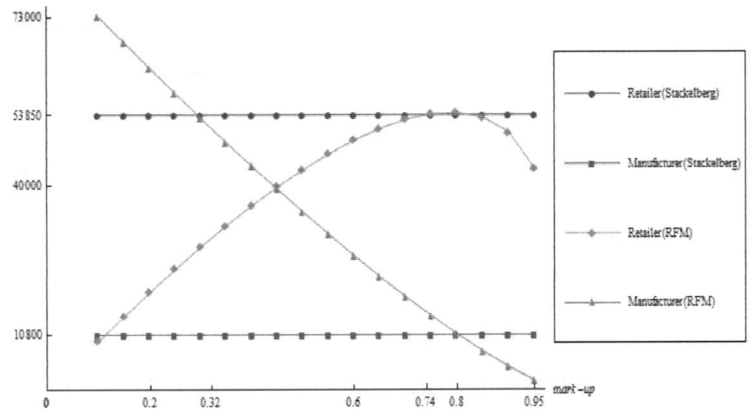

Figure 3. Variations of profit functions respect to α.

Figure 3 shows the manufacturer's and retailer's profit functions with respect to α under the Stackelberg and RFM policies. It can be seen that the retailer's profit function is concave and has a maximum value. However, the manufacturer's profit function is convex and decreasing in α. There exists a α_{max} such that if $\alpha \leq \alpha_{max} = 0.83$, the manufacturer can benefit from RFM policy comparing to Stackelberg policy. Similarly, there exist α_{min} and α_{max} such that if $0.74 = \alpha_{min} \leq \alpha \leq \alpha_{max} = 0.83$, the retailer can benefit from RFM policy comparing to Stackelberg policy. Furthermore, we can see that in the interval $(\alpha_{min}, \alpha_{max}) = (0.74,\ 0.83)$, both the manufacturer and the retailer will always prefer RFM policy to stackelberg policy. The interval $(0.74,\ 0.83)$ is a Pareto efficient strategy for the given data. Next, we investigate this Pareto-improving region numerically. The variation of this region with respect to retailer fixed ordering cost, A, and scaling factor of demand rate, γ, is illustrated in Figure 4. From the figure, the higher the ordering cost, the longer the interval of Pareto efficient strategy. However, the opposite results occurs in the scaling factor of demand rate.

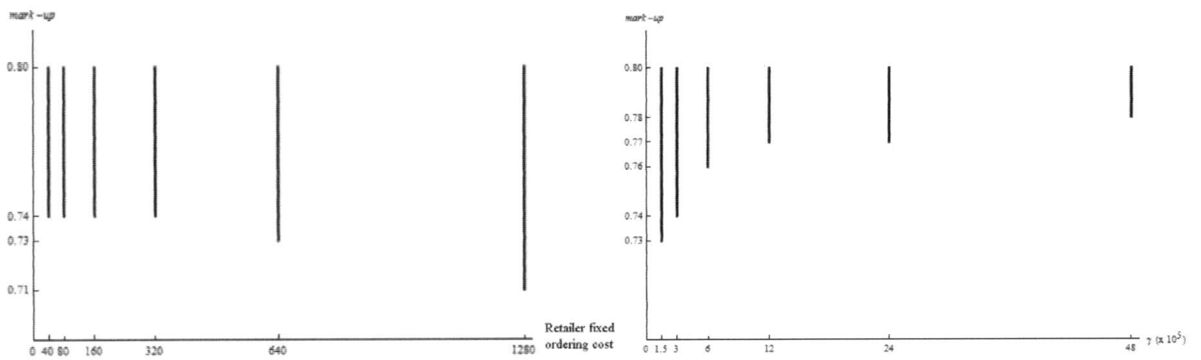

(a) (b)

Figure 4. Pareto-improving region in (a) retailer fixed ordering cost and (b) scaling factor of demand rate.

6. Sensitivity Analysis

The sensitivity analysis is performed by changing the values of major parameters. Table 2 illustrates the sensitivity analysis of the parameters and the competition penalty (ρ) for RFM policy with three different retail fixed mark-up rates. In the table, Δ_r is defined as the retailer's percentage improvement of RFM policy comparing to Stackelberg policy. Similarly, Δ_m is defined for the retailer's. Negative values shows that the retailer's/manufacturer's profit in RFM policy is lower than that of Stackelberg's. The RFM is not a Pareto efficient strategy if either Δ_r or Δ_m is negative. Moreover, the variations of decision variables for integrated, Stackelberg, and RFM policies with three different retail fixed mark-up rate are illustrated in Table 3.

Table 2. Two approach solutions under variation of major parameters

Solutions→		Stackelberg	RFM (0.74)			RFM (0.76)			RFM (0.78)		
Paramaters ↓		ρ	ρ	Δ_m	Δ_r	ρ	Δ_m	Δ_r	ρ	Δ_m	Δ_r
c	7	23	17	41.489	0.369	19	27.775	0.862	21	14.318	1.079
	18	22	17	40.664	-0.128	18	27.089	0.416	20	13.805	0.696
γ	150000	23	18	41.834	0.577	19	28.032	1.048	21	14.536	1.239
	500000	22	17	40.425	-0.272	18	26.896	0.288	20	13.656	0.586
β	1.08	15	5	294.158	-10.368	6	261.347	-8.567	7	228.762	-6.851
	1.6	29	41	-44.836	-6.219	44	-51.891	-8.951	47	-58.571	-12.168
A	40	23	18	40.88	0.073	20	27.172	0.525	21	13.762	0.698
	200	22	16	41.364	0.192	17	27.776	0.797	19	14.474	1.142
S	150	21	16	40.516	-0.265	17	27.029	0.339	19	13.826	0.689
	600	25	19	41.738	0.617	21	27.832	0.987	22	14.234	1.064
h	1	22	17	40.724	-0.0924	18	27.138	0.448	20	13.842	0.724
	32	24	18	42.212	0.807	19	28.338	1.255	21	14.768	1.417
H	0.5	22	17	40.915	0.022	18	27.292	0.551	20	13.96	0.812
	1.2	22	17	40.913	0.022	18	27.29	0.55	20	13.96	0.812
k_1	500	22	17	40.887	0.002	18	27.273	0.536	18	13.785	3.192
	2000	22	17	40.968	0.061	19	27.327	0.579	20	13.979	0.828
k_2	0.0001	22	17	40.915	0.022	18	27.292	0.551	20	13.96	0.812
	0.001	22	17	40.911	0.02	18	27.288	0.549	20	13.959	0.81
l	0.01	22	17	40.887	0.002	18	27.273	0.536	20	13.951	0.804
	0.1	23	18	41.128	0.177	19	27.431	0.664	21	14.031	0.877

We solve 101 problems and drive conclusions about the results. To evaluate the RFM policy, we set α to 0.74, but other parameters generated randomly from the intervals as below:

$$c \in [7,18], \gamma \in [150000,500000], \beta \in [1.08,1.6], A \in [40,200], S \in [150,600],$$
$$h \in [1,3], H \in [0.5,1.2], k_1 \in [500,2000], k_2 \in [0.0001,0.001], l \in [0.01,0.1].$$

Next, we compare the competition penalty for these problems. The penalty's histogram for RFM and Stackelberg policies can be seen in Figure 5. The figure shows that the highest and the lowest frequency of the competition penalty of Stackelberg policy are $\rho \in [22\%, 22.5\%]$ and $\rho < 21\%$, respectively. The highest frequency of competition penalty of RFM policy is $\rho \in [17\%, 18\%]$; whereas the lowest one is $\rho < 15\%$ and $\rho \in [15\%, 16\%]$.

Table 3. Two approach solutions under variation of major parameters.

Solutions→		Integrated		Stackelberg			RFM (0.74)			RFM (0.76)			RFM (0.78)		
Paramaters ↓		Q^*	p^*	Q^*	p^*	w^*	Q^*	p^*	w^*	Q^*	p^*	w^*	Q^*	p^*	w^*
c	7	1495	36.438	216	221.8	43.9	268	158.032	41.088	253	172.617	41.428	239	190.097	41.821
	18	836	92.495	125	530.26	105.414	153	386.84	100.578	145	421.481	101.155	137	462.824	101.82
γ	150000	717	67.892	102	423.511	83.918	127	298.895	77.712	120	326.819	78.436	113	360.346	79.276
	500000	1324	66.611	201	375.275	74.657	244	275.643	71.667	232	300.107	72.025	219	329.268	72.439
β	1.08	851	182.116	80	3286.44	242.437	175	771.221	200.517	167	840.384	201.692	159	922.895	203.04
	1.6	804	36.047	145	112.222	41.531	107	164.268	42.709	100	180.576	43.338	92	200.395	44.087
A	40	970	66.957	104	412.235	82.062	127	298.895	77.712	120	326.819	78.436	113	360.346	79.276
	200	1161	67.340	245	378.628	74.908	302	270.78	70.402	287	294.537	70.689	271	322.811	71.018
S	150	811	66.643	158	364.969	72.489	193	267.382	69.519	183	90.649	69.755	173	318.309	70.028
	600	1343	67.707	139	451.781	89.779	173	319.699	83.121	163	350.83	84.199	153	388.427	85.454
h	1	1121	66.886	168	385.118	76.647	205	280.477	72.924	194	305.65	73.355	183	335.701	73.854
	3	639	68.231	90	437.296	86.565	112	305.349	79.390	106	334.259	80.222	100	369.032	81.187
H	0.5	1022	67.034	152	391.34	77.74	186	283.473	73.703	176	309.093	74.182	166	339.71	74.736
	1.2	1021	67.108	152	391.536	77.779	186	283.618	73.740	176	309.251	74.220	166	1339.883	74.774
k_1	500	1009	67.037	153	389.642	77.402	186	282.46	73.439	177	307.925	73.902	167	338.345	67.743
	2000	1046	67.108	151	395.189	78.507	185	285.826	74.314	175	311.789	74.829	165	342.836	75.424
k_2	0.0001	1022	67.037	152	391.34	77.74	186	283.473	73.703	176	309.093	74.182	166	339.71	74.736
	0.001	1019	67.263	151	392.598	77.991	186	284.407	73.945	176	310.109	74.426	166	340.823	74.981
l	0.01	1009	67.059	152	389.781	77.429	186	282.564	73.466	177	308.037	73.929	166	338.468	74.462
	0.1	1115	67.413	148	407.484	80.956	182	293.381	76.279	172	320.431	76.903	162	352.847	77.626

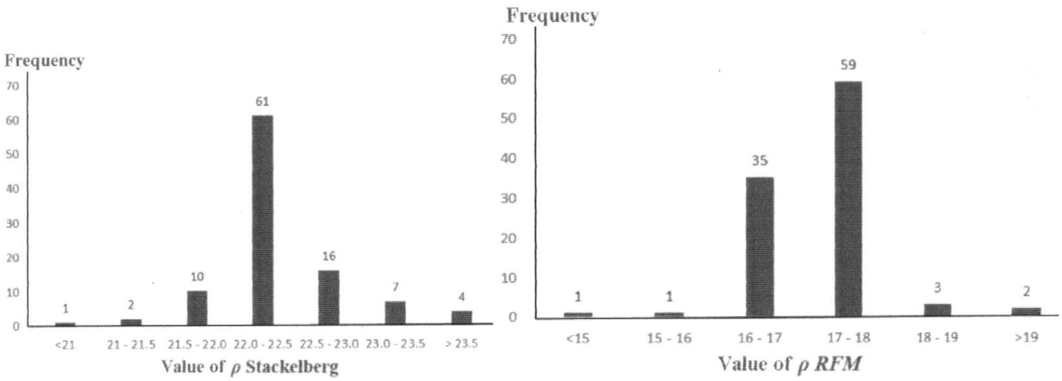

(a) (b)

Figure 5. Histogram of (a) Stackelberg policy's penalty and (b) RFM policy's penalty.

Figure 6 illustrates the variations of total profit, retailer's profit and manufacturer's profit respect to the retailer's fixed ordering cost. It can be seen that total profit of the RFM policy is higher than that of Stackelberg policy. Moreover, it is shown that greater value of α leads to lower profit for the manufacturer, but greater profit for the retailer. It is obvious that with $\alpha = 0.74$ is preferred by the retailer. However, manufacturer may choose $\alpha = 0.6$ to maximize his profit.

IOP Conf. Series: Materials Science and Engineering **166** (2017) 012014 doi:10.1088/1757-899X/166/1/012014

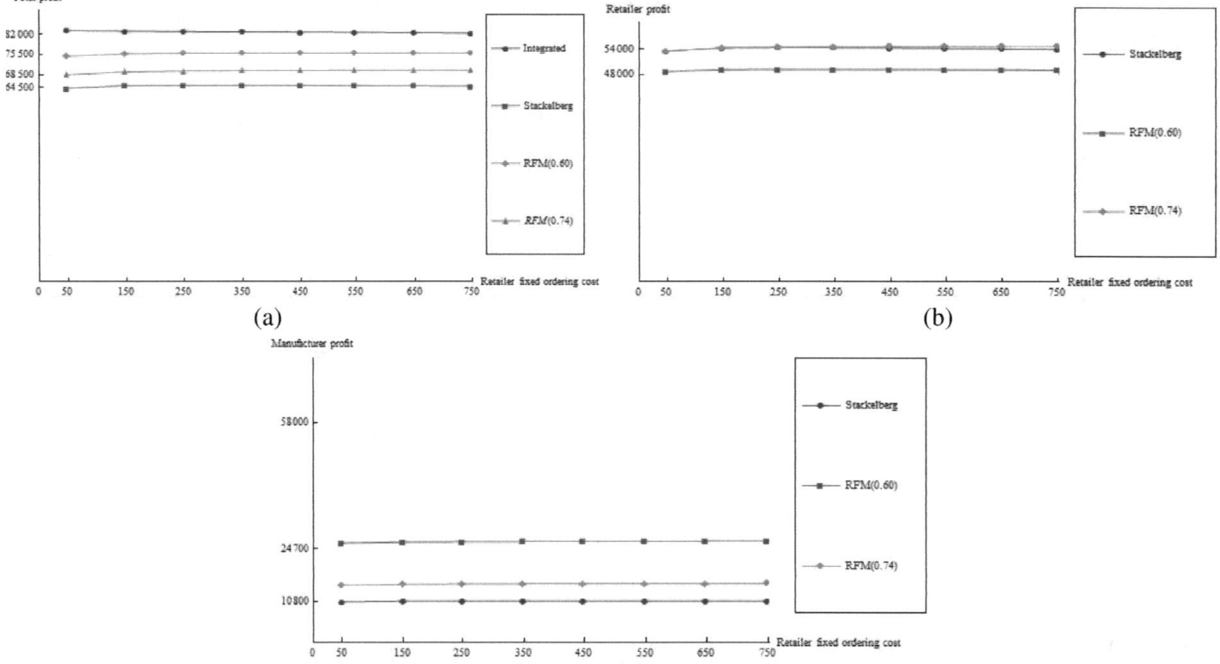

Figure 6. (a) Total profit, (b) retailer profit, and (c) manufacturer profit respect to retailer fixed ordering cost.

The variations of retail price and order quantity respect to the procurement cost are illustrated in Figure 7. We examine four different values of fixed mark-up, α. It can be seen that the greater the procurement cost, the greater the retail price, but the lower the order quantity. Moreover, the retail price of Stackelberg policy is greater than that of integrated policy. However, the order quantity of integrated policy is greater than those of Stackelberg and RFM policies.

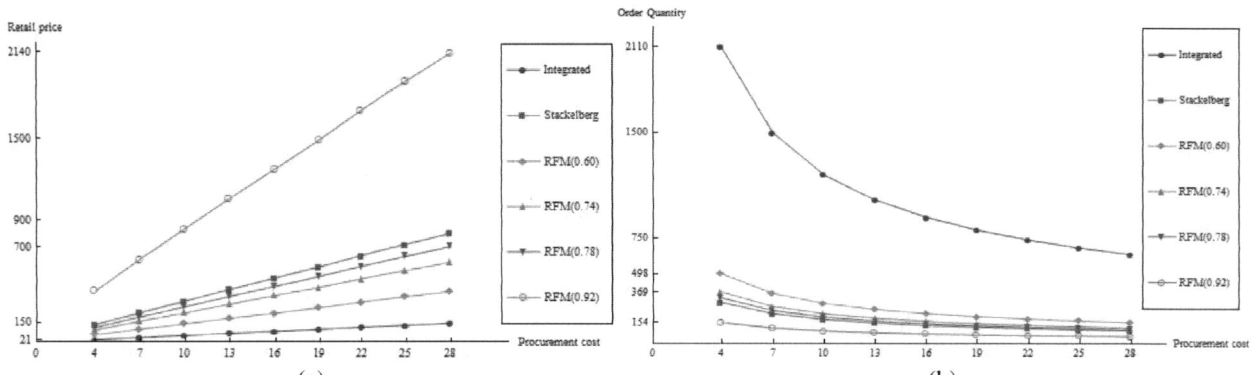

Figure 7. (a) Retail price and (b) order quantity respect to the procurement cost.

7. Conclusions

In this paper, we consider a production-inventory supply chain system with single-manufacturer and single-retailer. The decentralized policies (*i.e.* Stackelberg and RFM) are examined with the goal of coordinating the channel. There are many types of contract that guarantee the supply chain. However, the administrative costs of the contract are usually neglected in real situation. The additional gain from integration may not cover the extra administrative costs and may not be addressed to supply chain. Therefore, RFM policy is used it has minor administrative costs comparing to the other policies. With properly designed RFM policy, Pareto improvement is obtained over the Stackelberg policy. Although the RFM policy is not capable of coordinating the channel, it leads to considerable improvements over the channel. Possible extensions for future research could be by considering other general demand function with more efficient inventory models. Furthermore, it is also interesting to consider the administrative costs of the integration contracts in the objective function of the members.

Acknowledgment
This research is supported by PNBP Sebelas Maret University, Surakarta, Indonesia, year 2016.

References

[1] Alaei S, Behravesh M, and Karegar N 2014 Analysis of Production-Inventory Decisions In A Desentralized Supply Chain *International Journal of Production Economics* **2** pp 198-216.
[2] Chung K J and Liao J J 2011 The simplified solution algorithm for an integrated supplier–buyer inventory model with two-part trade credit in a supply chain system *European Journal of Operational Research* **213** pp 156–165.
[3] Chung C J and Wee H M 2008 An integrated production-inventory deteriorating model for pricing policy considering imperfect production, inspection planning and warranty-period-and stock-level-dependant demand *International Journal of Systems Science* **39** 8 pp 823–837.
[4] Hays F H and DeLurgio S A 2009 Where's the Beef?: Statistical demand estimation using supermarket scanner data *Journal of Case Research in Business and Economics* **1** 40–51.
[5] Khanra S, Sana S S, and Chaudhuri K 2010 An EOQ model for perishable item with stock and price dependent demand rate *International Journal of Mathematics in Operational Research* **2** 3 pp 320-335.
[6] Kim J, Hong Y, and Kim T 2011 Pricing and ordering policies for price-dependent demand in a supply chain of a single retailer and a single manufacturer *International Journal of Systems Science* **42** 1 pp 81-89.
[7] Lau A H L & Lau H S 2003 Effects of a demand-curve's shape on the optimal solutions of a multi-echelon inventory/pricing model *European Journal of Operational Research* **147** 3 pp 530-548.
[8] Li Q and Atkins D 2002 Coordinating Replenishment and Pricing in a Firm *Manufacturing & Service Operations Management* **4** 4 pp 241–257.
[9] Lin Y J and Ho C H 2011 Integrated inventory model with quantity discount and price-sensitive demand, *TOP* **19** 1 pp 177-188.
[10] Liou Y C, Schaible S, Yao J C 2006 Supply Chain Inventory Management via a Stackelberg Equilibrium *Journal of Industrial and Management Optimization* **2** 1 pp 81–94.
[11] Rad M A and Khoshalhan F 2011 An Integrated Production-Inventory Model with Backorder and Lot for Lot Policy *International Journal of Production Research and Industrial Engineering* **22** 2 pp 127-134.
[12] Rad M A, Khoshalhan F, and Setak M 2014 Supply Chain Single Vendor-Single Buyer Inventory Model with Price-Dependent Demand, *Journal of Industrial Enginering and Management* **7** 4 pp 733-748.
[13] Shah N H, Patel D G, and Shah D B 2014 Optimal Integrated Inventory Policy For Stock-Dependent Demand When Trade Credit is Linked to Order Quantity, *Revista Investigation Operational* **35** 2 pp 130-140.
[14] Whitin T M 1955 Inventory control and price theory *Management Science* **2** 1 pp 61-68.

A design of mathematical modelling for the mudharabah scheme in shariah insurance

R. Cahyandari[1], D Mayaningsih[2], Sukono[3]

[1]Faculty of Sciences and Technology, UIN Sunan Gunung Djati Bandung, Indonesia
[2]Faculty of Shariah and Law, UIN Sunan Gunung Djati Bandung, Indonesia
[3]Faculty of Mathematics and Natural Sciences, Universitas Padjadjaran, Indonesia

E-mail: rcahyandari@yahoo.com

Abstract. Indonesian Shariah Insurance Association (AASI) believes that 2014 is the year of Indonesian Shariah insurance, since its growth was above the conventional insurance. In December 2013, 43% growth was recorded for shariah insurance, while the conventional insurance was only hit 20%. This means that shariah insurance has tremendous potential to remain growing in the future. In addition, the growth can be predicted from the number of conventional insurance companies who open sharia division, along with the development of Islamic banking development which automatically demand the role of shariah insurance to protect assets and banking transactions.The development of shariah insurance should be accompanied by the development of premium fund management mechanism, in order to create innovation on shariah insurance products which beneficial for the society. The development of premium fund management model shows a positive progress through the emergence of *Mudharabah, Wakala*, Hybrid (*Mudharabah-Wakala*), and *Wakala-Waqf.* However, 'model' term that referred in this paper is regarded as an operational model in form of a scheme of management mechanism. Therefore, this paper will describe a mathematical modeling for premium fund management scheme, especially for Mudharabah concept. Mathematical modeling is required for an analysis process that can be used to predict risks that could be faced by a company in the future, so that the company could take a precautionary policy to minimize those risks.

1. Introduction

The development of shariah insurance started in 1994 with the establishment of PT Syarikat Takaful Indonesia (STI), initiated by the Indonesian Muslim Intellectuals Association (ICMI) through Abdi Bangsa Foundation, Bank Muamalat Indonesia, PT Asuransi Jiwa Tugu Mandiri, the Ministry of Finance, as well as some Indonesian muslim businessmen. [12]

Shariah Insurance as an alternative protection for Muslims needs to be managed in accordance to the Islamic law. Shariah insurance can also be an option for other faiths who consider that the shariah concept is fair for them, wherein shariah is a universal concept and can be applied by everyone.

National Shariah Board (MUI) determined the meaning of sharia insurance (*ta'min, takaful, tadhamun*) as an effort to protect and mutual assistance among persons or parties through investment funds, in form of assets or *tabarru*, which has a refund arrangement to face particular risks, based on a contract which corresponds to shariah (DSN MUI Fatwa No.21 / DSN-MUI / X / 2001 regarding General Guidelines for Shariah Insurance). [12]

Indonesian Sharia Insurance Association (AASI) believes that 2014 is the year of Indonesian Sharia Insurance, since its growth was above the conventional insurance. In December 2013, 43% growth was recorded for sharia insurance, while the conventional insurance was only hit 20%. This means that sharia insurance has tremendous potential to remain growing in the future. In addition, that growth can be predicted from the number of conventional insurance companies who open sharia division, along with the development of Islamic banking development which automatically demand the role of sharia insurance to protect assets and banking transactions.

The development of sharia insurance should be accompanied by the development of premium fund management mechanism, in order to create innovation on shariah insurance products which beneficial for the society. The mechanism of fund management on shariah insurance premium is divided into two, namely with saving portion and without saving portion [13]. The mechanism with saving portion divides the premium fund into two accounts, which are saving account (saving premium) and special account (*tabarru* premium). Both premiums are managed and invested by the company in accordance with the concept of sharia, in places that are lawful and do not contain usury. Special account (*tabarru*) is paid by participants with the aim to help (*taáwun*).[11]

The development of fund management mechanism model on sharia insurance shows an immense progress through the emergence of *Mudharabah, Wakala*, Hybrid (*Mudharabah-Wakala*), and *Wakala-Waqf* concepts [1, 6, 7, 14]. However, 'model' term that referred in this paper is regarded as an operational model in form of a scheme of premium fund management mechanism.

This paper will discuss about a mathematical modelling design for the scheme of premium fund management mechanism, particularly Mudharabah. Mathematical modelling is required for analysing process to predict what will happen in the future if certain changes occur in the system. This model can also be utilized to identify the most influential variable in the system. Firstly, the authors will discuss principles in shariah insurance which is applied as a base of the premium fund management mechanism. Afterward, the design of mathematical modelling will be discussed to translate operational model in form of schemes of premium fund management mechanism into mathematical language as a mathematical model.

The applied theory in sharia insurance calculation, among other, are:

a. Probability theory

Probability is defined as the extent to which an event is likely to occur, measured by the ratio of the favourable cases to the whole number of cases possible. For an illustration, an event occurred in m ways and failed to occur in n ways, then the probability of the event occurrence is: [11]

$$p = \frac{m}{m+n}$$

and the probability of the failed occurrence is:

$$q = \frac{n}{m+n}.$$

b. Mortality table

This table contains of probability of a person to die based on their age, in particular population, in this case the insured group. The number of people in age of x is stated in l_x symbol. While the number of people from the l_x group who died before they reach x+1 is stated in d_x symbol,

$$d_x = l_x - l_{x+1}.$$

A probability of a person with age of x died before they reach x+1 of age, or a probability of a person with age of x died between age of x and x+1 are stated in q_x symbol,

$$q_x = \frac{l_x - l_{x+1}}{l_x} = \frac{d_x}{l_x}$$

The mortality table is utilized to determine the amount of compensation from the probability of losses caused by the decease, and to predict a person life expectancy. Calculating the relation between age and time is useful to determine life-death probability.

1) Life probability
   A person's life expectancy with x of age is n years.

$$_nP_x = \frac{l_{x+n}}{l_x}$$

2) Death probability
   A person death probability with x of age is n years, or before they reach $x+n$ of age

$$_nQ_x = 1 - \ _nP_x$$

$$= 1 - \frac{l_{x+n}}{l_x}$$

$$= \frac{l_x - l_{x+n}}{l_x}$$

c. Gross Contribution

Gross Contribution is gross contribution that have to collect by the shariah insurance company in the form of insurance premium participants. In other words, this fund will be used as based of shariah insurance in determining the amount of the premium per participant.
General formula that used to count GC, defined as:

$$GC = RC + \lambda + E$$

where:
RC = Risk Contribution ; λ = Reserve Fund ; E = Dismissal of company

d. Risk Contribution

Risk Contribution is amount of fees that will be used in shariah insurance company to fulfill the needs of the insured claims.

e. The reserve fund

This reserve fund is collected by shariah insurance company as the fund that will be used in deficit situation *underwriting*.

f. Dismissal of company

Dismissal of Shariah insurance company consists of various cost as Fees from the company *wakala, operational cost, employee salaries, company profit, etc.*

2. Methodology

Two steps were taken in this research, namely literature study and drafting mathematical modeling

a. Literature Study
   In this step the researchers reviewed principles that related the shariah insurance as a foundation for premium fund management. The sources of references are books, papers, proceedings, and articles on the internet.

b. Drafting Mathematical Model
   This step was aimed to construct a plan to translate the Mudharabah scheme into mathematical language to produce a mathematical model for premium fund management mechanism.

3. Results and Discussion

3.1. The principles of Shariah Insurance [12]

1. Tawhid Principle
   Tawhid is a basic principle in sharia insurance. Fundamentally, every moslem should base their life and do every activity by tawhid (monotheism), including in muamalat. This means that the basic intention in performing shariah insurance is to obtain blessing form Allah SWT.
2. Justice Principle
   This principle means that the shariah insurance should be fair, impartial, especially in making the relationship pattern whether between participants, or between the shariah insurance companies and their customer, concerning their respective rights and obligations. Shariah insurance should not inflict disadvantage or detriment to customers.
3. Mutual Assistance Principle
   The spirit of mutual assistance is a very important aspect in the shariah insurance bussines. According to this principle, the concept of shariah insurance is fellow participants give charity for other participants who suffer from misfortune. Participants do not donate to the shariah insurance company, but they give it to fellow participants, while the companies merely act as fund managers. As a result, the companies are not entitled to claim or take participants' tabarru fund. The companies obtain ujrah (fee) upon the tabarru fund management, which is paid by the participants along with the contribution payment (premiums). The insurance companies manage the *tabarru* funds to be invested in shariah manner (*syar'iyyah*), and then allocate its result for other participants who suffer from misfortune.
4. Cooperation Principle
   Good cooperation should be established between participants and insurance companies. If both participants and the companies favorably fulfill their rights and responsibilities, it will establish a good and cooperative relationship pattern; which, in God's will, will bring blessing for both sides.
5. Amanah Principle
   Amanah (trustworthiness/reliability) is also a very important principle, because essentially this life is a mandate and everyone should responsible to Allah SWT. The companies are required to be reliable in managing the premium fund. Likewise, the participants should be true and fair about any occurred risk that happen to them. Do not let the participants lie or exaggerate and claim something that does not exist which will give disadvantages to other participants. Equally, the companies should not arbitrarily taking advantages, and detriment participants.
6. Mutual Willingness Principle
   Mutual willingness should accompany every transaction. The participants are willing their fund is managed by the insurance companies professionally and trustworthy, and the companies should solemnly perform the entrusted responsibility. In addition, the participants should agree and willing that their funds being allocated to other participants who suffered, to ease their burden. With this principle, shariah insurance has made mutual assistance broader and deeper meaning, since every party helping, cooperating, and transacting with each other in sincere manners.
7. Non-Riba Principle
   Riba (usury) is a form of transaction that should be avoided as far as possible, especially in insurance. Contributions (premiums) fund from the participants must be invested in shariah investment and has a clear halal status. Likewise, its operating system must applying risk sharing concept, which is based on *tabarru*contract, in order to eliminate the element of riba in participants' insurance benefits (claims).
8. Non-Maisir principle
   A conventional managed insurance has *maisir* (gambling) element, since participants may pay premium up to dozen times but never claim. On the other hand there is a new participant

who only pays premium once but claims immediately. This happens because the basic concept in conventional insurance is the transfer of risk. According to this concept, whenever the company receives premiums, it automatically becomes the company's possession. Any claim will be paid directly from the company's account. The company will gain a huge profit if they receive high premiums fund but low claims. Vis a versa, they will lose whenever the amount of claims is high while the collected premiums fund is low.

9. Non-Gharar Principle
   Gharar is obscurity, because the risk could happen or not. In conventional insurance, the participants do not know whether they get a claim or not, since it depends on risk occurrence. A participant can claim whenever a particular risk happens, but otherwise the participant will not be able to claim. This condition becomes *gharar* because the applied concept is the transfer of risk. Meanwhile, if the company applies the sharing of risk concept, the obscurity will not become *gharar*, because any claim will be carried together with other participant based on *tabarru* concept.
10. Non-*Risywah* Principle
   In its operational, both shariah insurance companies and its participants must avoid *risywah* (bribery) aspect. For example, a participant should not bribe an insurance official in order to get benefits (claims), likewise the company should not bribe the participant in order to obtain insurance premiums (contributions).

3.2. Design Modeling

The stages of mathematical modeling of a system in general can be illustrated through the following figure:

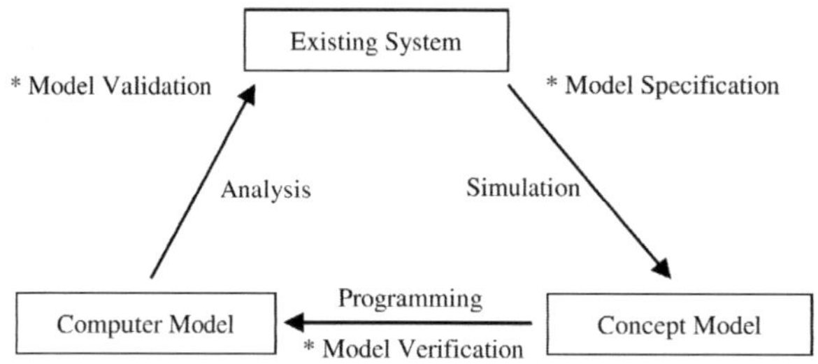

Figure 1. The Stages of Mathematical Modelling of a System

It shows that the first step that needs to be performed is to collect data in form of information about the existing system. The next step is to model the system or translate the existing system into mathematical language through mathematical equations. Afterward, the mathematical equations are transformed into programming language such as syntax. A computer simulation is performed to test, verify and validate the obtained mathematical models.

The Mudharabah scheme in the mechanism of premium-fund management in shariah insurance can be described as follows:

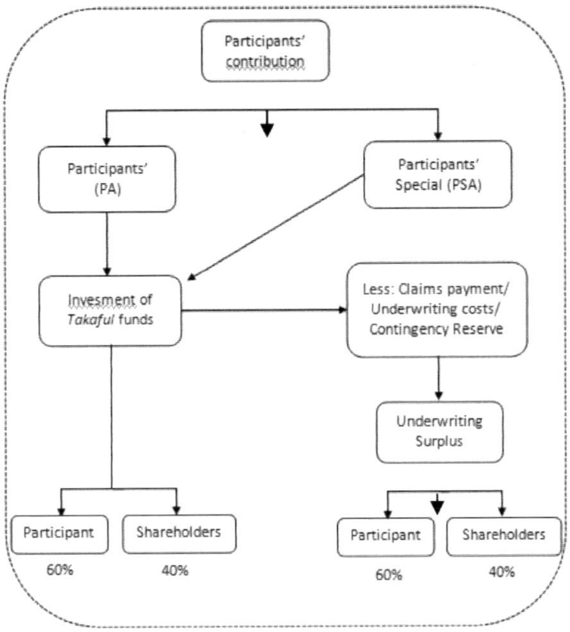

Figure 2 Mudharabah scheme in the shariah insurance mechanisms

The design of mathematical modeling for Mudharabah scheme in Figure 2 is started with determining research variables, as a basis to create mathematical equations, namely:

	Mudharabah Scheme
Research Variable	*Participant's Contribution* *Participant's* *Participant's Special (Tabarru)* *Invesment of Takaful Fund* *Claims paymentContingency Reserve* *Underwriting costs* *Contingency Reserve* *Underwriting Surplus* *Participant* *Shareholders*

Afterward, defining the research variables into following symbols:

a. *Participant's Contribution* = $Z = X + Y$
b. *Participant's* = $X = (100-p)\%*Z$
c. *Participant's Special (Tabarru)*= $Y=p\%*Z$
d. (investment level)
   Participant's = U
   Participant's Special (Tabarru) = V
e. *Claims payment* = W
f. *Underwriting costs* = T
g. *Contingency Reserve* = R
h. *Underwriting Surplus*= $S = a\%*T$ *(if available)*

i. Profit Sharing Invesment of Takaful (example)
Participant = 60%
Shareholders = 40%

j. Profit Sharing of Underwriting Surplus (if available)
Participant = 60%
Shareholders =40%

The next step is to build a mathematical modeling formulation by arranging the said symbols into mathematical equations and a system of mathematical equations.

Table 1.Mathematical Modeling Formulation

	polis year t				
	t =1	t =2	t =3	…	t =n
Number of collected Participant's Contribution = Z	Z(1)=1*Z	Z(2)=2*Z	Z(3)=3*Z	…	Z(n)=n*Z
Number of collected Participant's Special (Tabarru) = Y	Y(1)=p%*Z(1)	Y(2)=p%*Z(2)	Y(3)=p%*Z(3)	…	Y(n)=p%*Z(n)
Number of collected Participants = X	X(1)=(1-p%)*Z(1)	X(2)=X(1)+(1-p)%*Z	X(3)=X(2)+(1-p)%*Z	…	X(n)=X(n-1)+(1-p)%*Z
Collected profit sharing from Y	M(1)= q%*V*Y(1)	M(2)=q%*V*(Y(2)+M(1))+M(1)	M(3)=q%*V*(Y(3)+M(2))+M(2)	…	M(n)=q%*V*(Y(n)+M(n-1))+M(n-1)
Collected profit sharing from X	N(1)= r%*U*X(1)	N(2)= r%*U*(X(2)+N(1))+N(1)	N(3)=r%*U*(X(3)+N(2))+N(2)	…	N(n)=r%*U*(X(n)+N(n-1))+N(n-1)
Cash value	L(1)=X(1)+N1)	L(2)=X(2)+N(2)	L(3)=X(3)+N(3)	…	L(n)=X(n)+N(n)

Afterward, is to translate the mathematical model into syntax, a programming language, which will be utilized in the final stage, which is performing a computer simulation to verify and validate the obtained model.

4. Conclusion

Mudharabah scheme in the perspective of mathematics in shariah insurance is translating the scheme into the language of mathematics to produce mathematical models. Generally, mathematical modeling stage consists of: 1) learn the real system 2) translate the real system into the language of mathematics through the model 3) translate the mathematical models into computer programming language 4) do simulations for testing, verification and validation of the model by comparing the result of modeling with reality at the system. In this paper, research is still in the second stage and will be continued until the final stage.

Acknowledgment

We thank LP2M Universitas Islam Negeri Sunan Gunung Djati Bandung, which has been providing facilities for research and publication. We would also like to show our gratitude to the all who have helped both directly and indirectly.

References

[1] Akhter, W. "*Takaful Models and Global Practices*". MPRA Paper No. 40010, http://mpra.ub.uni-muenchen.de/40010/. 2012.
[2] Antonio, M.S . "Prinsip Dasar Operasi Asuransi Takaful dalam Arbitrase Islam di Indonesia". Jakarta, Badan Arbitrase Muamalat Indonesia, 1994.
[3] Boland, P.J . "Statistical and Probabilistic Methods in Actuarial Sciense". Chapman & Hall. 2015.
[4] Borowiak, D.S, Shapiro, A.F. "*Financial and Actuarial Statistics (An Introduction)*".

Chapman & Hall. 2014.
[5] Futami, T. "*Matematika Asuransi Jiwa* (*Bagian* I). Incorporated Foundation Oriental Life Insurance Cultural Development Centre.
[6] Ghazali, P.L.B, et.al. "*Mathematical Modelling in Family Takaful*". Journal of Applied Sciences 11(19): 3381-3388, ISSN: 1812-5654. 2011.
[7] Ghazali, P.L.B, et.al. "*Implementation of Integration Model for All*". Journal of Applied Sciences Research 8 (3): 1802-1812, ISSN: 1819-544X. 2012.
[8] Masud, H. "Takaful: An Innovative Approach to Insurance and Islamic Finance". U. Pa. J. Int'l L. Vol 32:4
[9] Nuryanto, A. "Pelaksanaan Prinsip Taáwun dalam Pengelolaan Premi Tabarru' Asuransi Syariah Menurut Hukum Positif Indonesia". Jurnal Ilmiah (fh.unram.ac.id). 2015.
[10] Puspitasari, N. "Sejarah dan Perkembangan Asuransi Islam serta Perbedaannya dengan Asuransi Konvensional". JEAM Vol X No.1, ISSN: 1412-5366. 2011.
[11] Rasyidah. "Model Matematika dan Aplikasinya dalam Masalah Bagihasil pada Asuransi Takaful Keluarga". Poli Rekayasa Vol.5, No.1. 2009.
[12] Sadeghi, M. "*The Evolution of Islic Insurance – Takaful*: *a Literature Survey*. Insurance Markets and Companies: Analyses and Actuarial Computations Vol.1, Issue2. 2010.
[13] Sujarwanto.staff.gunadarma.ac.id (accessed inMarch 2016)
[14] www.wikipedia.org (accessed in May 2016)
[15] www.takaful.co.id (accessed in June 2016)

Integer programming model for optimizing bus timetable using genetic algorithm

F.D. Wihartiko[1*], A. Buono[2], B.P. Silalahi[3]

[1]Department of Computer Science, Pakuan University, Indonesia
[2]Department of Computer Science, Bogor Agricultural University, Indonesia
[3]Department of Mathematic, Bogor Agricultural University, Indonesia

E-mail: fajardelli@gmail.com[1], pudesha@yahoo.co.id[2], bibparuhum1@yahoo.com[3]

Abstract. Bus timetable gave an information for passengers to ensure the availability of bus services. Timetable optimal condition happened when bus trips frequency could adapt and suit with passenger demand. In the peak time, the number of bus trips would be larger than the off-peak time. If the number of bus trips were more frequent than the optimal condition, it would make a high operating cost for bus operator. Conversely, if the number of trip was less than optimal condition, it would make a bad quality service for passengers. In this paper, the bus timetabling problem would be solved by integer programming model with modified genetic algorithm. Modification was placed in the chromosomes design, initial population recovery technique, chromosomes reconstruction and chromosomes extermination on specific generation. The result of this model gave the optimal solution with accuracy 99.1%.

1. Introduction

Bus timetable is the information which is provided by transportation provider for the passenger to give a service certainty time in term of departure/arrival bus schedule. Bus transportation service provider which is usually use timetable in giving the schedule information to the customer is the road based mass transit (*bus rapid transit*/BRT), shuttle bus and the other scheduled bus service transportation.

The consequences of using timetable is the bus should be deployed on schedule without waiting for a full passenger on the bus. As a result, service providers must be able to determine the number of optimum bus departure/frequency from the place of origin to the destination (optimal trip number) so that a timetable is made in accordance with passenger demand conditions and existed constraints, with regard to passenger service. It means in the off-peak time, the amount of trip will be less compared with the amount of trip during rush hour and vice versa. If it's not be calculated well, when the trips frequency is too small, there will be accumulation of passengers (unserved passenger) or vice versa, when the trips frequency is too much, it will waste the operating costs.

Making a timetable with the optimal number of trips (optimal timetable) without the computer programming, requires high accuracy and long processing time. The time required to achieve optimal working timetable is 2-7 working days, depends on the availability of data, tools, precision and expertise of the timetable maker. In providing services, service providers often face the changes of passenger demand and operational constraints that make the existing timetable to be improved. The long processing time in making timetable would create delays to anticipate the situation changing which will affect the services quality or operational costs.

Some studies related to the development of optimal bus scheduling model already made, some of them are Ceder *et al.* who already described timetables synchronization problems at transit stops by maximizing the number of arrival bus simultaneously [1]. Fu *et al.* modelling real time model nonlinear programming with using GPS. This model has the advantage in term of giving departure information to the passengers [2]. Kidwai develops genetic algorithm to solve the bus scheduling. On his article, scheduling models was made in two phases, the first phase is finding out the minimum frequency and the other phase is to find out the optimization model for the assignments in the services route [3]. Chuanjio *et al.* focus in developing model to find out the optimal headway and solve it using genetic algorithm. The developed model that has been made was in the form of nonlinear programming and also with considering the departure time and arrival time among shelters in the route. Chuanjio also did sensitivity analysis of his developed model [4]. Kidwai *et al.* and Chuanjio *et al.* haven't considering the bus worker aspect on their model. Wihartiko already modelling the number of trip problem with using integer programming. His model already considered the working hours of employees but has not considered the minimum service standards and it's limitations in finding the optimal number of trips on one route in one day [5]. Furthermore, Wagale *et al.* have developed Demand Travel-time Responsive model for scheduling problems [6]. Most of the research developed in the scope of bus scheduling are done by modelling the problem to find the optimum headway. Headway can be sought from the amount of frequency/route has been found. The differences amongst models were in the term of assumptions, indicator that be used and the method of solving problem.

This research is the development of integer programming model by Wihartiko [5], which will be designed by modelling a model that can accommodate a minimum service standards and overcome the limitations of the previous model by using genetic algorithm. Integer programming model can be solved using genetic algorithms [7]. Genetic algorithm is an evolutionary and numerical technique to find the optimal solution by using the principles of genetics theory [8]. Modifications with using genetic algorithm in this article was taken place in the chromosomes design which is the representations of bus operational time, chromosome generation techniques in feasible area, chromosome reconstruction techniques and population extermination on specific generation. The objective of this research is to model the optimal timetable problem into the form of integer programming, which then be solved by using a genetic algorithm. This research is limited to the manufacture of optimal timetable which contains the bus departure time schedule at the starting point and at the end point. Timetable explanations, bus arrival time schedule, departure schedules in each shelter along the trip, the making of "conditional" timetable, schedules synchronization at the transit, determination of the optimal number of officers, determination of the optimal number of buses, timetable evaluation method to predict the number of passengers and the bus travel time is not part of this research

2. Research Method

2.1. Problem formulation in the form of Integer Programming

In achieving optimal bus timetable, it be done by complementing the optimum trip/rit determination model [5] by considering minimum service standard and also by finding out the optimum rit for every route in the specific time period. The assumptions that be used:

1. Passengers distributed uniformly in one scheduling partition.
2. The same headway for each partition.
3. Make the bus journey happened from the starting point to the final destination repeatedly. Thus, the bus departure frequency (bus trip number) from the starting point to the final destination equals with a number of bus trips from the final destination to the starting point.

Bus timetable can be made according to the available routes, the passengers movement pattern in a time period, and the rush hour in a day. For example defined:

- $R = \{1,2,\dots,r|r \in \mathbb{N}\}$, is a set of routes, paths or corridors served by the service provider.
- $V_r = \{1,2,\dots,v_r|v \in \mathbb{N}, r \in R\}$, is a set of schedule variations imposed in a specific period on a route r∈R.

- $P_{r,v} = \{1,2,\dots,p_{v,r} | p \in \mathbb{N}, v \in V, r \in R\}$, is a set of partition or time division of the applicable timetable in a schedules variations $v \in V$ dan $r \in R$.

To formulate the problem of determining the optimal number of trips in integer programming, defined:

- $q^{psgr}_{r,v,p}$: Average number of passengers on the route $r \in R$ in schedule variations $v \in V$ on partition $p \in P$.
- $q_{bus,r}$: The numbers of vehicles that can be operated at route $r \in R$
- q_{cty} : Bus capacity in the term of passengers
- q_{emp} : Number of drivers employed
- $t_{r,v,p}$: Departure time of the bus n on route $r \in R$ in schedule variations $v \in V_r$ on partition $p \in P_{r,v}$
- $T_{r,v,p}$: Bus travel time (minutes) in one rit/trip at route $r \in R$ in schedule variations $v \in V_r$ on partition $p \in P_{r,v}$
- $h_{r,v,p}$: Bus *headway*, is the time differences (minutes) between one departure and the next departure ($h_{r,v,p} = t_{r,v,p,n+1} - t_{r,v,p,n}$). Headway can be found using the following equation:

$$h_{r,v,p} = \frac{2T_{r,v,p}}{q_{bus,r}} \tag{1}$$

- $h_{std,r,v,p}$: *headway based on the minimum service standard*
- LF : *Load Factor* defined as the percentage between the numbers of passengers and the bus capacity.
- K_{emp} : Maximum working hour of the bus operation employee per day, added by the maximum overtime hour in minutes along 1 variation periods.
- K_{emp} : Minimum working hour of the bus operation employee per day, in minutes along 1 variation periods.
- $k_{LF\,r,v,p}$: Number of passengers that be expected to be served on route $r \in R$ in schedule variations $v \in V_r$ on partition $p \in P_{r,v}$
- $k_{r,v}$: Number of schedule repetition during $v \in V_r$ on route $r \in R$
- $M^{Rit}_{r,v,p}$: Maximum rit/trip number that should be fulfilled for journey on route $r \in R$ in schedule variation $v \in V_r$ on partition $p \in P_{r,v}$

$$M^{Rit}_{r,v,p} = \frac{\Delta t_{r,v,p}}{h_{r,v,p}} = \frac{t_{r,v,p,n_{max}} - t_{r,v,p,n_0}}{h_{r,v,p}} \tag{2}$$

- $m^{Rit}_{r,v,p}$: Minimum rit/trip that possible to be fulfilled for journey on route $r \in R$ in schedule variation $v \in V_r$ on partition $p \in P_{r,v}$

$$m^{Rit}_{r,v,p} = \frac{\Delta t_{r,v,p}}{h_{std,r,v,p}} \tag{3}$$

- $x_{r,v,p}$: Amount of Rit or bus departure frequency on route $r \in R$in schedule variations $v \in V_r$ on partition $p \in P_{r,v}$
- $inc_{r,v,p}$: Average bus income for every rit/trip on route $r \in R$ in schedule variations $v \in V_r$ on partition $p \in P_{r,v}$
- $outc_{r,v,p}$: Average operational costs for every rit/trip on route $r \in R$in schedule variations $v \in V_r$ on partition $p \in P_{r,v}$
- $c_{r,v,p}$: Average profit/loss in every rit/trip on route $r \in R$ in schedule variations $v \in V_r$ on partition $p \in P_{r,v}$

$$c_{r,v,p} = inc_{r,v,p} - outc_{r.v,p} \tag{4}$$

The result of developing the optimum rit/trip determination model with using integer programming is as Model (5) follows:

Maximize $$z(x) = \sum_{r\in R}\sum_{v\in V_r}\sum_{p\in P_{v,r}} c_{r,v,p}.x_{r,v,p} \tag{5.1}$$

Subject To $$k_{emp}.\frac{q_{emp}}{2} \le \sum_{r\in R}\sum_{v\in V_r}\sum_{p\in P_{r,v}} T_{r,v,p}k_{r,v}.x_{r,v,p} \le K_{emp}.\frac{q_{emp}}{2} \tag{5.2}$$

$$m^{Rit}_{r,v,p} \le x_{r,v,p} \le M^{Rit}_{r,v,p} \qquad ;\forall r \in R, \forall v \in V_r, \forall p \epsilon P_{r,v} \tag{5.3}$$

$$x_{r,v,p} \ge \frac{q^{psgr}_{r,v,p}}{k_{LF_{r,v,p}}} \qquad ;\forall r \in R, \forall v \in V_r, \forall p \epsilon P_{r,v} \tag{5.4}$$

$$x_{r,v,p} \in \mathbb{Z}^+ \cup \{0\} \qquad ;\forall r \in R, \forall v \in V_r, \forall p \epsilon P_{r,v} \tag{5.5}$$

Objective function (5.1) states to find out the rit/trip numbers in order to maximize profits or minimize losses. Constraint (5.2) ensures that all of departure can be served by employees of the company with also considering the minimum employee working hour. Constraint (5.3) ensures that the number of trips being served does not exceed the amount of the existing fleet and in accordance with the minimum service standards. Constraints (5.4) to ensure that passengers can be served according to expectations of the company. Constraint (5.5) ensures that the solution result is nonnegative integers.

2.2 Optimal timetable creation with using genetic algorithm

John Holland introduced the concept of Genetic Algorithm in the early decades of 1970. Genetic Algorithm is a numerical optimization method inspired by the theory of natural selection and genetics. This algorithm contains stages sequential procedure that processes a population of artificial chromosome (artificial) into other new populations [8]. There are five main processes used in the genetic algorithm such as the initial population, the fitness value calculation, selection, crossover and mutation [9]

Optimal timetable creation by using GA was done by modifying the representation model of Huy *et al.* [7], especially in the process of the chromosomes design which is the representations of bus operational time, chromosome generation techniques in feasible area, chromosome reconstruction techniques and population extermination on specific generation. Result of optimal timetable representation to GA is as follows:

1. Establishment of Initial Population Chromosomes.
   The established chromosomes are the representation of the time on all variables $x_{r,v,p}$.
   Each variables is coded with using genes with the length of time (in minutes) on route $r \in R$ in schedule variations $v \in V_r$ on partition$p \in P_{r,v}$. In this case, the numbers of genes for variable $x_{r,v,p}$is $\Delta t_{r,v,p}$ (or equal$\Delta t_{r,v,p} + 1$ if $x_{r,v,p}$ placed on the end of operational time). Each genes on chromosomes has values:
   $$\text{gene}(n) = \begin{cases} 1 & \text{, if } t_{r,v,p,n} \text{ selected on timetable} \\ 0 & \text{, if } t_{r,v,p,n} \text{ unselected on timetable} \end{cases} \tag{6}$$
   Below is the ilustration of genes & chromosomes from Model (5).

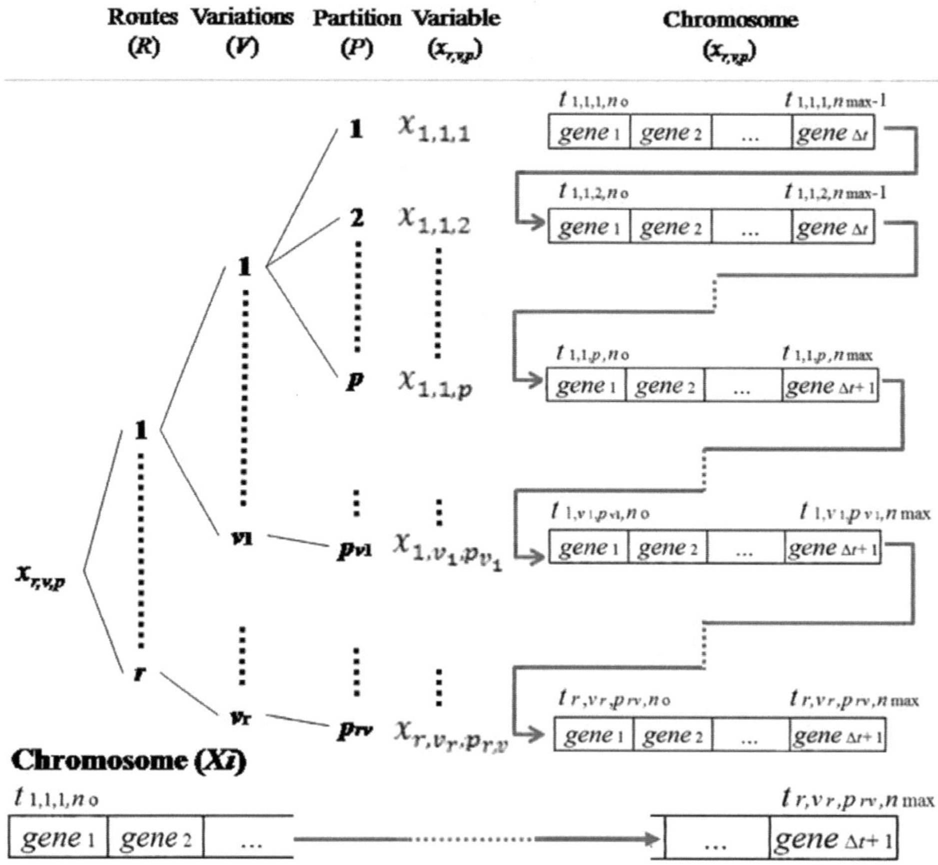

Figure 1. Design of chromosome

Each chromosome represent all variables on Model (5). Numbers of chromosomes was determined with the amount of N that will be raised randomly, limited, and arranged well to become first population. The early chromosomes was formed based on the random establishment from the value that is possible as result from one variable. In this case, the randomized variable value should fulfilled the constraints in Model (5), which is the integer ranging between Max $\left(\frac{q^{psgr}_{r,v,p}}{k_{LF_{r,v,p}}}, m^{Rit}_{r,v,p}\right)$ until $M^{Rit}_{r,v,p}$ and fulfilled Constraints (5.1).

The value that be obtained randomly then will be represented in the form of chromosomes which is also the representations of timetable.

2. Fitness function.
   To find out the fitness values for each chromosomes (Xi), it is done with using formula below:

$$f(X_i) = z(X_i) - \sum_{j=1}^{m} |MC_j . VC_j| ; i = 1,2, \dots N \tag{7}$$

Where $z(X_i)$ is the objective functions value from model (8) for chromosome X_i. The number of variable $x_{r,v,p}$ is as large as all genes in part of chromosomes which representing the variable $x_{r,v,p}$.

MC_j value is the constanta to measures the constraints interest *j*. $VC_j$is the variable value which is not suitable with constraints *j*. In this case, the value of VC_j for individu *I is as follows:*

$$VC_j = \begin{cases} A(X_i) - b & ; A(X_i) > b \\ 0 & ; \text{otherwise} \end{cases} \quad j = 1,2, \dots m \tag{8}$$

Assumption that be used in calculating the fitness values is each chromosome which is the result of genetic algorithm operator is an irregular chromosome and there is no checking for the chromosome regularity.

3. Reconstruction, Extermination of Chromosome and Elitism.
   Chromosome with the bess fitness value will be reconstructed, so it will produce the regular chromosome. The regularity of chromosomes was needed in order to create the timetable with the well-ordered headway. Extermination of chromosomes done when the population is considered to be no longer able to produce a better generation. In this extermination process, only the chromosomes with the best values ($X_{i_{best}}$) that will be maintained and will be included with the new population (elitism process [9]). For example, $g \in \mathbb{Z}^+$ is the generation in one genetic algorithm. In this case, the extermination of chromosomes done when the best fitness values in generation *g* equals the best fitness values on generation *g+w* ($f(X_{i_{best,g}}) = f(X_{i_{best,g+w}})$). It means the extermination process will happen if the best fitness values equals along *w* generations consecutively (*w* = 50). The new population that will be used to replace the exterminated population, is the early population when first time raised. Every *W* generation, the new population will be raised, will and replace the early population.
4. Selection, crossover and mutation.
   On this research, it will use rank selection in the selection process [10], *crossover* 2-points on *crossover process* and mutation operator to find the optimal solutions.
5. Chromosome representation on timetable
   The best chromosomes will be combined with departure time vector and then it will be arranged becomes 10 columns matrices.
6. Parameter choice
   The number of chromosomes (*n* = 50), the probability of mutation (p_m = 0.001), the probability of crossover (p_c = 0.6) [11], the number of generations (*g* = 6000), the extermination of chromosomes (*w* = 50) and the new population (*W* = 0, 2000, 4000).

2.3 Error Analysis
Absolute Error (E) is the absolute value of the difference approximation solution with the exact solution ($E = |z - \hat{z}|$). While the relative error (E_R) is the value of Absolute Error (E) compared with the absolute value of the exact solution, in this case $E_R = |z - \hat{z}|/|z|$ [12]. Accuracy is calculated based on the proximity to the optimal value which is defined as:

$$\text{Accuracy} = (1 - E_R) \times 100\% \tag{9}$$

3. Results and Analysis
Based on the experiment result that has be done (Table 1), the initial population generation techniques in the feasible area, regular chromosome arrangement techniques, and chromosomes extermination techniques in the specific generation could improve the genetic algorithm performance in finding out the optimal solution in the case of arranging timetable. On each table, it can be seen that in this case, the improvement of crossover possibility 0.1 and mutation possibility (from 0.001 becomes 0.01) will not significantly increases the accuracy of fitness values.

IOP Conf. Series: Materials Science and Engineering **166** (2017) 012016 doi:10.1088/1757-899X/166/1/012016

Table 1. Preliminary Experiment Results Search Solutions with Genetic Algorithm

No	Treatment								Best Position	accuracy
	Initial Population		Pc	Pm	reconstruction	extermination		generation		
	Random	*Feasible*				w=50	W=2000			
1	√		0.6	0.001	x	x	x	1000	70	-388.10%
2		√	0.6	0.001	x	x	x	1000	103	87.52%
3		√	0.7	0.001	x	x	x	1000	103	88.27%
4		√	0.6	0.01	x	x	x	1000	29	83.38%
5		√	0.7	0.01	x	x	x	1000	39	89.04%
6		√	0.6	0.001	√	x	x	1000	108	90.38%
7		√	0.6	0.001	√	√	x	1000	732	96.81%
8		√	0.6	0.001	√	√	x	3000	1418	97.98%
9		√	0.6	0.001	√	√	√	4000	3524	99.41%

Settlement of the Model (5) be done with using initial population generation techniques in feasible areas, chromosomes arrangement techniques and chromosomes extermination techniques in specific area. The Model (5) is implemented on the existing BRT service in Bogor, Indonesia. Recapitulation of the optimal solution search using genetic algorithms to 9 repetitions can be seen in Table 2. From the solution results,it shows that the fitness value that is obtained equals with the objective functions values for every repetition. Based on Formula (7), it means that the entire solution produced by the genetic algorithm is a feasible solution.

Table 2. Optimal Solutions Search Repetition Results

Repetition	Variable																		Best Position	Fitness Value	Objective Value	Condition	Accuracy
	X_{111}	X_{112}	X_{113}	X_{114}	X_{115}	X_{121}	X_{122}	X_{123}	X_{124}	X_{131}	X_{132}	X_{133}	X_{134}	X_{211}	X_{212}	X_{311}	X_{312}	X_{313}					
verification	4	31	16	41	5	6	15	60	6	15	15	8	35	18	18	7	56	3			-73334054		
1	4	32	18	41	5	6	17	60	6	15	15	8	35	18	18	7	50	3	5186	-74029219	-74029219	*feasible*	99.05%
2	4	31	16	43	6	6	15	60	6	15	15	8	35	18	18	7	50	3	4560	-73489285	-73489285	*feasible*	99.79%
3	4	32	16	44	5	7	17	60	6	15	15	8	35	18	18	7	48	3	5645	-74267919	-74267919	*feasible*	98.73%
4	4	32	17	41	5	7	15	60	6	15	15	8	35	18	18	7	52	3	4640	-73753009	-73753009	*feasible*	99.43%
5	5	31	16	41	5	8	16	61	7	15	15	8	35	18	18	7	53	3	5175	-74636165	-74636165	*feasible*	98.22%
6	4	34	16	42	5	6	17	60	6	15	15	8	35	18	18	7	48	3	5222	-73935336	-73935336	*feasible*	99.18%
7	4	31	16	41	5	6	15	60	6	15	15	8	35	18	18	7	56	3	5342	-73334054	-73334054	*optimal*	100.00%
8	4	31	16	42	5	8	15	60	6	15	15	8	35	18	18	7	53	3	5625	-73918832	-73918832	*feasible*	99.20%
9	4	32	16	44	5	8	17	60	6	15	15	8	35	18	18	7	48	3	5422	-74586206	-74586206	*feasible*	98.29%
																		Average	5201.89		-73994447		99.10%

The examination results of solution optimization show that genetics algorithm could find the optimum solution on the 7th repetition with the fitness values -73,334,054 on the 5,342 generation in accordance with Figure 2. Chromosome resulting from such solution is presented in the form of timetable as in the Table 3. Analysis result of towards the generated error shows that the solution which is produced by using genetic algorithm has the optimal value with the average of errors (0.009) or having the accuracy 99,10%. Total fitness value calculation performed in the genetic algorithm is $(50\times 6000) \leq 2^{19}$ which is much smaller when compared with the search space for 2^{l} (l = the chromosome length, in this case $l = 4340$). Efficiency of the genetic algorithm usage towards the search space reaches 99.99%.

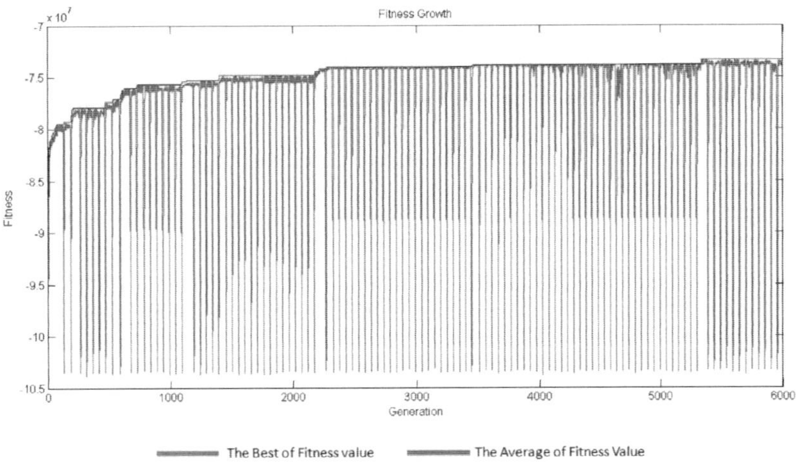

Figure 2. Fitness Values of Graphical Implemented Model (5)

Table 3. The Optimal *Timetable* on BRT service in Bogor, Indonesia.

Bubulak - Baranang Siang									
5.00	5.08	5.15	5.23	5.30	5.38	5.46	5.54	6.02	6.10
6.18	6.26	6.35	6.43	6.51	6.59	7.07	7.15	7.23	7.31
7.39	7.47	7.55	8.03	8.11	8.19	8.27	8.35	8.44	8.52
9.00	9.08	9.16	9.24	9.32	9.40	9.52	10.04	10.16	10.28
10.39	10.51	11.03	11.15	11.27	11.39	11.51	12.03	12.14	12.26
12.38	12.50	12.59	13.08	13.17	13.26	13.35	13.44	13.53	14.02
14.11	14.20	14.29	14.38	14.47	14.56	15.05	15.14	15.23	15.32
15.41	15.50	16.00	16.09	16.18	16.27	16.36	16.45	16.54	17.03
17.12	17.21	17.30	17.39	17.48	17.57	18.06	18.15	18.24	18.33
18.42	18.51	19.00	19.15	19.31	19.46	20.00	-	-	-

Baranang Siang - Bubulak									
5.31	5.39	5.46	5.54	6.14	6.22	6.30	6.38	6.46	6.54
7.02	7.10	7.19	7.27	7.35	7.43	7.51	7.59	8.07	8.15
8.23	8.31	8.39	8.47	8.55	9.03	9.11	9.19	9.28	9.36
9.44	9.52	10.00	10.08	10.16	10.22	10.34	10.46	10.58	11.10
11.21	11.33	11.45	11.57	12.09	12.21	12.33	12.45	12.56	13.08
13.20	13.40	13.49	13.58	14.07	14.16	14.25	14.34	14.43	14.52
15.01	15.10	15.19	15.28	15.37	15.46	15.55	16.04	16.13	16.22
16.31	16.40	16.50	16.59	17.08	17.17	17.26	17.35	17.44	17.53
18.02	18.11	18.20	18.29	18.38	18.47	18.56	19.05	19.14	19.23
19.32	19.41	19.50	20.05	20.21	20.36	20.50	-	-	-

Bubulak - Baranang Siang									
5.00	5.20	5.40	6.00	6.20	6.40	7.00	7.14	7.28	7.42
7.56	8.10	8.24	8.38	8.52	9.06	9.20	9.34	9.48	10.02
10.16	10.30	10.39	10.47	10.56	11.04	11.13	11.21	11.30	11.38
11.47	11.55	12.04	12.12	12.21	12.29	12.38	12.46	12.55	13.03
13.12	13.20	13.29	13.37	13.46	13.54	14.03	14.11	14.20	14.28
14.37	14.45	14.54	15.02	15.11	15.19	15.28	15.36	15.45	15.53
16.02	16.10	16.19	16.27	16.36	16.44	16.53	17.01	17.10	17.18
17.27	17.35	17.44	17.52	18.01	18.09	18.18	18.26	18.35	18.43
18.52	19.00	19.12	19.24	19.37	19.49	20.00	-	-	-

Baranang Siang - Bubulak									
5.35	5.55	6.15	6.35	6.55	7.15	7.40	7.54	8.08	8.22
8.36	8.50	9.04	9.18	9.32	9.46	10.00	10.14	10.28	10.42
10.56	11.20	11.29	11.37	11.46	11.54	12.03	12.11	12.20	12.28
12.37	12.45	12.54	13.02	13.11	13.19	13.28	13.36	13.45	13.53
14.02	14.10	14.19	14.27	14.36	14.44	14.53	15.01	15.10	15.18
15.27	15.35	15.44	15.52	16.01	16.09	16.18	16.26	16.35	16.43
16.52	17.00	17.09	17.17	17.26	17.34	17.43	17.51	18.00	18.08
18.17	18.25	18.34	18.42	18.51	18.59	19.08	19.16	19.25	19.32
19.36	19.41	19.53	20.05	20.18	20.30	20.41	-	-	-

Bubulak - Baranang Siang									
5.00	5.19	5.39	5.58	6.17	6.37	6.56	7.15	7.35	7.54
8.13	8.33	8.52	9.11	9.31	9.50	10.01	10.11	10.22	10.33
10.43	10.54	11.05	11.15	11.26	11.37	11.47	11.58	12.09	12.19
12.30	12.39	12.48	12.56	13.05	13.14	13.23	13.31	13.40	13.51
14.02	14.14	14.25	14.36	14.47	14.58	15.10	15.21	15.32	15.43
15.54	16.06	16.17	16.28	16.39	16.51	17.02	17.13	17.24	17.35
17.47	17.58	18.09	18.20	18.31	18.43	18.54	19.05	19.16	19.27
19.39	19.50	20.00	-	-	-	-	-	-	-

Baranang Siang - Bubulak									
5.40	5.59	6.19	6.38	6.57	7.17	7.36	7.55	8.15	8.34
8.53	9.13	9.32	9.51	10.11	10.34	10.45	10.55	11.06	11.17
11.27	11.38	11.49	11.59	12.10	12.21	12.31	12.42	12.53	13.03
13.17	13.26	13.35	13.43	13.52	14.01	14.10	14.18	14.28	14.39
14.50	15.02	15.13	15.24	15.35	15.46	15.58	16.09	16.20	16.31
16.42	16.54	17.05	17.16	17.27	17.39	17.50	18.01	18.12	18.23
18.35	18.46	18.57	19.08	19.19	19.31	19.42	19.53	20.04	20.15
20.27	20.38	20.48	-	-	-	-	-	-	-

Baranang Siang - Ciawi									
6.00	6.20	6.40	7.00	7.20	7.40	8.00	8.20	8.40	9.00
9.20	9.40	10.00	10.20	10.40	11.00	11.20	11.40	12.00	12.21
12.42	13.04	13.25	13.46	14.07	14.29	14.50	15.11	15.32	15.54
16.15	16.36	16.57	17.19	17.40	18.00	-	-	-	-

Ciawi - Baranang Siang									
6.36	6.56	7.16	7.36	7.56	8.16	8.36	8.56	9.16	9.36
9.56	10.16	10.36	10.56	11.16	11.36	11.56	12.16	12.52	13.13
13.34	13.56	14.17	14.38	14.59	15.21	15.42	16.03	16.24	16.46
17.07	17.28	17.49	18.11	18.32	18.52	-	-	-	-

Baranang Siang - Sentul City (Bellanova)									
5.45	6.04	6.24	6.43	7.02	7.21	7.41	8.00	8.13	8.26
8.39	8.51	9.04	9.17	9.30	9.43	9.56	10.09	10.21	10.34
10.47	11.00	11.13	11.26	11.39	11.51	12.04	12.17	12.30	12.43
12.56	13.09	13.21	13.34	13.47	14.00	14.13	14.26	14.39	14.51
15.04	15.17	15.30	15.43	15.56	16.09	16.21	16.34	16.47	17.00
17.13	17.26	17.39	17.51	18.04	18.17	18.30	18.43	18.56	19.09
19.21	19.34	19.47	20.00	20.31	21.00	-	-	-	-

Sentul City (Bellanova) - Baranang Siang									
6.02	6.21	6.41	7.00	7.19	7.38	7.58	8.18	8.31	8.44
8.57	9.09	9.22	9.35	9.48	10.01	10.14	10.27	10.39	10.52
11.05	11.18	11.31	11.44	11.57	12.09	12.22	12.35	12.48	13.01
13.14	13.27	13.39	13.52	14.05	14.18	14.31	14.44	14.57	15.09
15.22	15.35	15.48	16.01	16.14	16.27	16.39	16.52	17.05	17.18
17.31	17.44	17.57	18.09	18.22	18.35	18.48	19.01	19.14	19.27
19.39	19.52	20.05	20.16	20.47	21.16	-	-	-	-

4. Conclusion

This research has successfully developed a model of optimal timetable by using a genetic algorithm. Model is prepared using timetable variation data, route, partition/time division, number of passengers, number of fleets, capacity of the bus, departure time, travel time, headway, load factor, employees working hours, bus revenues and bus operating costs. Solutions with using genetic algorithm is done by representing the chromosomes as the time in timetable. Searching for the optimal solution is done by using several operators in the genetic algorithm, such as selection, mutation, crossover and elitism. Several techniques are added to enhance the fitness value, such as initial population generation techniques in the feasible area, regular chromosome arrangement techniques and chromosomes extermination techniques in a particular generation. The output generated from the genetic algorithm is the timetable with optimum number of trips. It is shown that the developed model has an accuracy of 99.10% in terms of proximity to the optimal solution. The efficiency of genetic algorithms usage for this problem reaches 99.99%. Research related to the preparation of optimal timetable can be developed further by developing an application that starts from the stage of the evaluation, optimization of the number of trips, preparation of optimal timetable, optimization of the employee number, optimization of the number of buses, and the preparation of schedules and bus driver. In addition, the model can be improved by taking into account the distribution of passengers as well as its implementation for the other bus rapid transit. For the model with genetic algorithm, it can be improved by adding a regularity constraint in calculating the value of the fitness function. Examination of chromosomes regularity can be done by calculating the deviation of the distance of a gene (with value 1) with another gene (with value 1) in a partition. In addition, it is also necessary to repair or add another technique, so genetic algorithm able to find the optimal solution in each repetition.

References

[1] Cedar A, Golany B, Tal O. Creating Bus Timetables with Maximal Synchronization. *Transportation Research.* 2001; Part A: 913-928.

[2] Fu, L. Liu Q, Calamai P. Real-Time Optimization Model for Dynamic Scheduling of Transit Operations. *Transportation Research Record 1857.* 2003; Paper No. 03-3697.

[3] Kidwai FA. *A Genetic Algorithm Based Bus Scheduling Model for Transit Network.* Eastern Asia Society for Transportation Studies. Bangkok. 2005; 05: 477–489.

[4] Chuanjiao S, Wei Z, Yuanqing W. Scheduling Combination and Headway Optimization of Bus Rapid Transit. *Journal of Transportation Systems Engineering and Information Technology.* 2008; 8(5): 61–67.

[5] Wihartiko FD. Pengoptimuman Jumlah Rit pada Pelayanan Jasa Bus Trans Pakuan Kota Bogor. *Komputasi-Scientific Journal of Computer Science and Mathematics.* 2012; 9 (1): 58-73.

[6] Wagale M, Singh AP, Sarkar AK, Arkatkar S. Real-Time Optimal Bus Scheduling for a City using A DTR Model. *Procedia - Social and Behavioural Sciences.* 2013; 104: 845–854.

[7] Huy PNA, San CTB, Triantaphyllou E. *Solving Integer Programming Problems Using Genetic Algorithms.* ICEIC. Ha-Noi. 2004.

[8] Coley DA. An Introduction to Genetic Algorithms for Scientists and Engineers. Singapore: World Scientific Publishing. 1999.

[9] Engelbrecht AP. Computational Intelligence. Chichester: John Wiley & Sons Ltd. 2007.

[10] Kumar. Blending Roulette Wheel Selection & Rank Selection in Genetic Algorithms. *International Journal of Machine Learning and Computing.* 2012; 2(4): 365-370.

[11] Eiben AE, Hinterding R, Michalewicz Z. Parameter Control in Evolutionary Algorithm. *IEEE Transactions On Evolutionary Computation.* 1999; 3(2): 124-141.

[12] Mathews JH, Kurtis DF. Numerical Methods using Matlab. New Jersey: Pearson Education International. 2004.

Implementation of Self Organizing Map (SOM) as decision support: Indonesian telematics services MSMEs empowerment

E.T. Tosida, S. Maryana, H. Thaheer, Hardiani

Computer Science Department, Pakuan University, Jl. Pakuan PO Box 452, Indonesia

E-mail: enengtitatosida@unpak.ac.id

Abstract. Information technology and communication (telematics) is one of the most rapidly developing business sectors in Indonesia. It has strategic position in its contribution towards planning and implementation of developmental, economics, social, politics and defence strategies in business, communication and education. Aid absorption for the national telecommunication SMEs is relatively low; therefore, improvement is needed using analysis on business support cluster of which basis is types of business. In the study, the business support cluster analysis is specifically implemented for Indonesian telecommunication service. The data for the business are obtained from the National Census of Economic (Susenas 2006). The method used to develop cluster model is an Artificial Neural Network (ANN) system called Self-Organizing Maps (SOM) algorithm. Based on Index of Davies Bouldin (IDB), the accuracy level of the cluster model is 0.37 or can be categorized as good. The cluster model is developed to find out telecommunication business clusters that has influence towards the national economy so that it is easier for the government to supervise telecommunication business.

1. Introduction

Information technology and communication (telematics) is one of the most rapidly developing business sectors in Indonesia. The sector has strategic position in its contribution towards planning and implementation of national developmental, economics, social, politics and defence strategies. Studies that focus on development of the national telecommunication service are pretty limited despite the annual growth of the sector that reaches nearly 4% [1]. A study conducted by [1] of which focus is telecommunication service reveals that the most preferable products of telecommunication for the public as the end-users of telecommunication products are ones related to computer and internets, either direct use of computer and internets or in the form of information. Indonesia ranks the fifteenth in terms of countries with the largest internet users even though level of penetration is still pretty low (1.7%). Shifting to digital communication is launched as public information policy initiatives [2]. It is expected that the initiatives become the key contributor to spread internet access in every household. Public policy has an active role in expanding and promoting information technology diffusion that emphasizes on quality of service.

One method to prepare telecommunication industry as national priority is developing model for classifying the national telecommunication service [1]. In order to develop the classification, qualitative study that compares and harmonizes types of business groups should be conducted. The classification model has yet been completed with detail characteristics related to potential of each area and its contribution towards developing factors for telecommunication service. The data related to Indonesian telecommunication service mapping is relatively large. The approach used to deal with large data is

Knowledge Discovery in Database (KDD) approach and one of its components is widely known as Data Mining. According to [3], data mining is more meaningful as a means of gaining knowledge (data) from interdisciplinary subject. One of data mining techniques implemented to explore characteristics of particular business sector is clustering.

Implementation of clustering related to business development strategy has been conducted by some researchers for example clustering-based strategic formulation to explore business characteristics [4]. Information and Communication Technology (ICT) sector is closely related to innovations, clustering towards types of existing technology based on its growing patterns as well as correlations between one technological information [5] and another can be used as strategic decision support help for the sector [6]. It is expected that the clustering can help stakeholders understanding characteristics of the industry or existing business groups more particularly telecommunication sector as well as analyzing process of innovations within the business groups.

Studies related to identification of Indonesian telecommunication service characteristics and potentials have been conducted by [1]. The studies are limited to qualitative and descriptive approach. Thus, deeper exploration that involves data mining is of necessity. The concept should be equipped with visualization of business developing factor inter-correlation as well as potentials of each area using clustering process with Artificial Neural Network (ANN) with SOM algorithm.

2. Indonesian Telematics Services MSMEs Competences

Indonesian telematics businesses are grouped into the hardware industry; the software industry; and industrial and non-industrial telematics services. The results of Multi Sectoral Qualitative Analysis (MSQA) concluded the level of interest of the development of the telematics industry types in Indonesia [7]. Specifically the study of the classification of telematics services business in Indonesia resulted in three major groups namely business services, communications services and education services [1]. This classification model is obtained through a comparative study and followed by expert acquisition using the Delphi method. The basis of the development of this model is taken from the document of the World Trade Organization / WTO [8], as it relates to the preparation of Indonesia towards the liberalization of the field of telematics. These fields have a subfield and in detail derived by following the numbering systematically arranged following the numbering system. The numbering system is known as Group of business scope standard in Indonesia (ISIC).

Related to the development business of telematics services group in Indonesia, has been arranged competence map services telematics group [9]. The existing competence map remains dominated by a group of MSMEs telematics service, which are have the conditions that still needed help in the development of their business. One of the government's efforts in empowering SMEs telematics through the ICT Fund. ICT Fund which was directed to improve business ecosystems, including through venture capital assistance, with the framework of Evolution Concept of USO (Universal Service Obligation).

Indonesian telematics service business competence can be identified through the value chain framework. The results of this framework indicates that the computer consulting business is a business that has the support of both internal and external, so that it becomes easier in its development. Field of telematics education efforts will still need affirmative action to strengthen the competence of external support. Publishing Software, Computer Programming, Information Technology and other market potential is still vast, but the internal capabilities need to be improved. Computer Repair Enterprises, Web Portal, Hosting, and Special Design, began to show saturation. This matter may be attributable to the increasing number of businessmen, the increasingly fierce competition, and the carrying capacity of the human resources inadequate [10].

3. Method

This research resulted a model that includes empowering of national telematics services MSMEs, as one of the strategies to increase competitiveness, which is done through clustering models. This model is supported by Data Mining process. Research carried out in stages using the stage as shown in Figure 1. The main data used are economic census sample data from the Central Bureau of Statistics (BPS), with the scope of entities and attributes that will be tailored to the needs of stakeholders.

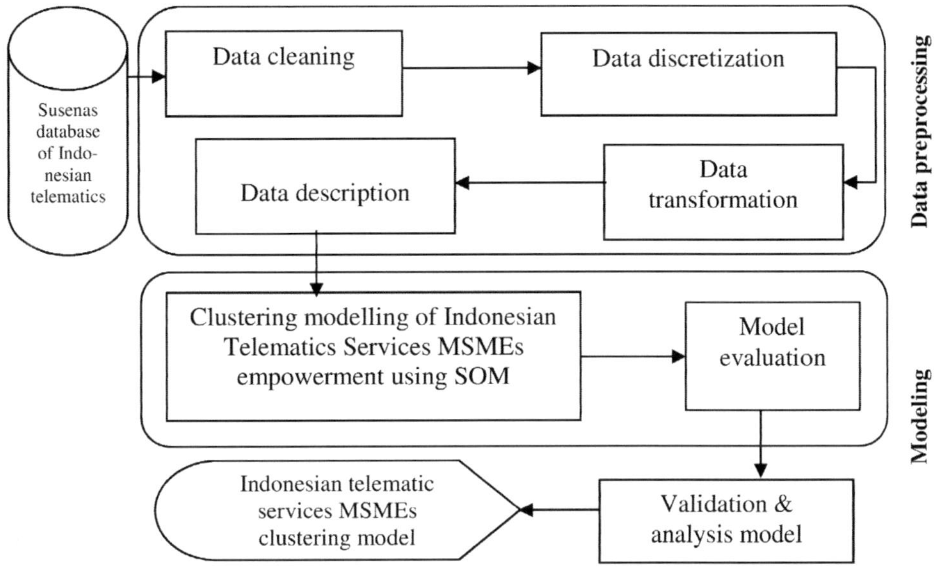

Figure 1. Research method

4. Result and Discussion

4.1. Analysis and Design Clustering Model used SOM method

Analysis of system and network architecture SOM algorithm performed in order to determine the work process SOM algorithm. This was done to evaluate and give an overview of clustering models Indonesia telematics services MSMEs to be created. SelfOrganizing Map (SOM) or what is often called a Kohonen neural network topology is one form of Unsupervised Artificial Neural Network (ANN Unsupervised, where in the training process does not require surveillance (target output). Kohonen network / SOM is used for grouping (clustering) data based on the characteristics / features of the data [11]. SOM algorithm is shown in Figure 2.

SOM architecture consists of two layers that are the input and output. Inter- layers are connected by a certain weight are often referred to as representation vector or vector code book or reference vector [11]. In the one-dimensional architecture, input layer (X) will take the form of linear and similarly the output (Y). In this architecture, the adjacent unit would have been more different than the farther unit. Weights (W) is used as one of the components to determine the distance of the output layer. The shortest distance will become references clustering data. If there is a distance to the second output is the closest then the vectors are included will be placed in the second output class, and then the weight that is to be updated by considering the inclusion of the input vector to the output layer of the second class. This new weight then the next will be a reference for further input.

Architecture of SOM in general is shown in Figure 3. Case of Indonesian telematics services MSMEs clustering model based on the architecture as shown in Figure 4. The results of interviews with experts and refers to the data Susenas 2006, determined 10 factors that affect the empowerment of SMEs telematics services Indonesia [12]. Therefore, the inputs used in the architecture of SOM in these cases totaled 10. Determination of the number of clusters based on the type of Indonesian telematics services

IOP Conf. Series: Materials Science and Engineering **166** (2017) 012017 doi:10.1088/1757-899X/166/1/012017

MSMEs that was consist of 10 types. Type of MSMEs encompasses Rental Services Computer and VCD, Repair and Computer Service, Printing Services, Photo Copy, Training Service and Private Computers, Service Setting Computer and Internet Services Data Processing and Analysis Statistics, Consulting Services Software, Services Procurement and Assembling Computer, Buy-Sell Computers and accessories and Computer Services other. Thus the output of architecture SOM for this case amounts to 10. The SOM architecture for Indonesian telematics services MSMEs clustering models shown in Figure 4 .

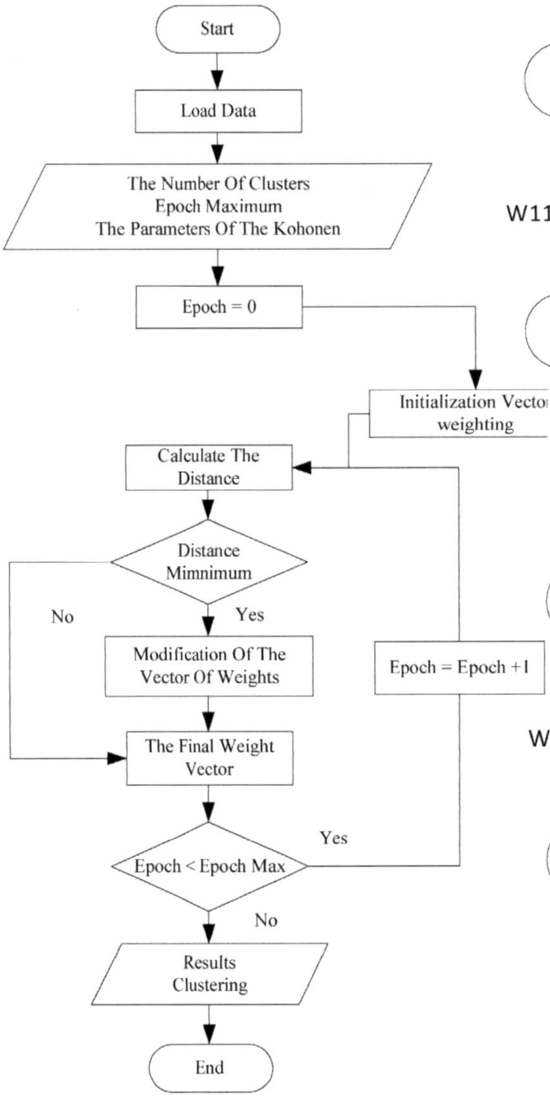

Figure 2. SOM algorithm

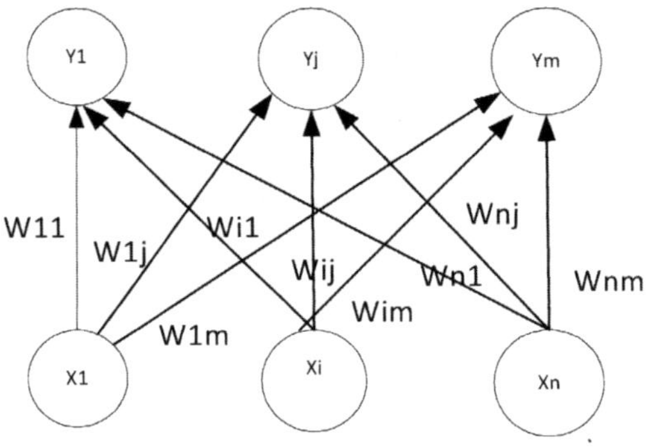

Figure 3. SOM architecture in general

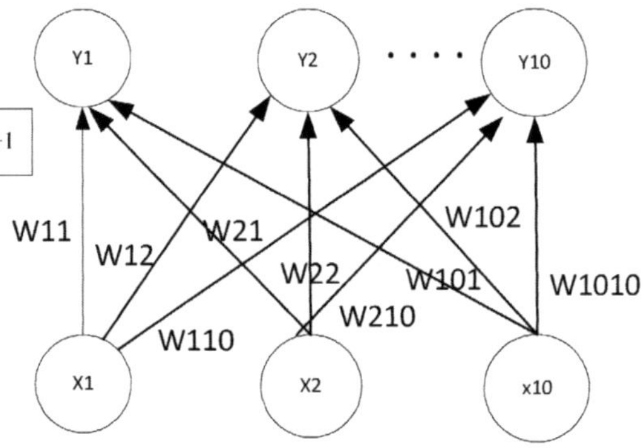

Figure 4. SOM architecture of Indonesian telematics services MSMEs clustering model

4.2. Clustering Indonesian Telematics Services MSMEs Model

Model of clustering is developed base on the data from the 2006 National Economic Census. The number of samples are 2704 that is divided into two, 70% for training data and the remaining 30% for testing data. Clustering model outlook is developed using *plotsomhits* (Matlab 2010a). The outlook makes it easier for users to find out total number in each cluster. Testing was conducted using different clusters, namely 10, 7, and 5 clusters. The basis of using the clusters in the testing related to [5]. Epoch used are 100, 80, 50, 30 while learning rate is 0.01, 0.03, 0.04, 0.05, 0.07, 0.08 and 0.09. The clustering model is conducted using the SOM algorithm and is evaluated using Index of Davies Bouldin (IDB). Index of Davies Bouldin aims at maximizing inter cluster and minimizing intra cluster in one cluster. The index is used to measure validity from result of clustering in order to obtain the most suitable IDB, which is minimum IDB (Zaki dan Meira, 2013).

The results of modelling using training data is shown in Figure 5, while the results of modelling using testing data shown in Figure 6 .

Figure 5. Experimentation result of clustering models using training data

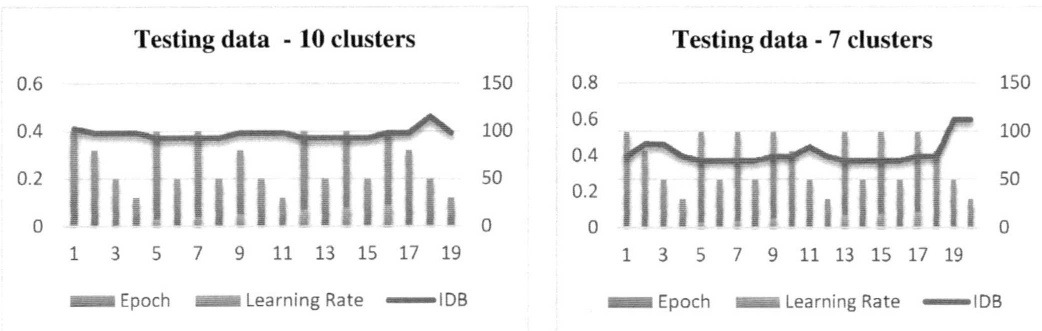

Figure 6. Experimentation result of clustering models using testing data

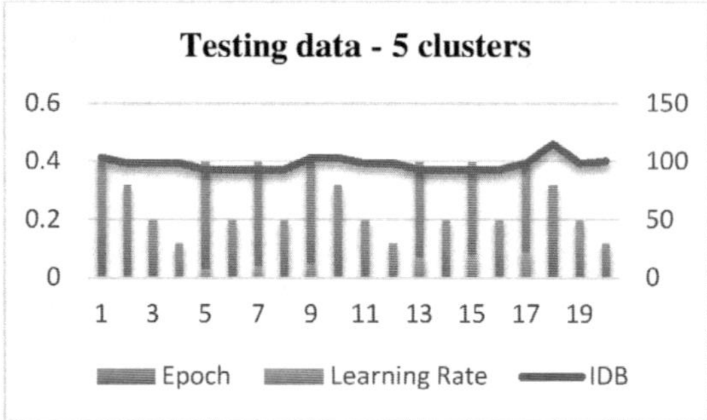

Figure 6. Experimentation result of clustering models using testing data (continued)

Based on Figure 5 experimentation clustering models using training data IDB has produced a minimum of 0.35. The experimental results of the training data indicated that the number of clusters and Learning Rate (LR) did not affect the achievement of the minimum IDB. The epoch affect the achievement of the IDB. The experimental results from the training data, shows that the minimum value of IDB was in the epoch 80. These results form the basis for the learning process is done for modelling using testing data (based on Figure 6). Experiments to testing data showed similar characteristics with the results of training data, which changes the number of clusters and LR did not affect the achievement of the minimum IDB. IDB minimum value obtained at 0.37 value and occurs in epoch 80. Value IDB experimental results of a more stable testing data at each change of LR and epoch, when compared with the experimental results of the training data. Clustering models obtained in this study as a whole has achieved fairly good accuracy. This is demonstrated by the achievement of IDB relatively small value nearly zero.

Further analysis of the experimental results show interesting findings. One example in cluster 5 the modelling results using 10 clusters with optimal condition of the IDB, has found the characteristics of MSMEs conditions unique telematics services. The condition in question include the following: the amount of capital has an average of 4 million, with an average spending of 1 million, and the average income of over 2 million. Business entities that exist in this cluster dominant following the guidance or counselling and the state of the company and the company's prospects were very good. Based on this information, the model of clustering using SOM method can be used as a decision support for the government in the process of empowerment of Indonesian telematics services MSMEs. Based on 2006 Susenas, cluster 5 is referred to the types of Computer Settings Services and Internet businesses. In 2006 those are telematics service business that are growing with excellent business prospects. Based on the model results clustering exposure conditions, the government can direct the kind of empowerment that is more suitable for this type of business. These telematics services business is dominated by businesses that have received guidance and counselling, as well as the conditions were relatively good profit margins. Therefore, empowerment which was more suitable for these business type was increase the competence of human resources or marketing competitive regulation. So that in the future these types of telematics services business is expected to have a stronger competitiveness.

Exposure to other conditions may be identified through this clustering models. It shows that through the Susenas database, SOM method can be used as decision support for the government in determining the type of empowerment of SMEs telematics services. Disadvantages of this study include data used still have data that is not balanced on a number of factors or parameters used. Therefore, further research requires the balancing process the data, so the expected value obtained IDB will become more optimal.

5. Conclusion

Model clustering of empowerment of Indonesian telematics services MSMEs using the Susenas database has performed quite well, with the achievement of a minimum value of IDB which revolves around the value of 0.36. The experiments show that the diferent number between cluster and Learning Rate (LR) did not have any influence towards optimum IDB scores. Optimum clustering model takes place when epoch is 80, with IDB score of 0.37. Different occurence happens towards the training data resulting in IDB score of 0.35. The clustering characteristics of testing data is similar to the model of training data clustering. It shows that the model works well. Prominent clustering model takes place in cluster 5 that shows dominating number of telecommunication MSMEs that less worthy to get grants compared to other clusters. Such domination causes bias in the result of cluster as the effect of imbalance number of data in one of the types of telecommunication MSMEs. However, the model results clustering using SOM has been able to provide information of interest and can be used as decision support for the government in determining the type of empowerment of MSMEs telematics services. This information was obtained from the results of further analysis of the characteristics of the cluster formed from the model . Disadvantages of this study are associated with the data that is not balanced is expected to be overcome with the preprocessing of data that involves the balancing process of the data. Thus the model obtained will be more accurate with the IDB a more optimal value.

Acknowledgment

1. DP2M DIKTI Educational and Cultural Ministry as the main sponsor, which gives us Competitive Grants Scheme
2. Computer Science Department, Mathematics and Natural Science Faculty, Pakuan University, and Research Institute Pakuan University, for supporting, coordinating and facilitating to achieve this grants.
3. Indonesian Communication & Information Ministry, Indonesian Cooperation and SMEs Ministry and Bandung Technopark for active participation in the activities of interviews and user requirement.

References

[1] Tosida ET, Harsani P, Hermawan, Setyaningsih S. 2012. Classification models of information technology services bussiness in Indonesia. Proceeding International Seminar on Science and Technology Innovations 2012. University of Al Azhar Indonesia, October 2-4th 2012, ISBN 978-602-95064-5-7, UAI Press. 304-309.

[2] Sourbati M. 2011. The digital switchover as an information society initiative: The role of public policy in promoting access to digital ICTs. Journal of Telematics and Informatics, Volume 28, Issue 4, November 2011, Pages 295–30.

[3] Han J, Kamber M, Pei J. 2012. Data mining: Concepts and techniques. Third Edition. Morgan Kaufmann is an imprint of Elsevier, 225 Wyman Street, Waltham, MA 02451, USA.

[4] Hadighi SA, Sahebjamnia N, Mahdavi I, Shirazi MA. 2013. A framework for strategy formulation based on clustering approach: A case study in a corporate organization. Knowledge-Base Systems, 49 (2013), 37-49. http://dx.doi.org/10.1016/j.knosys.2013.04.008.

[5] Lee H, Lee S, Byungun Y. 2011. Technology clustering based on evolutionary patterns: The case of information and communications technologies. J. Technological Forecasting & Social Change, 78 (2011) 953-967. DOI: 10.1016/j.techfore.2011.02.002.

[6] Tsui Chia-jung, Ping W, Fleischann KR, Oard DW, Sayeed AB. 2010. Exploring the relationships among ICTs: A scalable computational approach using KL divergence and hierarchical clustering. Proceedings of the 43rd Hawaii International Conference on System Sciences. 978-0-7695-3869-3. Added to IEEExplorer. DOI: 10.1109/HICSS.2010.203

[7] Setyaningsih S, Hermawan, Tosida ET. 2013. Competency mapping of human resources in the field of Indonesian telematics industry: As an investment policy. Proceedings of the National Conference of Mathematics and Natural Sciences: MIPA as Platform Creation and Technological Innovation. IPB International Convention.251-258.

[8] WTO. 1991. Services sectoral classification list [Internet]. [Downloaded 2012 Feb 10]. Available at https://www.wto.org/english/tratope/serve/mtn_gns_w_120_e.doc.
[9] Tosida ET, Maryana S, Thaheer H. 2014. The Potencies of Indonesian Telematics Services Enterprises Group. Proceedings of National Conference of Information, Communication & Management Proceeding. Bina Darma University, Palembang, August, 2014, 45-52.
[10] Tosida ET, Thaheer, Maryana S. 2015. Strategy to enhance competitiveness through the value chain of framework for the Indonesian telematics business service competence. J. Postal and Informatics Research. Vol. 5(1) 2015. 1-18.
[11] Fausett L. 1994. Fundamentals of neural networks, architectures, algorithms, and applications. Prentice-Hall, Inc. Upper Saddle River, NJ, USA, ISBN: 0-13-334186-0.
[12] Tosida ET, Maryana S, Thaheer H, Damin FA. 2015. Visualization model of Small Medium Enterprises (SMEs) telematics services potentiality map in Indonesia. Published in: International Conference on Information & Communication Technology and Systems (ICTS), 2015. Added to IEEExplorer. January 14th 2016. DOI: 10.1109/ICTS.2015.7379890.
[13] Zaki MJ, Meira Jr W. 2013. Data mining and analysis: Fundamental concepts and algorithms.

Tsukamoto fuzzy implementation to identify the pond water quality of koi

A. Qur'ania and D.I. Verananda

Computer Science Department, Pakuan University, Indonesia.

E-mails: qurania@yahoo.com and dwiibnuvz@gmail.com

Abstract. The colour quality of koi was affected by the water quality in the pond. Koi fish have a diversity of types differentiated based on the body colour groups, such as one colour pattern, two colour patterns, three colours patterns and even more. Each colour characteristic of the koi have different handling, particularly in the handling of water quality, this is because the colour pigments in the body was affected by the composition of water quality include temperature, pH, TDS, do and salinity. The data of koi fish used were *sanke*, *sowa*, *kohaku*, *shiro*, *yamabuki*, *ogon* and *chagoi*. The aim of this study is to make an application to inform the condition of the pool water quality that can help breeders to know the water quality that will improve the handling strategies through water media. Tsukamoto Fuzzy method used to produce the three outputs namely water quality, water grade, and water conditions. The output of water quality consists of four categories, namely optimal, moderate, poor, and very poor. The output of water grade consists of grade A to D, while the output of water conditions consist of an excellent, good, bad, and very bad. Input to the application consists of five parameters, namely water temperature, pH, TDS, do and salinity.

1. Introduction

Koi was the most commodity of Indonesian ornamental fish exports. According to statistics the *Ministry of Marine Affairs and Fisheries* (KKP) Republic of Indonesia 2014, the average ornamental fish total exports was 30%. The quality of the koi that exported should meet several criteria, based on the quality and the sharpness of colour. Each colour characteristic of the koi have different handling, particularly in the handling of water quality, this is because the colour pigments in the body was affected by the composition of water quality include temperature, pH, TDS, do and salinity. Koi have a diversity of types differentiated based on the body colour groups, such as one colour pattern, two colour patterns, three colours patterns and even more. Each colour characteristic of the koi have different handling, particularly in the handling of water quality, this is because the colour pigments in the body was affected by the composition of water quality include temperature, pH, TDS, do and salinity. Water is a factor that support up to 20% to improve the colour quality of the koi besides 70% of quality factor and 10% of other factors (Takeo, 1986) [8]. The previous study as a reference is the determination of the expert system for freshwater fish farming based on the location and quality of water [2]. Research on the use of Tsukamoto Fuzzy method for the determination of water quality aims to determine the condition of the river water with four categories: good condition, lightly polluted, medium polluted and heavy polluted [1]. The aim of this study is to make an application to inform the condition of the pool water quality that can help breeders to know the water quality that will improve the handling strategies through water media. Tsukamoto Fuzzy method will produce three outputs namely water quality, water grade, and water conditions. The output of water quality consists of four categories, namely optimal, moderate, poor, and very poor. The output of water grade consists of grade A to D, while the output of water conditions consist of an excellent, good, bad, and very bad. Input to the application consists of five parameters, namely water temperature, pH, TDS, do and salinity.

2. Methodology

The research methodology used to identify water quality of Koi using an expert system approach [6] with Tsukamoto fuzzy method [3] through several stages as shown in Figure 1.

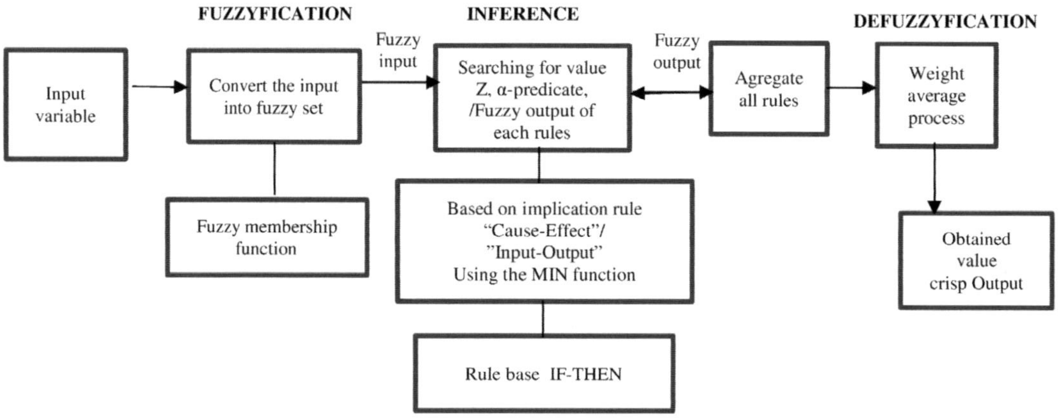

Figure 1. Stages of Fuzzy Tsukamoto.

The optimal water quality can improve and boost the quality of Koi, while poor water quality can resulting in decreased quality of Koi. Data types of water used in this study is a kind of water for Koi species by colour pattern, namely the Koi one colour, two colours, three colours and mix. These fishes are Sanke, Sowa, Kohaku, Shiro, Ogon, Yamabuki, and mix. All four categories of Koi have different needs and characteristics almost the same but different. The parameters used are water parameters that independent from each other and influence on the quality of Koi. Water parameters include temperature, pH, TDS (solutes in the water), do (dissolved oxygen) and salinity (salt content) [5] [7]. The tools needed to measure the parameters of water can use the pH meter tools, TDS meter, DO meter, Salinity meter, and Thermometer. Objects were observed in this study is the quality of water by reference to the type of water used by Koi champion [4] and ordinary koi pet, by categorized into several qualities include: Optimal (A), Moderate (B), Poor (C) and Very Poor (D). Data of Koi fish pond water quality shown in Table 1.

2.1 Fuzzification process

Fuzzification was done to divide the variables into fuzzy sets that will be used in making the rules. Water parameters were developed on the basis of fuzzy system that uses a Fuzzy Inference System (FIS) as a process to determine the final results as shown in Table 2.

Table 1. Water quality of Koi pond

Koi Type	Water Parameter	Unit	[A] Optimal	[B] Moderate	[C] Poor	[D] Very Poor
Mix	Temperature	Celsius	23-25	20-28	15-20 ; 28-30	<=10 ; >=32
	pH	Ph	7,2 – 7,5	7,0 -7,8	5,5-6,5 ; 8,0 -8,5	0 - 4,0 ; >8,5
	TDS	Ppm	20-80	0-150	200-300	>400
	DO	Ppm	>7,0	5.0 – 6,0	3,0-4,0	< 2
	Salinity	Ppt	0,05-0,15	0,00- 0,2	0,25 - 0,3	>0,4
Chagoi Ogon Yamabuki	Temperature	Celsius	25-29	20-30	15-20 ; 30-32	<=10 ; >=35
	pH	pH	7,5 – 8,5	7,0 -8,7	5,5-7 ; 8,7-8,8	0 - 5,0 ; >9
	TDS	Ppm	75-100	0-150	200-300	>400
	DO	Ppm	>7,0	5.0 – 6,0	3,0-4,0	< 2
	Salinity	Ppt	0,05-0,15	0,00- 0,2	0,25 - 0,3	>0,4
Shiro	Temperature	Celsius	18-22	15-28	13-15; 28-30	<=10 ; >=32
	pH	pH	7,5 – 8,5	7,0 -8,7	5,5-7 ; 8,7-8,9	0 - 5,0 ; >9
	TDS	Ppm	300-350	0-400	450-500	>550
	DO	Ppm	>7,0	5.0 – 6,0	3,0-4,0	< 2
	Salinity	Ppt	0,05-0,15	0,00- 0,2	0,25 - 0,3	>0,4
Showa Sanke	Temperature	Celsius	25-26	20-28	15-20 ; 28-30	<=10 ; >=32
	pH	pH	6,8 – 7,4	6.5-;-7.5	5.5-6.5; 7,5-8.0	0 - 5,0 ; >8.5
	TDS	ppm	250-300	0-350	400-450	>500
	DO	ppm	>7,0	5.0 – 6,0	3,0-4,0	< 2
	Salinity	ppt	0,05-0,15	0,00- 0,2	0,25 - 0,3	>0,4
Kohaku	Temperature	Celsius	25-28	20-30	15-20 ; 30-32	<=10 ; >=35
	pH	pH	7,5 – 8,5	7,0 -8,7	5,5-7 ; 8,7-8,9	0 - 5,0 ; >9
	TDS	ppm	75-100	0-150	200-300	>400
	DO	ppm	>7,0	5.0 – 6,0	3,0-4,0	< 2
	Salinity	ppt	0,05-0,15	0,00- 0,2	0,25 - 0,3	>0,4

Table 2. Water parameter

Input Parameter	Unit Value	Fuzzy
Temperature	Celsius	Fuzzy
pH	Acid (< 7), normal (7), base (> 7)	Fuzzy
TDS	ppm	Fuzzy
DO	ppm	Fuzzy
Salinity	ppt	Fuzzy

2.2 Membership function

Membership functions were used to assign values to fuzzy set and domain for each water quality parameters, as shown in Table 3.

2.3 Membership function of water quality output

Output identification of water quality represented by the triangle curve shown in Figure 2. The output value rated 100 to 500, scoring of 100 to 500 was to determine the range of each output, so it can be classified on the calculation result of water quality output was optimal (A), moderate (B), poor (C), or very poor (D) as shown in Table 4.

Table 3. The fuzzy set and water parameter domain

Parameter	Fuzzy set	Domain
Temperature	Excellent	[22.5 25 26 27]
	Good	[17.5 20 22.5 25], [26 27 28 29]
	Bad	[12.5 15 17.5 20], [28 29 30 31]
	Very Bad	[0 10 12.5 15], [30 31 32 32]
Ph	Excellent	[6.65 6.8 7.4 7.7]
	Good	[6 6.5 6.65 6.8], [7.4 7.7 8 8.2]
	Bad	[6.25 5.5 6 6.5], [8 8.25 8.5 8.6]
	Very Bad	[0 0 5.25 5.5], [30 31 32 32]
TDS	Excellent	[125 250 300 325]
	Good	[0 0 125 250], [300 325 350 400]
	Bad	[350 400 450 475]
	Very Bad	[450 475 500 500]
d.o	Excellent	[6 7 7 7]
	Good	[4 5 6 7]
	Bad	[3 3.5 4 5]
	Very Bad	[0 0 3 3.5]
Salinity	Excellent	[0.025 0.05 0.15 0.175]
	Good	[0.15 0.175 0.2 0.225]
	Bad	[0.2 0.225 0.3 0.35]
	Very Bad	[0.3 0.35 0.4 0.4]

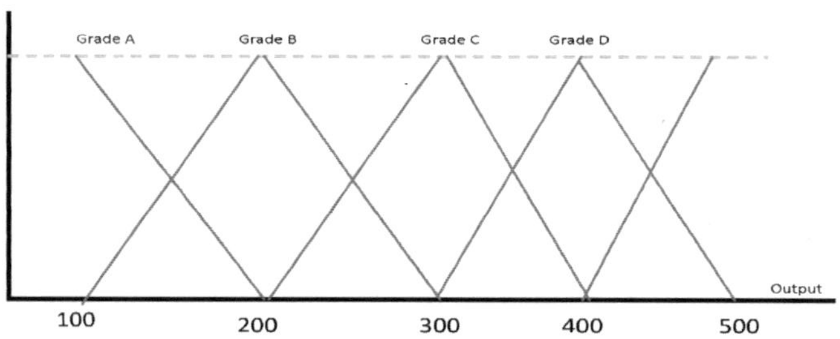

Figure 2. Membership function output of water quality

Table 4. Fuzzy set and output parameter domain

Parameter	Fuzzy set	Domain
Output (water quality)	Optimal (A)	[0 100 200]
	Moderate (B)	[100 200 300]
	Poor (C)	[200 300 400]
	Very Poor (D)	[300 400 500]

2.4 Making Rules and Fuzzy Reasoning

The rules applied represented in form "if-then" rules and connect between the premises with the operator "and" while the degree of membership selected on each parameter was the biggest value because using the operator "or". The fuzzy reasoning method based on an evaluation of each rule before it can be concluded (defuzzification), at this stage the degree of membership value parameter of water that has been sought to be used as input information of each rule that was created, of course, the rule used was the rule that only has the relationship between the fuzzy membership with the rule. Then, this process will produce output values based on the output membership function of each rule. The rule made was two rules that have relationship as the following example:

[R1] IF temperature quality EXCELLENT AND ph quality EXCELLENT AND TDS quality GOOD AND DO quality EXCELLENT AND Salinity quality GOOD THEN pond water quality is Grade B

$\alpha 1$ =µtemperatureEXCELLENT[temperature] ∩ µphEXCELLENT[ph] ∩ µtdsGOOD[tds] ∩µdoEXCELLENT[do] ∩ µsalinityGOOD[salinity]

= Min (µtemperatureEXCELLENT[25.5], µphEXCELLENT[7.4], µtdsGOOD[93], µdoEXCELLENT[8.4], µsalinityGOOD[0.00])

= Min (0.52 , 1, 1, 1, 1)

= 0.52

Based on the water quality set Grade B, then:

$\mu\, gB = \{(300 - x)/(300 - 200)$

Z1= 300 - (0.52 * (300 -200)) = 248

[R2] IF temperature quality GOOD AND ph quality EXCELLENT AND TDS quality GOOD AND DO quality EXCELLENT AND Salinity quality GOOD THEN pond water quality is grade B

$\alpha 2$ =µtemperatureGOOD[temperature] ∩ µphEXCELLENT[ph] ∩ µtdsGOOD[tds] ∩µdoGOOD[do] ∩ µsalinityGOOD[salinity]

= Min (µtemperatureGOOD[25.5], µphEXCELLENT[7.4], µtdsGOOD[93], µdoEXCELLENT[8.4], µsalinityGOOD[0.00])

= Min (0.48 , 1, 1, 1, 1)

= 0.48

Based on the water quality set Grade B, then:

$\mu\, gB = \{(300 - x)/(300 - 200)$

Z2= 300 - (0.48 * (300 -200)) = 252

2.5 Defuzzification

The output process marked by defuzzification stage to produce a crisp values of some output fuzzy rule evaluation results to the knowledge base. Defuzzification method used in this study was the weighted average method, these results will be used as criteria to assess water quality, crisp values obtained as in equation (1).

$$Z = \frac{(\alpha 1 * z1) + (\alpha 2 * z2) + \ldots + (\alpha n * zn)}{\alpha 1 + \alpha 2 + \ldots + \alpha n} \qquad (1)$$

The final result based on sample of test data by using equation (1) then,

$$Z = \frac{(248*0,52)+(252*0,48)}{0,52 + 0,48}$$

Z= 249,92

Based on the Tsukamoto fuzzy values obtained = 249.92, because it is in the range between 200 and 300 then this value is categorized as Grade B, Moderate types of water quality.

2.6 Validation Test of Water Quality System Assessment on Expert

The testing process carried out on 21 test data to ensure the compatibility between the systems and expert. The test results by grouping types of Koi fish are shown in Table 5.

Table 5. Testing result of water quality system between the output system and expert

Koi Pond type	tempe rature (⁰C)	pH	TDS (ppm)	d.o (ppm)	Salinity (ppt)	Tsukamoto value	Wáter quality (Grade)	Expert validation
Mixed	26.3	7.54	74	8.2	0.11	**238.59**	B	Valid
Mixed	23.7	7.5	87	7.2	0.16	**185.71**	A	Valid
Mixed	28.7	7.1	146	4.2	0.16	**347.5**	C	Valid
Mixed	32.7	6.1	354	4.2	0.16	**457.14**	D	Valid
Mixed	24.2	7.3	67	7	0.16	**148**	A	Valid
Kohaku	28.2	7.7	105	8.9	0.15	**`188.57**	A	Valid
Kohaku	24.2	7.3	67	7	0.16	**260.26**	B	Valid
Kohaku	24.2	6.3	164	4	0.16	**367.38**	C	Valid
Kohaku	32.2	8.9	209	7	0	**418.24**	D	Valid
Sanke	23.8	7.4	93	8.4	0	**249.92**	B	Valid
Sanke	28.2	7.9	209	7	0	**334.226**	C	Valid
Sowa	32.2	8.9	209	7	0	**444.08**	D	Valid
Sowa	25.2	7.4	239	7	0	**124.85**	A	Valid
Shiro	28.2	7.7	105	8.9	0.15	**252**	B	Valid
Shiro	22	7.5	249	7	0.15	**178.88**	A	Valid
Shiro	29	7.5	456	5	0.15	**249.28**	C	Valid
Shiro	29	6.7	590	4	0.2	**448**	D	Valid
Ogon	23.7	8.1	110	7.2	0.16	**231.13**	B	Valid
Ogon	29	6.7	590	4	0.2	**448**	D	Valid
Yamabuki	27.5	7.4	87	6	0.1	**140**	A	Valid
Yamabuki	25.2	7.8	356	4	0	**400**	C	Valid

Based on Table 5, the precision accuracy can be calculated using the Confusion Matrix, by comparing the 21 test data to classified water quality grade A (+) and grade BCD (-). The final result was shown in Table 6 with the accuracy up to 100%.

Table 6. The Final Result

Testing		Expert (Actual)	
		Negative	Positive
Fuzzy Tsukamoto (Predict)	Negative	15	0
	Positive	0	6

3. Conclusion

Research on the application of Fuzzy Tsukamoto for the identification of Koi water quality pond aims to create a system that can assess the quality of water in four criteria, ie optimal (grade A), moderate (grade B), poor (grade C), and very poor (grade D). The quality of water was determined based on five parameters: temperature, pH, Tds, Do Kyung-soo and salinity. The method used was the development of expert systems with Tsukamoto Fuzzy application.

The testing process carried out on 21 test data of Koi species based on colour pattern namely Koi one colour, two colours, three colours and mix. The koi species was Sanke, Sowa, Kohaku, Shiro, Ogon, Yamabuki and mix. The test result compared the calculation of the system and expert assessment, which produced an accuracy of 100%.

References

[1] Dewi C. 2014. Implementasi Fuzzy Inference System (Fis) Metode Tsukamoto Pada Sistem Pendukung Keputusan Penentuan Kualitas Air Sungai. Campus Brawijaya. Journal. Malang.
[2] Haslinda S. 2013. Sistem Pakar Penentuan Jenis Budaya Ikan Air Tawar Berdasarkan Lokasi dan Kualitas Air. Campus IPB Bogor. Journal. Bogor.
[3] Kusumadewi S dan Purnomo H. 2004. Aplikasi Logika Fuzzy Untuk Mendukung Keputusan. Graha Ilmu, Yogyakarta.
[4] Koi Owners of Indonesia-Society. 2010. KOISHI AWARDS. www.koi-s.org.com 06 April 2015.
[5] Malokoi. 2012. Oksigen Terlarut (DO) dan temperatur. www.malokoi.com. 04 April 2015.
[6] Marimin. (2005). Teori dan aplikasi sistem pakar dalam tehnologi manajerial. IPB – Press, Bogor.
[7] Purwakusuma, W. 2012. Parameter umum air (pH). www.o-fish.com. 04 April 2015.
[8] Takeo K.1986. The Latest Manual to Nishikigoi. Shin Nippon Kyoiku ToshoCo. Japan

Web-based application on employee performance assessment using exponential comparison method

S. Maryana, E. Kurnia, and A. Ruyani

Computer Science Department, Pakuan University, INDONESIA.

E-mails: sufiatul.maryana@unpak.ac.id, emakurnia@gmail.com, ahmadruyani10@gmail.com

Abstract: Employee performance assessment is also called a performance review, performance evaluation, or assessment of employees, is an effort to assess the achievements of staffing performance with the aim to increase productivity of employees and companies. This application helps in the assessment of employee performance using five criteria: Presence, Quality of Work, Quantity of Work, Discipline, and Teamwork. The system uses the Exponential Comparative Method and Weighting Eckenrode. Calculation results using graphs were provided to see the assessment of each employee. Programming language used in this system is written in Notepad++ and MySQL database. The testing result on the system can be concluded that this application is correspond with the design and running properly. The test conducted is structural test, functional test, and validation, sensitivity analysis, and SUMI testing.

1. Introduction

Employee Performance Assessment in modern organizations provides an important role for the management mechanism to be used in explaining the objectives and standards of performance and to motivate individual performance. Currently, the employee performance assessment of the company still using manual system by calculating the assessment of each employee one by one and discussing with the managers with insufficient data. This is considered as less effective and less efficient to determine the employee performance assessment. This situation can be resulted into unfavourable decisions.

Therefore it is needed to create a system that could help in fast data processing, exact, and accurate. To support in the calculation process this application uses Exponential Comparison Method (ECM). This method is one of the decision making method that quantifies the opinion of one person or more in a certain range. In principle, it is a scoring method for the existing choices. By using the exponential calculation, the difference value between criteria can be distinguished depend on the capability of the person in evaluating. The aim of this study is to develop a web-based application on employee performance assessment using exponential comparison method. The scope of this study including the employee data, calculation process, report of the assessment and rank.

2. Methodology

2.1 Model Base Application

The system used the Exponential Comparative Method for calculating performance assessment and Eckenrode weighting method. The criteria which have been determined from the interviews for employee performance assessment include: login, the presence, quantity of work, quality of work, initiative, teamwork, integrity, discipline, work ethics, health and safety. The calculation used Exponential Comparative Method, by power all the criteria and look for the highest score results. The results of the analysis will be able to assist the company in determining who is worthy and appropriate to be the first rank and also used as a promotion in accordance provided by the company so that the selection of promotions is no longer selected in the kinship system but with reference to employee performance assessment application [5].

2.2 User Interface

User interface was designed simple so that can allow the user to use and provide the information required. The interface generated in this system is home page, login admin, Eckenrode criteria, weighted value, expert assessment, employee, MPE criteria, employee assessment, and employee assessment results. Applications used in this interface is phpMyAdmin and MySQL database [6]. The method used in this research is System Development Life Cycle (SDLC). It is a series of activities carried out by professionals and users of information systems, to develop and implementing information systems [2]. The stages of research can be seen in Figure 1.

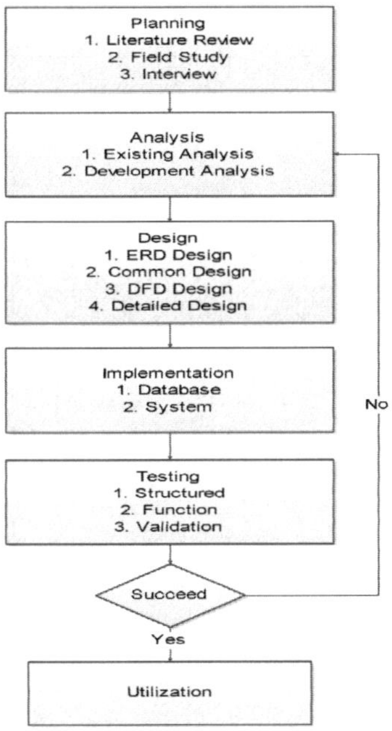

Figure 1. Stages of research

3. Results and Discussion

The determination of criteria was done by interviews with personals and manager that adjusted with the condition of the Company. The determination resulted in 5 criteria as explained in Table 1.

Table 1. Type of criteria

Criteria		Indicator
K1	Presence	Employees presence on work day
K2	Quality of Work	Work quality of low mistake
K3	Quantity of Work	Target achievement
K4	Discipline	Obedience
K5	Teamwork	Willingness to work together in teams

From the results of the selection criteria by three experts summed every choices based on the order or the scale of assessment in which the order of 1 or first is the order of the highest importance degree and value for the multiplicative also getting bigger, then the total selection obtained for each criterion [1]. The next process was calculated by weigh the value on the order of 1 = 5, 2 = 4, ..., 5 = 1. The weighted grades can begin by any number with the proviso sequentially from the largest to the smallest, where if the weigh calculation process getting smaller, then the smaller its results. The total selection results can be seen in Table 2.

Table 2. Total Selection

No	Criteria	Order				
		1	2	3	4	5
1	K1	0	1	2	0	0
2	K2	2	1	0	0	0
3	K3	1	1	1	0	0
4	K4	0	0	0	3	0
5	K5	0	0	0	0	3
Value		5	4	3	2	1

Calculation of weighted criteria [4]:

$$We = \frac{\sum_j^n = 1\,\lambda_{ej}}{\sum_k^e = 1\;\lambda_{ej}\,\sum_j^n = 1\,e_{ej}} \qquad \text{where} \qquad \sum_{e=1}^{k} We = 1 \qquad \text{or total of weight} = 1$$

$$K1 = \frac{[(0x5)+(1x4)+(2x3)+(0x2)+(0x1)]}{[\,(0x5)+(1x4)+(2x3)+(0x2)+(0x1)+(2x5)+(1x4)+(0x3)+(0x2)+(0x1)+(1x5)+(1x4)+(1x3)+(0x2)+(0x1)+(0x5)+(0x4)+(0x3)+(3x2)+(0x1)+(0x5)+(0x4)+(1x3)+(0x2)+(3x1)\,]}$$

= 0,22

K2 = 0,31

K3 = 0,27

K4 = 0,13

K5 = 0,07

We = K1+K2+K3+K4+K5
= 0,22+0,31+0,27+0,13+0,07
= 1

After the Eckenrode calculation process was done, then obtained the weight of each predefined criteria. The weight of each criteria as shown in Table 3.

Tabel 3. Weighting criteria

No	Criteria	Weight	Refinement weight
1	Presence	22%	0.22
2	Quality of work	31%	0.31
3	Quantity of work	27%	0.27
4	Discipline	13%	0.13
5	Teamwork	7%	0.07
	Total	100%	1

Calculations sample using exponential comparison method manually can be seen in Tables 4 and 5. The calculation is: Alternative = Assessment ^ weight.

Determining the value for employees

Name : Saeful Bahri
NIK : 262012

Table 4. Input Criteria

Criteria	Value
Presence	80
Quality of work	90
Quantity of work	75
Discipline	80
Teamwork	70

Value (Saeful Bahri)
=80^0.22+90^0.31+75^0.27+80^0.13+70^0.07
=12.986

Table 5. Employee performance assessment result matrices

No	Name	Criteria					Value	Ranking
		C1	C2	C3	C4	C5		
1	Saeful Bahri	80	90	75	80	70	12.986	4
2	Sela Agus	90	80	85	90	75	13.053	3
3	Hadi Wiratmono	100	90	90	85	80	13.305	1
4	Bibing Sublisolih	90	85	80	85	80	13.066	2
	Weight	0.22	0.31	0.27	0.13	0.07	1.0	

From the above calculation, obtained:
Saeful Bahri = 12.986
Sela Agus = 13.053
Hadi Wiratmono = 13.305
Bibing Sublisolih = 13.066

It is known that the order of priority ranking employee assessment, namely:
1st Ranking Hadi Wiratmono = 13.305
2nd Ranking Bibing Sublisolih = 13.066
3rd Ranking Sela Agus = 13.053
4th Ranking Saeful Bahri = 12.986

It can be concluded that the first rank of all the data that is Hadi Wiratmono with total value = 13.305.

Employee assessment results page displayed showing the results of calculation analysis using exponential comparison method, the display showed based on employee data input and the ranking process showed are also in graphic form. The display shown in Figures 2 and 3.

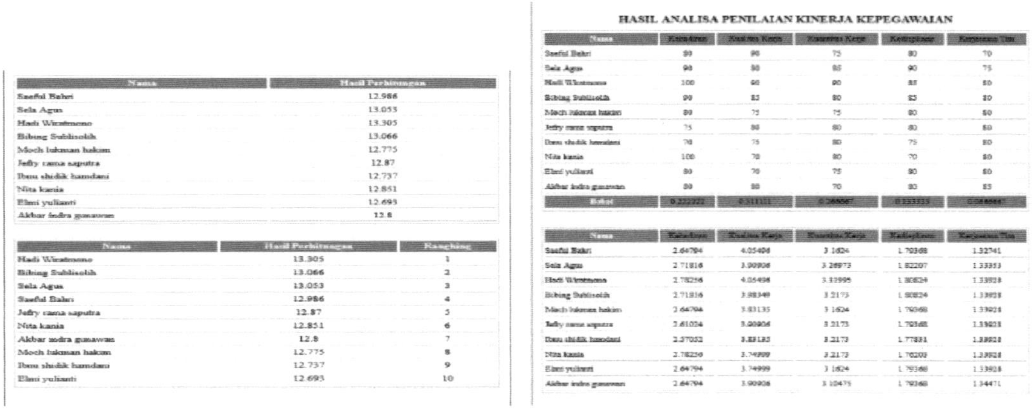

Nama	Hasil Perhitungan
Saeful Bahri	12.986
Sela Agus	13.053
Hadi Wiratmono	13.305
Bibing Sublisolih	13.066
Moch lukman hakim	12.775
Jefry rama saputra	12.87
Ibnu shidik hamdani	12.737
Nita kania	12.851
Elmi yulianti	12.693
Akbar indra gunawan	12.8

Nama	Hasil Perhitungan	Rangking
Hadi Wiratmono	13.305	1
Bibing Sublisolih	13.066	2
Sela Agus	13.053	3
Saeful Bahri	12.986	4
Jefry rama saputra	12.87	5
Nita kania	12.851	6
Akbar indra gunawan	12.8	7
Moch lukman hakim	12.775	8
Ibnu shidik hamdani	12.737	9
Elmi yulianti	12.693	10

HASIL ANALISA PENILAIAN KINERJA KEPEGAWAIAN

Nama	Kehadiran	Kualitas Kerja	Kuantitas Kerja	Kedisiplinan	Kerjasama Tim
Saeful Bahri	90	90	75	80	70
Sela Agus	98	80	85	90	75
Hadi Wiratmono	100	90	90	85	80
Bibing Subliolih	90	85	80	85	80
Moch lukman hakim	80	75	75	80	80
Jefry rama saputra	75	80	80	80	80
Ibnu shidik hamdani	70	75	80	75	80
Nita kania	100	70	80	70	80
Elmi yulianti	80	70	75	80	80
Akbar indra gunawan	80	80	70	80	85
Bobot	0.222222	0.111111	0.166667	0.333333	0.066667

Nama	Kehadiran	Kualitas Kerja	Kuantitas Kerja	Kedisiplinan	Kerjasama Tim
Saeful Bahri	2.64794	4.05406	3.1624	1.79368	1.32741
Sela Agus	2.71816	3.90906	3.26973	1.82207	1.33353
Hadi Wiratmono	2.78256	4.05498	3.33995	1.80824	1.33928
Bibing Subliolih	2.71816	3.98349	3.2173	1.80824	1.33928
Moch lukman hakim	2.64794	3.83135	3.1624	1.79368	1.33928
Jefry rama saputra	2.61024	3.90906	3.2173	1.79368	1.33928
Ibnu shidik hamdani	2.57052	3.83135	3.2173	1.77831	1.33928
Nita kania	2.78256	3.74999	3.2173	1.76209	1.33928
Elmi yulianti	2.64794	3.74999	3.1624	1.79368	1.33928
Akbar indra gunawan	2.64794	3.90906	3.10475	1.79368	1.34471

Figure 2. Display of employee assessment result

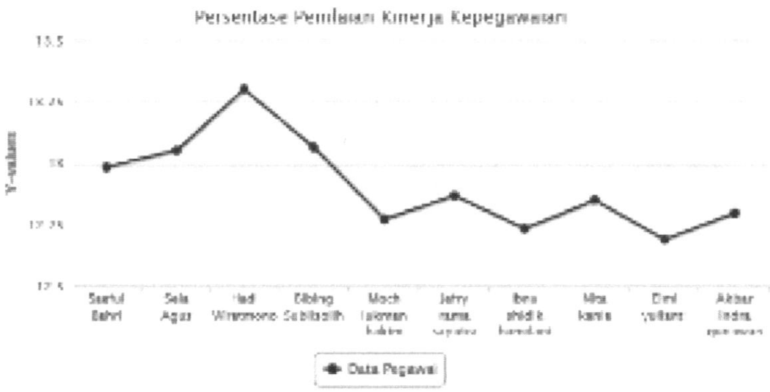

Figure 3. Employee assessment graph

4. Conclusion

Application of Employee Performance Assessment is an improvement of the previous system that still use manual systems in its data processing. The criteria which have been determined from the interviews for employee performance assessment include: the presence, quantity of work, quality of work teamwork, discipline. The calculations used to build applications of employee performance assessment by using methods and techniques weighting Eckenrode MPE. By power the value of the results of the system between all the criteria and look for the highest score.

The system development used the research stages of SDLC (System Development Life Cycle). The system was implemented using Adobe Dreamweaver CS6, Notepad++ and database using Xampp 1.6.7. This application cover of two stages of the calculation, the first process was calculated the Eckenrode weighting where the admin should input the Eckenrode criteria, the weight value, and input the selection order of criteria by experts. After the process was done then the calculation of weighted criteria Eckenrode will be entered automatically into the MPE criteria weights. The second MPE calculation process was the admin should input employee's data, criteria MPE, and employee assessment, then it can be resulted the calculation analysis of employees, ranking employees, and chart ranking employees. In this system, all the calculations was done dynamically by adding criteria, alternatives as well as edit or delete data.

Reference

[1] Eriyatno. 1999. Ilmu Sistem: Meningkatkan Mutu dan Efektivitas Manajemen. IPB Press. Bogor.
[2] Fathansyah. 2004. Basis Data. Bandung: Informatika
[3] Mangkuprawira S. 2002. Manajemen Sumber Daya Manusia Strategik. Jakarta: Ghalia Indonesia.
[4] Marimin. 2005. Teknik dan Aplikasi Pengambilan keputusan dengan Kriteria majemuk, volume 2, Jakarta: Grasindo.
[5] Simamora and Henry. 2001. Manajemen Sumber Daya Manusia. Sekolah Tinggi Ilmu Ekonomi YKPN.
[6] Sugiri and Saputro. 2008. Pengelolaan Database MySql dengan PhpMyadmin. Yogyakarta: Graha Ilmu.

Comparison of simple additive weighting (SAW) and composite performance index (CPI) methods in employee remuneration determination

L. Karlitasari[1], D. Suhartini[2] and Benny[3]

[1,3]Computer Science Study Program, Pakuan University, Bogor, INDONESIA
[2]Informatics Management D3 Study Program, Pakuan University, Bogor, INDONESIA

E-mail: rawit77@yahoo.com, dini.suhartini04@gmail.com, benny.acolyte@gmail.com

Abstract. The process of determining the employee remuneration for PT Sepatu Mas Idaman currently are still using Microsoft Excel-based spreadsheet where in the spreadsheet there is the value of criterias that must be calculated for every employee. This can give the effect of doubt during the assesment process, therefore resulting in the process to take much longer time. The process of employee remuneration determination is conducted by the assesment team based on some criterias that have been predetermined. The criteria used in the assessment process are namely the ability to work, human relations, job responsibility, discipline, creativity, work, achievement of targets, and absence. To ease the determination of employee remuneration to be more efficient and effective, the Simple Additive Weighting (SAW) method is used. SAW method can help in decision making for a certain case, and the calculation that generates the greatest value will be chosen as the best alternative. Other than SAW, also by using another method was the CPI method which is one of the calculating method in decision making based on performance index. Where SAW method was more faster by 89-93% compared to CPI method. Therefore it is expected that this application can be an evaluation material for the need of training and development for employee performances to be more optimal.

1. Introduction

The advances in technology cannot be separated from computerization, one of which is to support in decision making. The utilization of existing technology is expected to provide convenience in every problem faced. One of the problems is when companies are faced with several options for taking a decision.

The process of determining the remuneration of employees at PT Sepatu Mas Idaman currently still using Microsoft Excel-based spreadsheet where in the spreadsheet there is the value of criteria that must be calculated for each employee. This can give the affect of doubt during the assesment process, therefore resulting in the process to take much longer time. The process of employee remuneration determination is conducted by the assesment team based on some criterias that have been predetermined. The criteria used in the assessment process are the ability to work, human relations, job responsibility, discipline, creativity, achievement of targets, and presence. Based on this, it is necessary to design a system that can simplify the determination of remuneration of employees so that the process can run more efficiently and effectively. Therefore the aim of this research is to compare the Simple Additive Weighting (SAW) with the Composite Performance Index (CPI) in employee remuneration determination (Case Study at PT Sepatu Mas Idaman Bogor). The development of application is also needed in order to support the decision making and evaluation needs in training and development for the employee to be more optimal in performance and quality.

2. Methodology

The research methods applied of the application development using SDLC (System Development Life Cycle). SDLC approach consists of six stages, namely planning, analysis, design, implementation, testing and utilization as in Figure 1.

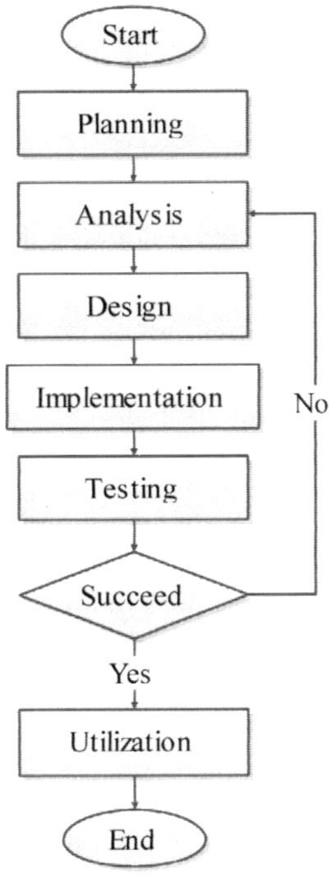

Figure 1. SDLC Approach

1.1 Planning stage

Planning stage was a process of collected some of the material presented as an initial basis to supplement the definition of the problem, including: interviews, and literature review.

1.2 Analysis stage

The analysis stage began with the data collection process. The data collected will be used as materials in the process of system analysis. Furthermore, analyzed the existing system and analyzed the ongoing system. The results analysis of the existing system will be the basis for system development.

Max value of each criterion
Ability to work: 6
Human relations: 6
Job responsibilities: 6
Creativity: 6
Discipline: 6
Achievement of targets: 8
Presence: 12

a. The first way with Simple Additive Weighting (SAW)

The process of normalization (dividing the value of the alternative on each criterion with a max value of each criteria)

Andi :

Ability to work		:
	C11	: 5 / 6 = 0.8333
	C12	: 5 / 5 = 1
	C13	: 3 / 5 = 0.6
Human relation		:
	C21	: 4 / 6 = 0.6667
	C22	: 2 / 5 = 0.4
	C23	: 4 / 5 = 0.8
	C24	: 3 / 5 = 0.6
	C25	: 3 / 5 = 0.6
Job responsibility		:
	C31	: 4 / 5 = 0.8
	C32	: 4 / 4 = 1
	C33	: 4 / 4 = 1
Creativity		:
	C41	: 5 / 5 = 1
	C42	: 5 / 5 = 1
	C43	: 4 / 5 = 0.8
	C44	: 3 / 6 = 0.5
	C45	: 3 / 5 = 0.6
Discipline		:
	C51	: 4 / 5 = 0.8
	C52	: 3 / 5 = 0.6
Achievement of target		:
	C61	: 6 / 7 = 0.8571
Presence		:
	C71	: 6 / 10 = 0.6

Ranking process was done by multiplied the normal alternatives values with the weight of each criterion.

Andi :

Ability to work		:
	C11	: 0.8333 * 0.17 = 0.1417
	C12	: 1 * 0.17 = 0.17
	C13	: 0.6 * 0.17 = 0.102
Human relations		:
	C21	: 0.6667 * 0.11 = 0.0733
	C22	: 0.4 * 0.11 = 0.044
	C23	: 0.8 * 0.11 = 0.088
	C24	: 0.6 * 0.11 = 0.066
	C25	: 0.6 * 0.11 = 0.066
Job responsibility		:
	C31	: 0.8 * 0.1 = 0.08
	C32	: 1 * 0.1 = 0.1
	C33	: 1 * 0.1 = 0.1

Creativity	:		
C41	: 1	* 0.12	= 0.12
C42	: 11	* 0.12	= 0.12
C43	: 0.8	* 0.12	= 0.096
C44	: 0.5	* 0.12	= 0.06
C45	: 0.6	* 0.12	= 0.072
Discipline	:		
C51	: 0.8	* 0.14	= 0.112
C52	: 0.6	* 0.14	= 0.084
Achievement of targets	:		
C61	: 0.8571	* 0.16	= 0.1371
Presence	:		
C71	: 0.6	* 0.2	= 0.012

The ranking result was:

Andi : 0.1417 + 0.17 + 0.102 + 0.0733 + 0.044 + 0.088 + 0.066 + 0.066
+ 0.08 + 0.1 + 0.1 + 0.12 + 0.12 + 0.096 + 0.06 + 0.0.72 + 0.112
+ 0.084 + 0.1371 + 0.12 = 1.9521

Percentage of performace = 1.9521 / 0.026 = 75.08 %
Percentage of late rise = 75.08 x 0.05 = 3.75 %

Nominal rise = 3.75 x basic salary
= 3.75 x 2.500.000
= 93.850,96

New salary = total salary + nominal rise
= 2.500.000 + 93.850.96
= 2.593.851

b. The second way with Composite Performance Index (CPI)
With the same value, convert the smallest alternative value into number 100, then the other alternative value except the smallest value that had been converted into number 100 divided by the smallest value. The result of each alternative value was multiplied by the weight.

Andi :

Ability to work	: 13 = 100
Human relations	: 16 = 100
Job responsibility	: 12 = 100
Creativity	: 14 = 100
Discipline	: 8 = 100
Achievement of targets	: 6 = 6
Presence	: 6 = 100

Process of alternative value except the smallest value divided by the smallest value of each criteria.

Andi :

Ability to work	: 13 = 100
Human relations	: 16 = 100
Job responsibility	: 12 = 100

Creativity	: 14 = 100
Discipline	: 8 = 100
Achievement of targets	: 6 = (6/5) * 100) = 120
Presence	: 6 = 100

The result of each alternative value multiplied by the weight.

Andi :

Ability to work	: 100 * 0.17 = 17
Human relations	: 100 * 0.11 = 11
Job responsibility	: 100 * 0.10 = 10
Creativity	: 100 * 0.12 = 12
Discipline	: 100 * 0.14 = 14
Achievement of targets	: 120 * 0.16 = 19.20
Presence	: 100 * 0.20 = 20

The result was:

Andi : 17 + 11 + 10 + 12 +14 + 19.20 + 20 = 103.20

1.3 Design stage

The design stage of the system is to provide an overview of the system design will be built, the design includes the ERD (Entity Relationship Diagram), database, relationships tables, flowchart systems, and design process of interface. This design is based on the user interface design to fit with the goals and the system requirements, such as in Figure 2.

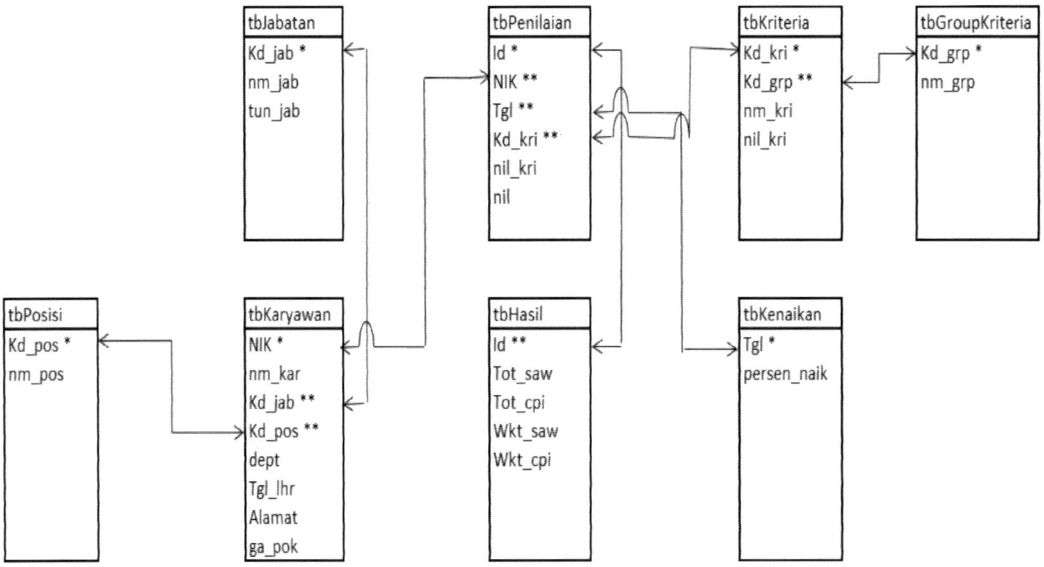

Figure 2. Relationship tables

In the system design also required a flowchart that illustrates the work flow process structured into several sections that will facilitate the implementation of the system, as shown in Figure 3.

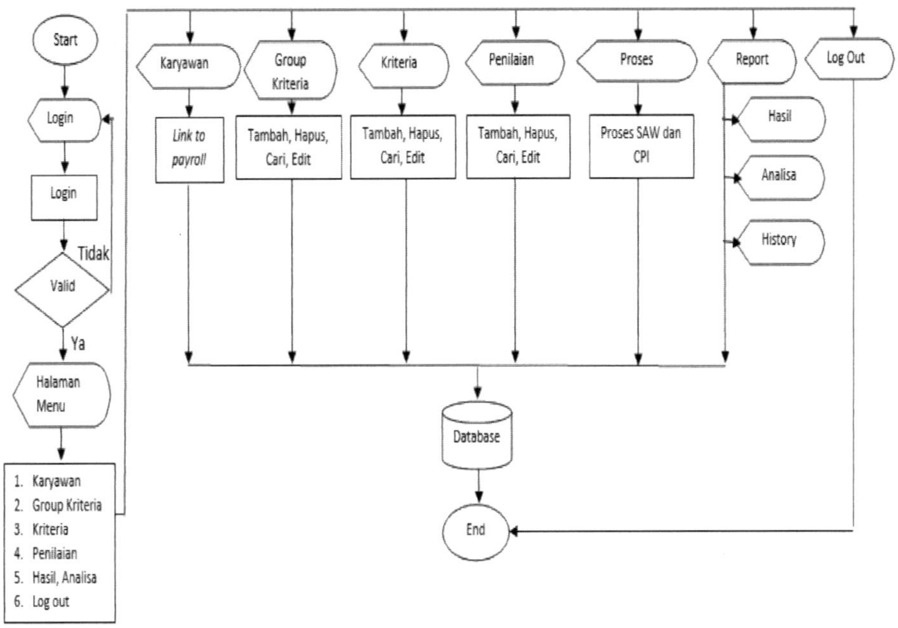

Figure 3. Flowchart Menu

3. Result and Discussion

Results of the application for SAW and CPI Comparison in the determination of employee remuneration shown in Figure 4.

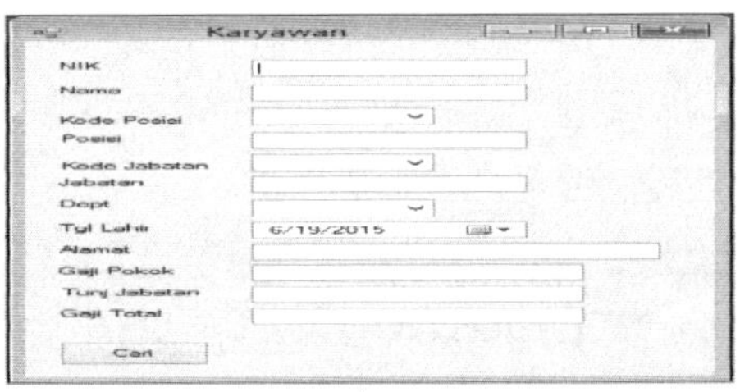

4a

4b

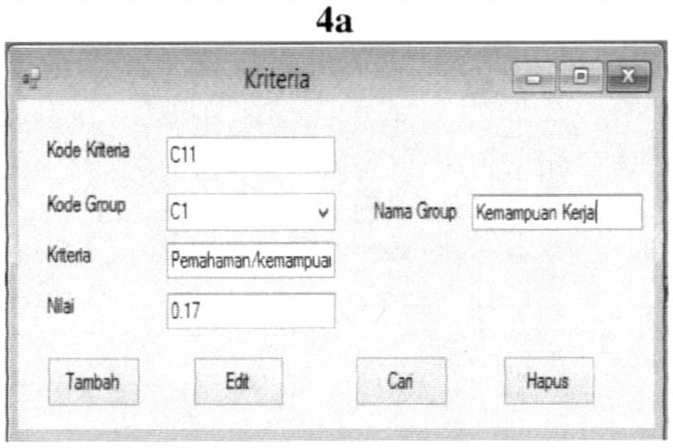

4c

4d

Figure 4. Display of aplication : 4a. Display of employee, 4b. Display of group criteria, 4c. Display of criteria, and 4c. Display of assessments

The process of History rise and Reports shown in Figure 5.

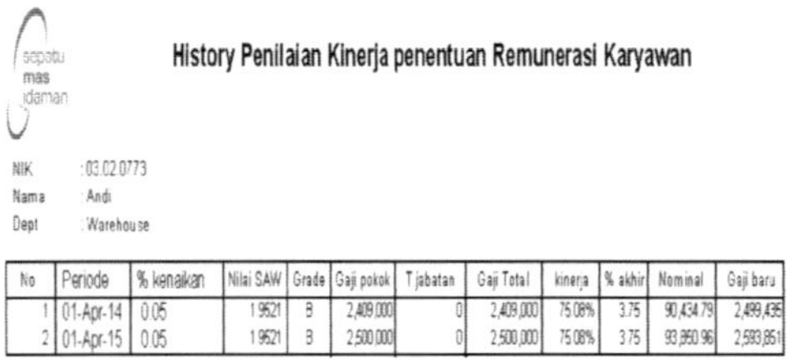

History Penilaian Kinerja penentuan Remunerasi Karyawan

NIK :03.02.0773
Nama :Andi
Dept :Warehouse

No	Periode	% kenaikan	Nilai SAW	Grade	Gaji pokok	T jabatan	Gaji Total	kinerja	% akhir	Nominal	Gaji baru
1	01-Apr-14	0.05	1.9521	B	2,409,000	0	2,409,000	75.08%	3.75	90,434.79	2,499,435
2	01-Apr-15	0.05	1.9521	B	2,500,000	0	2,500,000	75.08%	3.75	93,850.96	2,593,851

5a

Penilaian Kinerja untuk penentuan Remunerasi Karyawan

Perode 4/1/2015 3:01:53PM
% kenaikan tahunan 0.05

No	NIK	Nama	Jabatan	Dept	Nilai SAW	Grade	Gaji pokok	T jabatan	Gaji Total	kinerja	% akhir	Nominal	Gaji baru
1	04.07.6700	Dimas	Operator	Produksi	2.5477	A	3,000,000	0	3,000,000	97.99 %	4.90	146,982.69	3,146,983
2	07.08.9283	Guntur	Operator	Warehouse	2.2987	A	2,600,000	0	2,600,000	88.41 %	4.42	114,936.00	2,714,936
3	05.14.8557	Siti	Operator	Produksi	2.0726	B	2,600,000	0	2,600,000	79.72 %	3.99	103,630.00	2,703,630
4	03.02.0773	Andi	Operator	Warehouse	1.9521	B	2,500,000	0	2,500,000	75.08 %	3.75	93,850.96	2,593,851

5b

Figure 5. Display of result : 5a. History of performance assessments, 5b. Performance assessments of employee renumeration determination

While the analysis process of SAW and CPI shown in Figure 6.

ANALISA PROSES MENGGUNAKAN SAW DAN CPI

Satuan waktu : detik

No	NIK	Nama	Dept	Nilai (SAW)	Nilai (CPI)	Waktu (SAW)	Waktu (CPI)	Hasil
1	03.02.0773	Andi	Warehouse	1.9521	103.2000	0.0020	0.0286	93.00 %
2	04.07.6700	Dimas	Produksi	2.5477	138.9300	0.0030	0.0286	89.50 %
3	05.14.8557	Siti	Produksi	2.0726	108.0500	0.0030	0.0286	89.50 %
4	07.08.9283	Guntur	Warehouse	2.2987	121.0100	0.0020	0.0286	93.00 %

Figure 6. Analysis menu

By using the two methods: Simple Additive weighting method (SAW) and the Composite Performance Index (CPI) for data processing assessment, then the ranking of these two methods produce different values but gave the same rank. In addition of accuracy in showing the ranking, this study was also compared the time process of both methods, as shown in Table 4.

Table 4. The difference between SAW and CPI

Amount of data	SAW	CPI
4	0.0200 second	0.1142 second
5	0.0250 second	0.1900 second
10	0.0440 second	0.5643 second
20	0.0860 second	2.7286 second

The ranking process was done under computer of only one open application. From the time of two tested methods, SAW method was faster data processing time compared with the CPI method. From the overall processing time of SAW and CPI, there was no significant difference in time. From the rate average sample data used of SAW and CPI, SAW method was faster of 89-93 % than the CPI method. The time difference will increasingly appear to be associated with the amount of data processed, as shown in Figure 7.

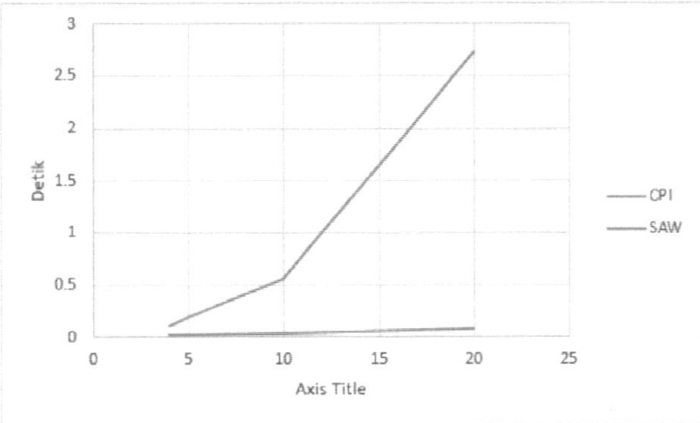

Figure 7. Time comparison chart

From both tests conducted, it is known that SAW method was the most superior in data processing as compared with the CPI method. The ranking process of SAW and CPI generate a sequence of the same ranking, but the more data processed then the CPI method will increasingly require a longer time in proceed the data because of many minimum value alternative of each criteria, so that it needed extra time to convert into the unit of 100 .

4. Conclusions

From this study was concluded that SAW and CPI methods produce the same grade, but the CPI method requires a longer time. SAW method was faster than the CPI method, because the CPI in data processing relies on the data processed, the more minimum value to be transformed into a value of 100 then the more need of extra time. SAW method does not depend on the data pattern, but the CPI method depends on the data pattern processed. The time difference will increasingly appear to be associated with the amount of data processed and the data patterns that must be processed when using the CPI method. The SAW method has more processing step instead of the CPI, but the CPI requires a longer time for processing because the processing should be completed in a structured and highly dependent on the amount of the minimum data pattern processed data.

References

[1] A Hidayat, H Mustafidah, and A Suyadi. 2015. *Penerapan Metode Simple Additive Weighting (Saw) Untuk Sistem Pendukung Keputusan Penilaian Kinerja Dosen Di Universitas Muhammadiyah Purwokerto.* Prosiding SENATEK 2015. Technique Faculty. Universitas Muhammadiyah Purwokerto. Purwokerto. ISBN 978-602-14355-0 -2, pp. 367-374.

[2] A Memariania, A Aminib and A Alinezhadc. 2009. Sensitivity Analysis of Simple Additive Weighting Method (SAW): The Results of Change in the Weight of One Attribute on the Final Ranking of Alternatives. *Journal of Optomization in Industrial Engineering*. Article 2. Volume 2. Issue 4. Page 13-18.

[3] G S Pandian, N Jawahar, and SP Nachiappan. 2013. Composite Performance Index for Sustainability. *IOSR Journal of Environmental Science, Toxicology and Food Technology (IOSR-JESTFT)* e-ISSN: 2319-2402, p- ISSN: 2319-2399.Volume 3, Issue 1. PP 91-102.

[4] Turban, Efrain, Aronson, and Jay. 2001. *Decision Support System and Intelligent System*, Prentice Hall, New Jersey.

[5] Valensia. 2012. *Sistem Pendukung Keputusan penilaian Kinerja Karyawan Dengan Menggunakan Metode Simple Additive Weighting (SAW).*

Decision support system for determining the contact lens for refractive errors patients with classification ID3

B.H. Situmorang, M.P. Setiawan, E.T. Tosida

Computer Science Department, Pakuan University, Indonesia

E-mails: boldson_skom@yahoo.com, paisalsetiawan16@gmail.com, enengtitatosida@unpak.ac.id

Abstract. Refractive errors are abnormalities of the refraction of light so that the shadows do not focus precisely on the retina resulting in blurred vision [1]. Refractive errors causing the patient should wear glasses or contact lenses in order eyesight returned to normal. The use of glasses or contact lenses in a person will be different from others, it is influenced by patient age, the amount of tear production, vision prescription, and astigmatic. Because the eye is one organ of the human body is very important to see, then the accuracy in determining glasses or contact lenses which will be used is required. This research aims to develop a decision support system that can produce output on the right contact lenses for refractive errors patients with a value of 100% accuracy. Iterative Dichotomize Three (ID3) classification methods will generate gain and entropy values of attributes that include code sample data, age of the patient, astigmatic, the ratio of tear production, vision prescription, and classes that will affect the outcome of the decision tree. The eye specialist test result for the training data obtained the accuracy rate of 96.7% and an error rate of 3.3%, the result test using confusion matrix obtained the accuracy rate of 96.1% and an error rate of 3.1%; for the data testing obtained accuracy rate of 100% and an error rate of 0.

1. Introduction
Most optical stores in Indonesia does not provide experts to assist in determining the right eye glasses or contact lenses for refractive errors patients by reason of inability to pay the expert eye. This research aims to develop a decision support system that can produce output on the right contact lenses for refractive errors patients with a value of 100% accuracy so as to replace the services of an eye expert in optical.

Iterative Dichotomize Three (ID3) classification methods (ID3) is a learning system algorithm that is often used by the developer to create an artificial intelligence system or expert systems in solving problems with single alternative/ fixed alternative output and not the best alternative to the lowest sort.

2. Material and Methods
Decision Support System for determining the contact lens for refractive errors is developed by following all phases of Software Development Life Cycle as shown in Figure 1. Iterative Dichotomize Three (ID3) classification method will generate entropy and gain values of attributes that has been collected from the UCI Learning Repository dataset and data experts of Salwi Farma Optical as shown in Table 1. The Attributes that include sample data, age of the patient, astigmatic, vision prescription,

IOP Conf. Series: Materials Science and Engineering **166** (2017) 012021 doi:10.1088/1757-899X/166/1/012021

the ratio of tear production, and classes will affect the outcome of the decision tree to determine the right contact lens for refractive errors patients.

The steps to create a decision tree using ID3 algorithm based are count the number of cases in each attribute classy hard lenses, soft contact lenses or not recommended to use contact lenses; determining the entropy value of each class by equation (1) and the gain value by equation (2). The calculation of the gain is repeated on each attribute and the largest gain value will be selected as the root in decision tree. The next step is making the decision tree. Do the repetitive calculation for determining the entropy value of each class by equation (1) and the gain value by equation (2) until all the attributes to be a node in the decision tree.

$$Entropy(S) = \sum_{i}^{c} - Pi\ Log2\ Pi \tag{1}$$

$$Gain(S,A) = Entropy(S) - \sum_{vEValues(A)} - \frac{|Sv|}{S} Entropy(Sv) \tag{2}$$

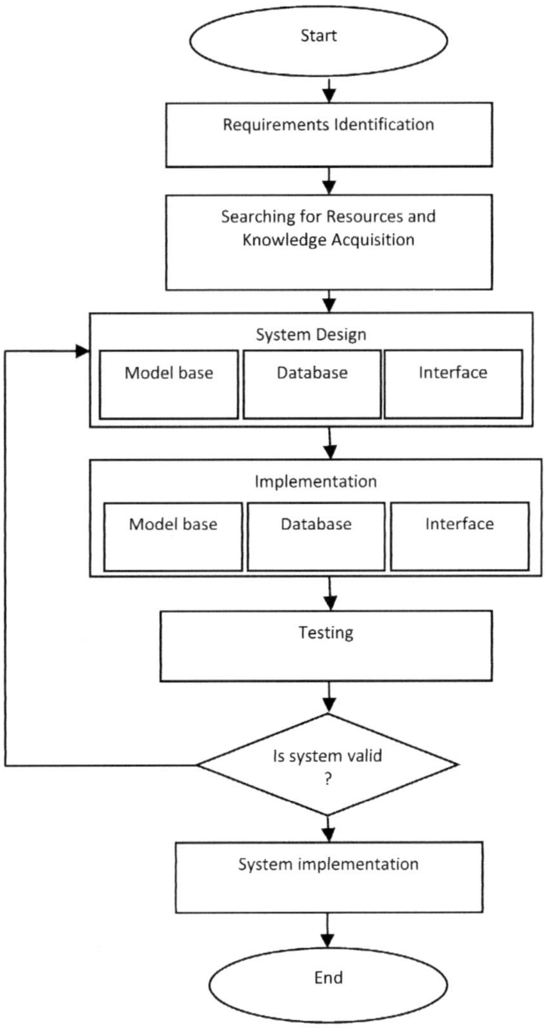

Figure 1. Software Development Life Cycle [2]

Table 1. The Attributes that will affect the outcome of the decision tree to determine the right contact lens for refractive errors patients

Attribute	Value	Description
Sample code number	-	Code is taken at random from the data of patients with disorders of the eye.
Age of patient	Young (1) , Pre-presbyopia (2), Presbyopia (3)	a. Young: 7-20 years old. b. Pre-presbyopia: 20-40 years old. c. Presbyopia: more than 40 years old.
Spectacle prescription	Myopia (1), Hypermetropia (2)	a. Myopia is the term used to define short sightedness. Light from a distant object forms an image before it reaches the retina. b. Hypermetropia means long sight and is where the image of a nearby object is formed behind the retina.
Astigmatic	Yes (1) , No (2)	Astigmatic is an eye condition with blurred vision as its main symptom. The front surface of the eye (cornea) of a person with **astigmatism** is not curved properly - the curve is irregular - usually one half is flatter than the other - sometimes one area is steeper than it should be.
Tear production rate	Reduced (1), Normally (2)	a. Reduced: eyes look soft and damp budge b. Normally: eyes look bright without water and not excessive.

The programming languages used to build decision support systems are PHP for DSS application interface and XAMPP MySql for database.

3. Result & Discussion

Based on 30 pieces of training sample data is obtained the number of classes of hard lens = 3 pieces, soft lens = 5 pieces, and not recommended to use a contact lenses = 16 pcs, the entropy value of each class:

$$\left(\frac{-3}{24}\ x\ log2\left(\frac{3}{24}\right)\right)+\left(\frac{-5}{24}\ x\ log2\ \left(\frac{5}{24}\right)\right)+\left(\frac{-16}{24}\ x\ log2\ \left(\frac{16}{24}\right)\right)=1.236440502$$

and the gain value of age of the patient attribute:

$$1.235440502-\left(\frac{8}{24}\ x\ 1.5\right)-\left(\frac{8}{24}\ x\ 1.2987991\right)-\left(\frac{8}{24}\ x\ 0.543564443\right)=0.122321$$

The result of the calculation of entropy and gain values shown in Table 1. Based on Table 1, astigmatic is the attribute with the largest gain value and will be selected as the root in decision tree. The result of the repetitive calculation of entropy and gain values shown in Table 2.

After designing the model base of ID3 algorithm, DSS interfaces , databases , and architectural, then the next stage is converting the result of design to coding, where one interface results is shown in Figure 3. The next step is to make a node to generate a decision tree as shown in Figure 2.

Table 1. The 1st result of the calculation of entropy and gain values

Atribut	Value	Class			Total	Entropy	gain
		Hard	Soft	Not Recommended			
Age of the patient	1	2	2	4	8	1.5	0.122321
	2	1	2	5	8	1.298794941	
	3	0	1	7	8	0.543564443	
Vision prescription	1	2	2	8	12	1.251629167	0.016265
	2	1	3	8	12	1.188721876	
Astigmatic	1	0	5	7	12	0.979868757	0.340867
	2	3	0	9	12	0.811278124	
The ratio of tear	1	0	1	12	13	0.391243564	0.303731
	2	3	4	4	11	1.572623664	

Table 2. The 2nd result of the calculation of entropy and gain values

Attribute	Value	Class			Total	Entropy	Gain
		Hard	Soft	Not Recommended			
Age of the patient	1	2	0	2	4	1	0.207519
	2	1	0	3	4	0.811278	
	3	0	0	4	4	0	
Vision prescription	1	2	0	4	6	0.918296	0.027119
	2	1	0	5	6	0.650022	
The ratio of tear production	1	0	0	6	6	0	0.311278
	2	3	0	3	6	1	

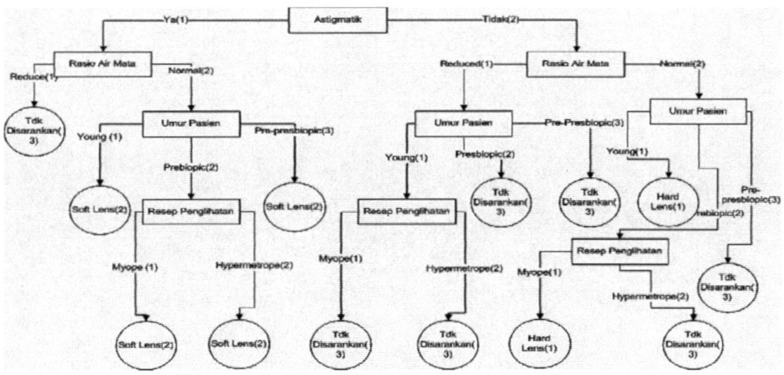

Figure 2. Final Results Decision Trees

System testing is performed to determine whether the application meets the expected needs and in accordance with its designs. Test using the confusion matrix as shown in Table 3.

Table 3. The result test using *Confusion Matrix.*

Data	Accuracy	Error Rate
Training	96,1%	3,9%
Testing	100%	0%

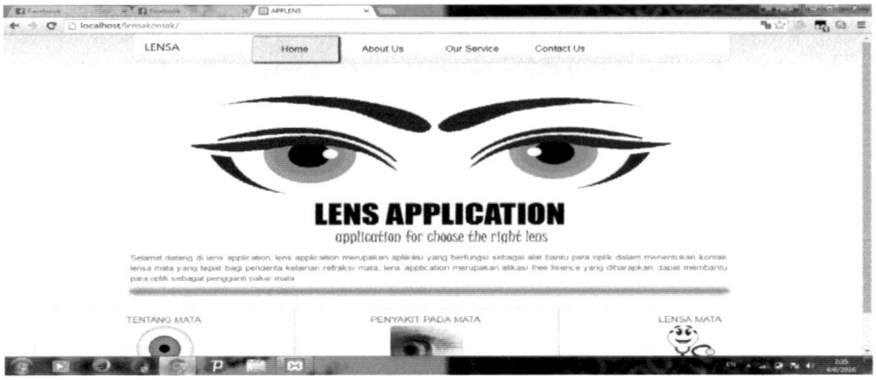

Figure 3. The main page of DSS application for determining the contact lens for refractive errors patients with Classification ID3

4. Conclusion

The eye specialist test result for the training data obtained the accuracy rate of 96.7% and an error rate of 3.3%, the result test using confusion matrix obtained the accuracy rate of 96.1% and an error rate of 3.1%; for the data testing obtained accuracy rate of 100% and an error rate of 0.

REFERENCES

[1] Bhattacharyya. 2009. Refraksi atau Kelainan Pada Mata, Jakarta.
[2] Kristanto. 2003. Konsep & Perancangan Database. Gava Media, Jakarta.

Estimation model of life insurance claims risk for cancer patients by using Bayesian method

Sukono*, **M. Suyudi**, **F. Islamiyati**, **S. Supian**

Department of Mathematics, Universitas Padjadjaran, Indonesia

*E-mail: sukono@unpad.ac.id

Abstract. This paper discussed the estimation model of the risk of life insurance claims for cancer patients using Bayesian method. To estimate the risk of the claim, the insurance participant data is grouped into two: the number of policies issued and the number of claims incurred. Model estimation is done using a Bayesian approach method. Further, the estimator model was used to estimate the risk value of life insurance claims each age group for each sex. The estimation results indicate that a large risk premium for insured males aged less than 30 years is 0.85; for ages 30 to 40 years is 3:58; for ages 41 to 50 years is 1.71; for ages 51 to 60 years is 2.96; and for those aged over 60 years is 7.82. Meanwhile, for insured women aged less than 30 years was 0:56; for ages 30 to 40 years is 3:21; for ages 41 to 50 years is 0.65; for ages 51 to 60 years is 3:12; and for those aged over 60 years is 9.99. This study is useful in determining the risk premium in homogeneous groups based on gender and age.

1. Introduction

Lately, people with critical illnesses are expanding. One of them cancer. Ironically, health care costs have also increased rapidly over the past few decades. Things need to be aware, unforeseen events can occur in cancer patients, resulting in cancer patients financially are in need of protection. Therefore, the role of insurance companies is very necessary [2], [8].

Critical illness insurance is devoted to protecting customers' patients with critical illnesses, such as cancer, kidney failure, and heart. Critical illness insurance is different with health insurance [9], [3]. Critical illness insurance provides a number of cash when the customer has been diagnosed with a critical illness [2], [5].

In insurance, the insurer (insurance company), provides financial assistance in the form of insurance money that is called a risk premium to the insured (customer). Determining the risk premiums be reckoned with that the company did not experience a loss [10]. With the increasing number of cancer patients, and the cost of cancer treatment is higher, the number of claims from year to year increasing [12]. This is a problem for the insurer in estimating future claims trends to determine the risk premium. Therefore, in this paper do research on the risk of claims, particularly in cancer patients.

2. Methodology

2.1. Bayesian methods

The person who filed a claim (X) distributed Binomial (n, θ_B) where θ_B is a probability of occurrence X. The values θ_B are estimated using Bayesian methods. Maximum likelihood estimation of the random variable X can be formulated with

$$est\theta = \frac{x}{n} \quad (1)$$

Then Bayes' theorem, the posterior distribution of the value obtained as follows.

$$f_{\theta_b}(\theta \mid x) \propto f(x \mid \theta_B) f(\theta_B) \quad (2)$$

The posterior distribution contains all the information about θ_B that can be used for Bayesian estimation of the parameters θ_B the binomial distribution [1]. To estimate the value θ_B of a single observation X (the person filing the claim) with prior distribution θ_B the binomial distribution, turned into the beta distribution with parameters α and β, we can examine the form of the posterior distribution of θ_B [1]:

$$f(\theta_B \mid x) \propto \theta_B^x (1-\theta_B)^{n-x} \theta_B^{\alpha-1}(1-\theta_B)^{\beta-1} = \theta_B^{\alpha+x-1}(1-\theta_B)^{\beta+n-x-1} \quad (3)$$

The value of θ_B given by a data sample $x = (x_1, x_2, \dots, x_n)$ has a loss function $g(x)$, which minimizes the loss estimate by observing its posterior distribution. The loss function commonly used is the quadratic loss function is defined as follows [11]:

$$L(g(x);\theta) = [g(x) - \theta]^2 \quad (4)$$

By minimizing the quadratic loss function, $\theta_B$value can be expressed as the average (mean) of the posterior distribution as follows:

$$\theta_B = \frac{\alpha + x}{(\alpha + x) + (\beta + n - x)} = \frac{\alpha + x}{\alpha + \beta + n} \quad (5)$$

Using the theorem credibility, we can declare the value of Bayesian estimation $\theta_B$in the form:

$$\theta_B = Z\theta + (1-Z)\mu = Z\frac{x}{n} + (1-Z)\mu \quad (6)$$

where credibility factor Z is as follows:

$$Z = \frac{n}{\alpha + \beta + n} \quad (7)$$

and μ is the average of the prior distribution of beta distributions declared by

$$\mu = \frac{\alpha}{\alpha + \beta} \quad (8)$$

In practice, there are some situations where a prior value is unknown. So, we use the value of non-informative priors. For example, if θ_B is the opportunity of a binomial distribution, and θ_B not have any information about the prior distribution, the distribution of which had been distributed prior to the hose Uniform (0,1) would seem appropriate. In this case, the prior distribution is the beta distribution with parameter α = 1 and β = 1. However, the interval that we take not the interval (0,1), but the interval is more realistic. Set the interval $(\theta_{min}, \theta_{max})$ to get a good estimate. We denote the value of s as the average of the prior distribution of beta which is the centre of this interval [1], [11].

$$s = \frac{x_{\min} + x_{\max}}{2} \quad (9)$$

We mark as θ_0 the more distant boundary from the value of 0.5 of the interval $(\theta_{min}, \theta_{max})$. Calculate the *error* h_B as follows:

$$h_B = |\theta_0 - s| \quad (10)$$

and calculate the number of claim q by using formula as follows:

$$q = \frac{2p\theta_0(1-\theta_0)}{ph_B^2 - \theta_0(1-\theta_0)} \quad (11)$$

where p is number of insured in the insurance company. Then, we estimate the value of the parameter α and β of beta prior distributions as follows:

$$\alpha = qs \tag{12}$$
$$\beta = q - qs \tag{13}$$

2.2. *Individual Risk Model*

To determine the risk premium in homogeneous groups according to age and gender, need to be estimated that many insurance policy issued in the following year by extrapolating the trend of the time series using Statistics program *Statgraphics Centurion*. To select the most appropriate trend function, we use the procedure Comparison of Alternative Models [9], [7]. As for estimating the value of n in the following period (in 2009), we use the procedure Forecast. In this research, there are two models of risk, namely the risk model of collective and individual risk models. In the collective risk model, we let $X_{1,}, X_2, X_3 \ldots\ldots X_N$ is a random variable that the variable determining the amount of the claim. The total of the amount of the claim, denoted by [4]:

$$S = X_{1,} + X_2 + X_3 + \ldots + X_N \tag{14}$$

While the individual risk model, the total of the amount of the claim can be denoted as S_n. So we can write as follows:

$$S_n = Y_1 + Y_2 + Y_3 + \ldots + Y_n \tag{15}$$

where Y_j indicates the number of claims for individual year j, and n indicates the period of observation. However, it is possible some risks will not give rise to a claim. Therefore, the value Y_j, $j = 1,2, \ldots, n$ may be 0. It will be given two assumptions, namely:

- The number of claims in the year to j, N_j is 0 or 1.
- Possible claims in the year to j is q_j.

Based on the above assumptions, $N_j \sim Bi(1; q_j)$, thus the distribution of Y_j is compound binomial with individual claims are denoted X_j. Then we can write it as follows:

$$E(Y_j) = q_j \mu_j \tag{16}$$

$$D(Y_j) = q_j(\sigma_j^2 + \mu_j^2) - q_j^2 \mu_j^2 = q_j \sigma_j^2 + q_j(1 - q_j)\mu_j^2 \tag{17}$$

Where $\mu_{j,}$ and σ_j^2 is the average and variance of X_j. Then, the average and variance of S_n are:

$$E(S_n) = E\left(\sum_{j=1}^{n} Y_j\right) = \sum_{j=1}^{n} E(Y_j) = \sum_{j=1}^{n} q_j \mu_j \tag{18}$$

$$D(S_n) = D\left(\sum_{j=1}^{n} Y_j\right) = \sum_{j=1}^{n} D(Y_j) = \sum_{j=1}^{n} \left[q_j \sigma_j^2 + q_j(1 - q_j)\mu_j^2\right] \tag{19}$$

In special cases, when Y_J, $j = 1,2,\ldots,n$ is a composite of several distribution and is a random variable. Based on the central limit theorem we can approach S_n distribution by the normal distribution. Therefore, in this case, we assign the value of the risk premium equal to 95% of the normal distribution with parameters$(\mu = E(S_n),\ \sigma^2 = D(S_n))$. Large risk premium (RP) can be calculated with the calculation below [6], [4]:

$$P(S_n \le RP) = 95\% \tag{20}$$

$$P\left(\frac{S_n - E(S_n)}{\sigma_{S_N}} \le \frac{RP - E(S_n)}{\sigma_{S_N}}\right) = 0.95$$

$$P\left(Z \le \frac{RP - E(S_n)}{\sigma_{S_N}}\right) = 0.95$$

From the standard normal distribution table, it is obtained that

$$\frac{RP-E(S_n)}{\sigma_{S_n}}=1.96 \Rightarrow RP=\left(1.96\left(\sigma_{S_n}\right)\right)+E(S_n)$$

$$RP=\left(1.96\left(\sqrt{D(S_n)}\right)\right)+E(S_n) \Rightarrow RP/\text{person}=\frac{\left(1.96\left(\sqrt{D(S_n)}\right)\right)+E(S_n)}{n_{2009}} \quad (21)$$

3. Results and Discussion

The data used is in the form of simulation data in the form of insured claims data that have been diagnosed with a critical illness at an insurance company for ten years, from 2005 to 2014.

- For insured men aged less than 30 years

Table 1.Bayesian estimation calculation for men aged less than 30 years.

year	n	x	$\frac{x}{n}$	α	β	θ_B	Z	%Z
2005	67	1	0.015	0.041	8.124	0.005	0.891	89%
2006	108	4	0.037	1.041	74.124	0.014	0.590	59%
2007	123	2	0.016	5.041	178.124	0.028	0.402	40%
2008	384	6	0.016	7.041	299.124	0.023	0.556	56%
2009	688	12	0.017	13.041	677.124	0.019	0.499	50%
2010	985	10	0.010	25.041	1,353.124	0.018	0.417	42%
2011	997	9	0.009	35.041	2,328.124	0.015	0.297	30%
2012	1456	12	0.008	44.041	3,316.124	0.013	0.302	30%
2013	1878	18	0.010	56.041	4,760.124	0.012	0.281	28%
2014	2146	12	0.006	74.041	6,620.124	0.011	0.243	24%
2015				86.041	8,754.124	0.010		

Based on the percentage of the value of Z obtained in Table 1, it can be concluded that the influence of information in previous years to take into account the value of the insured $\theta_B$men aged less than 30 years, namely in 2005, 2006, 2008, and 2009 are very large, in 2007 and in 2010 is large enough, whereas in 2011 until 2014 small.

- For insured women aged less than 30 years

Table 2. Calculation of Bayesian estimation for women aged less than 30 years.

year	n	x	$\frac{x}{n}$	α	β	θ_B	Z	%Z
2005	55	1	0.018	0.041	8.124	0.005	0.871	87%
2006	89	2	0.022	1.041	62.124	0.016	0.585	58%
2007	94	6	0.064	3.041	149.124	0.020	0.382	38%
2008	222	4	0.018	9.041	237.124	0.037	0.474	47%
2009	587	2	0.003	13.041	455.124	0.028	0.556	56%
2010	878	6	0.007	15.041	1,040.124	0.014	0.454	45%
2011	984	4	0.004	21.041	1,912.124	0.011	0.337	34%
2012	1165	6	0.005	25.041	2,892.124	0.009	0.285	29%
2013	1456	12	0.008	31.041	4,051.124	0.008	0.263	26%
2014	1987	18	0.009	43.041	5,495.124	0.008	0.264	26%
2015				61.041	7,464.124	0.008		

Based on the percentage of the value of Z obtained in Table 2, it can be concluded that the influence of information in previous years to take into account the value of the insured θ_B women aged less than 30 years, namely in 2005, 2006 and 2009 was very strong, while in 2007, 2008, 2010 and 2011 is quite large, but in the year 2012 to 2014 is small.

- For insured men aged 30 years to 40 years

Table 3. Calculation of Bayesian estimation for men aged 30 years to 40 years

year	n	x	$\frac{x}{n}$	α	β	θ_B	Z	%Z
2005	180	6	0.033	0.041	8.124	0.005	0.957	96%
2006	354	4	0.011	6.041	182.124	0.032	0.653	65%
2007	487	9	0.018	10.041	532.124	0.019	0.473	47%
2008	555	15	0.027	19.041	1,010.124	0.019	0.350	35%
2009	687	18	0.026	34.041	1,550.124	0.021	0.302	30%
2010	987	26	0.026	52.041	2,219.124	0.023	0.303	30%
2011	1165	34	0.029	78.041	3,180.124	0.024	0.263	26%
2012	1548	50	0.032	112.041	4,311.124	0.025	0.259	26%
2013	1856	49	0.026	162.041	5,809.124	0.027	0.237	24%
2014	1951	55	0.028	211.041	7,616.124	0.027	0.200	20%
2015				266.041	9,512.124	0.027		

Based on the percentage of the value of Z obtained in Table 3, it can be concluded that the influence of information in previous years to take into account the value of the insured $\theta_B$men aged 30 years to 50 years ie in 2005 and 2006 are very large, in 2007 and in 2008 is large enough, whereas in the year 2009 to 2014, small.

- For insured women aged 30 years to 40 years

Table 4. Calculation of Bayesian estimation for women aged 30 years to 40 years

year	n	x	$\frac{x}{n}$	α	β	θ_B	Z	%Z
2005	165	2	0.012	0.041	8.124	0.005	0.953	95%
2006	289	3	0.010	2.041	171.124	0.012	0.625	63%
2007	348	6	0.017	5.041	457.124	0.011	0.430	43%
2008	498	14	0.028	11.041	799.124	0.014	0.381	38%
2009	512	16	0.031	25.041	1,283.124	0.019	0.281	28%
2010	594	22	0.037	41.041	1,779.124	0.023	0.246	25%
2011	789	31	0.039	63.041	2,351.124	0.026	0.246	25%
2012	987	46	0.047	94.041	3,109.124	0.029	0.236	24%
2013	1254	43	0.034	140.041	4,050.124	0.033	0.230	23%
2014	1789	49	0.027	183.041	5,261.124	0.034	0.247	25%
2015				232.041	7,001.124	0.032		

Based on the percentage of the value of Z obtained in Table 4, it can be concluded that the influence of information in previous years to take into account the value of the insuredθ_B on women aged 30 years to 50 years ie in 2005 and 2006 are very large, in the year 2007 and the year 2008 is large enough, whereas in the year 2009 to 2014, small.

- For insured men aged 40 years to 50 years

Table 5. Calculation of Bayesian estimation for men aged 40 years to 50 years

year	n	x	$\frac{x}{n}$	α	β	θ_B	Z	%Z
2005	185	4	0.022	0.041	8.124	0.005	0.958	96%
2006	354	6	0.017	4.041	189.124	0.021	0.647	65%
2007	439	9	0.021	10.041	537.124	0.018	0.445	45%
2008	654	9	0.014	19.041	967.074	0.019	0.399	40%
2009	945	12	0.013	28.041	1,612.074	0.017	0.366	37%
2010	1145	18	0.016	40.041	2,545.074	0.015	0.307	31%
2011	1265	21	0.017	58.041	3,672.074	0.016	0.253	25%
2012	1548	24	0.016	79.041	4,916.074	0.016	0.237	24%
2013	1856	27	0.015	103.041	6,440.074	0.016	0.221	22%
2014	2156	36	0.017	130.041	8,269.074	0.015	0.204	20%
2015				166.041	10,389.074	0.016		

Based on the percentage of the value of Z obtained in Table 5, it can be concluded that the influence of information in previous years to take into account the value of the insured θ_B men aged 40 years to 50 years ie in 2005 and 2006 are very large, in 2007 until the year 2010 is quite large, whereas in 2011 until 2014, small.

- For insured women aged 40 to 50 years

Table 6. Calculation of Bayesian estimation for women aged over 50 years

year	n	x	$\frac{x}{n}$	α	β	θ_B	Z	$\%Z$
2005	176	1	0.006	0.041	8.124	0.005	0.956	96%
2006	326	1	0.003	1.041	183.124	0.006	0.639	64%
2007	399	2	0.005	2.041	508.124	0.004	0.439	44%
2008	587	9	0.015	4.041	905.124	0.004	0.392	39%
2009	878	12	0.014	13.041	1,483.124	0.009	0.370	37%
2010	1059	6	0.006	25.041	2,349.124	0.011	0.308	31%
2011	1159	16	0.014	31.041	3,402.124	0.009	0.252	25%
2012	1478	21	0.014	47.041	4,545.124	0.010	0.243	24%
2013	1784	30	0.017	68.041	6,002.124	0.011	0.227	23%
2014	2111	27	0.013	98.041	7,756.124	0.012	0.212	21%
2015				125.041	9,840.124	0.013		

Based on the percentage of the value of Z obtained in Table 6, it can be concluded that the influence of information in previous years to take into account the value of the insured θ_B on women aged 40 to 50 years, namely in 2005 and 2006 are very large, in the year 2007 to the year 2010 was quite large, whereas in 2011 until 2014, small.

- For insured men aged 50 years to 60 years

Table 7. Calculation of Bayesian estimation for men aged 50 years to 60 years

year	n	x	$\frac{x}{n}$	α	β	θ_B	Z	$\%Z$
2005	165	6	0.036	0.041	8.124	0.005	0.953	95%
2006	354	4	0.011	6.041	167.124	0.035	0.672	67%
2007	477	9	0.019	10.041	517.124	0.019	0.475	48%
2008	658	15	0.023	19.041	985.124	0.019	0.396	40%
2009	978	18	0.018	34.041	1,628.124	0.020	0.370	37%
2010	1523	26	0.017	52.041	2,588.124	0.020	0.366	37%
2011	1648	34	0.021	78.041	4,085.124	0.019	0.284	28%
2012	1853	50	0.027	112.041	5,699.124	0.019	0.242	24%
2013	2157	49	0.023	162.041	7,502.124	0.021	0.220	22%
2014	3245	55	0.017	211.041	9,610.124	0.021	0.248	25%
2015				266.041	12,800.124	0.020		

Based on the percentage of the value of Z obtained in Table 7, it can be concluded that the influence of information in previous years to take into account the value of the insured θ_B men aged 50 years to 60 years, namely in 2005 and 2006 are very large, in 2007 until the year 2010 is quite large, whereas in 2011 until 2014, small.

- For insured women aged 50 to 60 years

Table 8. Calculation of Bayesian estimation for women aged 50 to 60 years

year	n	x	$\frac{x}{n}$	α	β	θ_B	Z	%Z
2005	152	6	0.039	0.041	8.124	0.005	0.949	95%
2006	333	4	0.012	6.041	154.124	0.038	0.675	68%
2007	398	9	0.023	10.041	483.124	0.020	0.447	45%
2008	599	15	0.025	19.041	872.124	0.021	0.402	40%
2009	977	18	0.018	34.041	1,456.124	0.023	0.396	40%
2010	1485	26	0.018	52.041	2,415.124	0.021	0.376	38%
2011	1677	34	0.020	78.041	3,874.124	0.020	0.298	30%
2012	1789	50	0.028	112.041	5,517.124	0.020	0.241	24%
2013	2054	49	0.024	162.041	7,256.124	0.022	0.217	22%
2014	2987	55	0.018	211.041	9,261.124	0.022	0.240	24%
2015				266.041	12,193.124	0.021		

Based on the percentage of the value of Z obtained in Table 8, it can be concluded that the influence of information in previous years to take into account the value of the insured θ_B on women aged 50 to 60 years ie in 2005 and 2006 are very large, in the year 2007 to the year 2010 was quite large, whereas in 2011 until 2014, small.

- For insured men aged over 60 years

Table 9. Calculation Bayesian estimation for men aged over 60 years

year	n	x	$\frac{x}{n}$	α	β	θ_B	Z	%Z
2005	88	6	0.068	0.041	8.124	0.005	0.915	92%
2006	120	4	0.033	6.041	90.124	0.063	0.555	56%
2007	155	9	0.058	10.041	206.124	0.046	0.418	42%
2008	258	15	0.058	19.041	352.124	0.051	0.410	41%
2009	394	18	0.046	34.041	595.124	0.054	0.385	39%
2010	478	26	0.054	52.041	971.124	0.051	0.318	32%
2011	658	34	0.052	78.041	1,423.124	0.052	0.305	30%
2012	698	50	0.072	112.041	2,047.124	0.052	0.244	24%
2013	789	49	0.062	162.041	2,695.124	0.057	0.216	22%
2014	841	55	0.065	211.041	3,435.124	0.058	0.187	19%
2015				266.041	4,221.124	0.059		

Based on the percentage of the value of Z obtained in Table 9, it can be concluded that the influence of information in previous years to take into account the value of the insured θ_B men aged over 60 years, namely in 2005 and 2006 are very large, in the year 2007 to in 2010 large enough, whereas in 2011 until 2014, small.

- For insured men aged over 60 years

Table 10.Calculation -Eighteen Bayesian estimation for over 60 years

year	n	x	$\frac{x}{n}$	α	β	θ_B	Z	%Z
2005	74	6	0.081	0.041	8.124	0.005	0.901	90%
2006	95	4	0.042	6.041	76.124	0.074	0.536	54%
2007	111	9	0.081	10.041	167.124	0.057	0.385	39%
2008	198	15	0.076	19.041	269.124	0.066	0.407	41%
2009	242	18	0.074	34.041	452.124	0.070	0.332	33%
2010	298	26	0.087	52.041	676.124	0.071	0.290	29%
2011	397	34	0.086	78.041	948.124	0.076	0.279	28%
2012	874	50	0.057	112.041	1,311.124	0.079	0.380	38%
2013	897	49	0.055	162.041	2,135.124	0.071	0.281	28%
2014	985	55	0.056	211.041	2,983.124	0.066	0.236	24%
2015				266.041	3,913.124	0.064		

Based on the percentage of the value of Z obtained in Table 10, it can be concluded that the influence of information in previous years to take into account the value θ_B insured women over the age of 60 years, namely in 2005 and 2006 are very large, in the year 2007 to the year 2009 big enough, whereas in the year 2010 to 2014, small.

Table 11. Calculation of the risk premium for insured males by age group

Age	θ_B	n_{2009}	$E(Y_j)$	$D(Y_j)$	The risk premium / person
< 30 years	0,01	2.146	179.628	1,707.642	0,85
30 to 40 years	0.027	2.156	51,861.258	49,340,328.857	3.58
40 to 50 years	0.016	2.156	35,770.747	37,168,956.545	1.71
50 to 60 years	0.020	3.245	86,276.357	110,450,316.148	2.96
> 60 year	0.059	841	24,683.851	10,354,477.324	7.82

Table 12. Calculation of the risk premium for insured women by age group

Age	θ_B	n_{2009}	$E(Y_j)$	$D(Y_j)$	The risk premium / person
< 30 years	0.008	1.987	12,115.687	9,045,503.896	0.56
30 to 40 years	0.032	1.789	41,465.275	29,037,717.865	3.21
40 to 50 years	0.013	2.111	26,374.527	25,957,640.904	0.65
50 to 60 years	0.021	2.987	72,548.771	88,639,531.691	3.12
> 60 year	0.064	985	32,052.294	12,355,569.095	9.99

4. Conclusion

Bayesian estimation theory provides a great method for estimating the risk premium for the next period of the claim data information in the previous period. In this study, a large risk premium for insured males aged less than 30 years is 0.85, for those aged 30 to 40 years is 3:58, for those aged 40 to 50 years is 1.71, for those aged 50 to 60 years was 2.96, and for age over 60 years is 7.82. Meanwhile, for insured women aged less than 30 years is 0.56, for those aged 30 to 40 years is 3:21, for those aged 40 to 50 years is 0.65, for those aged 50 to 60 years was 3:12, and for those aged over 60 years is 9.99.

Acknowledgment

We would like to thank Rector of Universitas Padjadjaran and Dean of Faculty of Mathematics and Natural Sciences, which have provided the Academic Leadership Grant (ALG), under the coordination of Prof. Dr. Sudradjat, M.S., which is a means to increase research activities and publications at the Universitas Padjadjaran.

References

[1] Apsari, W., Yasin, H. dan Sugito.Estimasi Parameter Regresi Logistik Multinomial Dengan Metode Bayes.*Jurnal Gaussian*, Volume 2, Nomor 1, Tahun 2013, Halaman 79-88. Online di: http://ejournal-s1.undip.ac.id/index.php/gaussian

[2] Jindrova, P & Pacakova, V. (2015). "Actuarial Models for Valuation of Critical Illness Insurance Product". *International Journal of Mathematical Models and Method n Applied Sciences.* Volume 9, 2015

[3] Qu Yu, Guang. (2011). *Hierarchical Bayesian Modeling of Health Insurance Claims.* B.Com., the University of Melbourne.

[4] Dickson, D.C. (2005). *Insurance Risk and Ruin.* Cambridge University Press.

[5] Hamadu, D. and Adeleke, I. Model-Assisted Credibility Rating for Health Insurance Claims. *Journal of Mathematics and Technology*, ISSN: 2078-0257, Vol. 3. No. 2, 2012. 37-43. DOI: 10.7813/jmt.2012/3-2/7.

[6] Kirkby, N.F., Jefferies, S.J., Jena, R., and Burnet, N.G. (2007). A mathematical model of the treatment and survival of patients with high-grade brain tumours. *Journal of Theoretical Biology* 245 (2007) 112–124. www.elsevier.com/locate/yjtbi.

[7] Krishnamoorthy, K. (2006). *Handbook of Statistical Distribution with Applications*. London: Chapman & Hall/CRC.

[8] Lahallo, A.P.T.P. dan I Nyoman Widana Desak Putu Eka Nilakusmawati, I. N.D.P.E. (2013). Perumusan Premi Bulanan Asuransi Kesehatan Individu Perawatan Rumah Sakit (Anuitas Hidup Pembayaran Bulanan). E-Jurnal Matematika Vol. 2, No.4, Nopember 2013, 40-45. ISSN: 2303-1751.

[9] [9] Mesike, G.C., Adeleke, I.A., and Ibiwoye, A. (2012). Predictive Actuarial Modeling of Health Insurance Claims Costs. *International Journal of Mathematics and Computational,* Vol. 14; Issue No. 1; Year 2012. ISSN: 0974-570X (Online), ISSN: 0974-5718 (Print). Copyright @ 2011-12 by CESER Publications.

[10] Pham, M.H., Grinshpan, A., Kafle, R.C., and Chen, Y. (2014). Survival Analysis - Breast Cancer. *Undergraduate Journal of Mathematical Modeling: One + Two*. Volume 6 | 2014 Fall, Issue 1 | Article 4.

[11] Retnawati, H. (2015). Perbandingan Estimasi Kemampuan Laten Antara Metode Maksimum Likelihood Dan Metode Bayes. *Jurnal Penelitian dan Evaluasi Pendidikan.* Volume 19, No 2, Desember 2015 (145-155). Tersedia Online: http://journal.uny.ac.id/index.php/jpep

[12] Sheta, O.E. and Eldeen, A.N. (2012). Building a Health Care Data Warehouse for Cancer Diseases. *International Journal of Database Management Systems* (IJDMS) Vol.4, No.5, October 2012. DOI: 10.5121/ijdms.2012.4503.

IOP Conf. Series: Materials Science and Engineering **166** (2017) 012023 doi:10.1088/1757-899X/166/1/012023

Countries population determination to test rice crisis indicator at national level using *k*-means cluster analysis

Y. Hidayat[1*], T. Purwandari[2], Sukono[3], Y.D. Ariska[4]

[1,2]Department of Statistics, Universitas Padjadjaran, Indonesia

[3]Department of Mathematics, Universitas Padjadjaran, Indonesia

[4]Graduate School of Statistics, Faculty of Mathematics and Natural Sciences, Universitas Padjadjaran, Indonesia

[*]E-mail: yuyun.hidayat@unpad.ac.id

Abstract. This study aimed to obtain information on the population of the countries which is have similarities with Indonesia based on three characteristics, that is the democratic atmosphere, rice consumption and purchasing power of rice. It is useful as a reference material for research which tested the strength and predictability of the rice crisis indicators Unprecedented Restlessness (UR). The similarities countries with Indonesia were conducted using multivariate analysis that is non-hierarchical cluster analysis *k*-Means with 38 countries as the data population. This analysis is done repeatedly until the obtainment number of clusters which is capable to show the differentiator power of the three characteristics and describe the high similarity within clusters. Based on the results, it turns out with 6 clusters can describe the differentiator power of characteristics of formed clusters. However, to answer the purpose of the study, only one cluster which will be taken accordance with the criteria of success for the population of countries that have similarities with Indonesia that cluster contain Indonesia therein, there are countries which is sustain crisis and non-crisis of rice in 2008, and cluster which is have the largest member among them. This criterion is met by cluster 2, which consists of 22 countries, namely Indonesia, Brazil, Costa Rica, Djibouti, Dominican Republic, Ecuador, Fiji, Guinea-Bissau, Haiti, India, Jamaica, Japan, Korea South, Madagascar, Malaysia, Mali, Nicaragua, Panama, Peru, Senegal, Sierra Leone and Suriname.

1. Introduction

In 2007-2008 recorded as the world rice crisis [5]. It is marked with an extremely increasing of rice prices in the world market, it was reached 2.852 times higher between October 2007 and May 2008 [4]. Not a few countries in Asia and Africa affected by the rice crisis. Indonesian Government should be responsive to the situation that afflicted these countries. Why? Because the rice crisis can also afflict Indonesia. It is because majority of Indonesian consume the rice as one of their staple food.

Indonesian Government needs a predictor to look for the potential of rice crisis in Indonesia in the future. Therefore, some researchers do studies in order to know the strength and predictability of indicator called Unprecedented Restlessness (UR) to detect the occurrence of rice crisis in the future. However, there are still less of data to describe the economic situation In Indonesia to predict the rice crisis in the future, it creates fears of uncertainty to look for the dynamics of rice crisis in Indonesia. Therefore, the researchers used the data of countries that assumed have similar characteristic to

Indonesian's without do the testing first. The effect is not known prediction accuracy of the result of the studies about the occurrence of rice crisis in Indonesia [8].

Therefore, it needs a testing to form a group that consists the countries which have similarities to Indonesian's. In this study, the similarity of countries in Indonesian is measured by 3 characteristics (democratic atmosphere, rice consumption, and restlessness or rice purchasing ability) in each countries [7].

2. Literature Reviews

In this study, there are 3 based characteristics that formed a group or population of countries which is have similarities to Indonesian's. The importance of the measured similarities is basic formation to form the population of countries which is have similarities to Indonesian's based on 3 characteristics, so the formed population can be used as an object of study to test the indicator of occurrence of rice crisis in the future.

2.1. The Similarity of Characteristics

2.1.1 Democratic Atmosphere

The advanced study Polity IV Project always develop coding to know the characteristics of authority in countries (world's systems) to compare and do a quantitative analyse which is illustrated by score called Polity. Polity score that produced by this project has been proved by the researchers years by years. This score is the democratic measurement of each country. The scale of this score ranges between -10 and +10, which is -10 shows the very authoritative government's system and +10 shows the very democratized government's system [10].

2.1.2 Percent Rice Consumption

Rice is a staple food for almost half of population in the world. Besides, more than 90% of population in Asia consume rice as their staple or main food.

Percent Rice Consumption is a proportion of rice consumption to total consumption of staple food in a country [7].

$$\%C = \frac{Rc}{Tc}\,100\% \tag{1}$$

Note:
R=*Restlessness*; *P*=Rice Prices (US$/kg); *C*= Rice Consumption (kg/capita); *I*=Income per capita (US$/capita).

2.1.3 Restlessness

This characteristic measured with rice purchasing ability per capita in a country, this is the formula below [7]:

$$R = \frac{P}{I}\,C \tag{2}$$

Note:
R=*Restlessness*; *P*=Rice Prices (US$/kg); *C*= Rice Consumption (kg/capita); *I*=Income per capita (US$/capita).

2.2. Multivariate Analysis

Multivariate analysis is an application of methods that can be used when some measurement (1 and more) obtain from each individuals or observations. This measurement called variables. In multivariate analysis, all variables measured simultaneously in each sample unit [9]. Basically, in multivariate analysis there are 2 classifications of relationship between variables, that is dependency relationship and interdependency relationship. Dependencies technique occurs when there is a dependency relationship between variables, and Interdependencies occurs when there are no different between dependent variable and independent variable [2].

In this study, interdependencies technique was used, because there is no different of dependent or independent variables. Therefore, there are some possibilities of application of multivariate analysis that used to form a group from observations, such as multidimensional scaling analysis, correspondence analysis, biplot analysis, and cluster analysis [9].

2.2.1 Dependencies Technique

Dependencies technique of multivariate analysis is a method to explain or predict one or more criteria measurement based on a set of predictor variables (independent variables). Multivariate analysis with dependencies technique can be used when there are independent and dependent variables. Some methods that classified to dependencies technique are multiple regressions, discriminant analysis, canonic analysis, and others [2].

2.2.2 Interdependencies Technique

Interdependencies technique of multivariate analysis is a method to give a more evident description about data structure by simplify the complex data structure (data reduction). This technique is used when there is no different between dependent variables and independent variables (predictors). Multivariate method that consisted in this method are multidimensional scaling analysis, correspondence analysis, biplot analysis, and cluster analysis.

Multidimensional scaling analysis is a multivariate method that used to determine the position of an observation to another observation based on the similarities value. The advantage of this method, can be used for metric and non-metric data structure while the weakness of this method is plot data structure of observations became not evident when there is a data point with big dimension [9]. The advantage of correspondence analysis method is it flexible enough used for a big dimension of data structure, but the weakness of this method is it only can be used for non-metric data structure (nominal and ordinal) [9]. The advantage of biplot analysis, it is capable to show directly the dominant variables of group of observations formed [13]. However, the weakness of this method, when do a big dimension reduction to 2 dimensions, then it can be effect to decreasing of information in biplot.

2.3. Cluster Analysis

Cluster analysis is a troubleshooting methods based on multivariate data. This method used to determine the cluster to group from some observations that have high similarities inside the cluster, but they have high dissimilarities between the clusters. This method is different with another grouping technique like discriminant analysis, which in cluster analysis, it shows unknown groups so that the objective of this method is to form group (clusters). To form clusters using this method, the similarities technique is the first thing that mostly seen, that is based on the closeness among the pair of each observations. The other methods can also be used to determine the centroid of some clusters [14].

There are 2 methods of grouping in cluster analysis, hierarchical and non-hierarchical technique (partition). Hierarchical methods begin with grouping observations by the nearest distance among all pair of observations, it begin with the big number of clusters then become smaller as a clusters merge to another clusters. It is different from non-hierarchical (partition) method, that used by determine the number of clusters first, then it finds the observations that have the nearest distance to the centroid of clusters [14].

The closeness or distance that usually used in cluster analysis between 2 observations in p dimensions (variables) is Euclidean Distance [14].

$$d(x,y) = \sqrt{(x-y)'(x-y)} = \sqrt{\sum_{j=1}^{p}(x_j - y_j)^2} \quad (3)$$

where $x = (x_1, x_2, \dots\dots\dots, x_p)'$ and $y = (y_1, y_2, \dots\dots\dots, y_p)'$.

Hierarchical Method

Cluster analysis with this method can be done with 2 approaches, agglomerative and divisive approaches. Agglomerative approach begin with each observations as clusters then the nearest observations will be merged first and forward merge into another clusters that have the nearest distance each other, and then until the number of clusters become smaller and the member of each clusters become larger.

Whereas, in the divisive approach, it begin with dividing the population into 2 sub-groups so the observations consisted in each cluster have the furthest distance each other, then they will divided again into the next sub-groups, continued until it form the largest number of clusters but the smallest number of member of each clusters. The final result of this method formed a dendogram [9].

Non-Hierarchical Method (*k*-Means)

Cluster analysis with this method can be done with determine the number of clusters to be formed first. This method begins with one of two approaches that is initial partition of observations into groups and first determining of centroids of each clusters. *k*-Means method is done by grouping items or observations into cluster that have the nearest centroid [9].

There are some steps below that can be done in this cluster analysis with non-hierarchical methods:

- Partitioning the observations randomly into *k* initial clusters.
  The first thing that should be done is determining the number of initial clusters, that is *k* clusters, so we get the centroid of each clusters [14].
  The centroid value of each clusters calculated by below formula [15]:
  $$\mu_k = \frac{1}{N_k}\sum_{q=1}^{N_k} x_q \quad (4)$$
  Note:
  μ_k : Centroid of cluster-*k*
  N_k : The number of members of cluster-*k*
  x_q : Observations-q in cluster-*k*
- Groups the observations into clusters that have the nearest centroid.
  Distance that used in this analysis is Euclidean distance, both standardized and non-standardized. Recalculate the centroid value of each cluster that had a member change (addition or subtractions of cluster members) [9].
- Performs the repetition steps 2 until it impossible to occur the change of cluster members [9].

The advantage of *k*-Means cluster compared to hierarchical method is it can overcome the large number of data. *k*-Means cluster is method that used in this study because the similarities among the observations with metric data structure and interdependencies technique (there is no different between dependent and independent variables).

2.4. *Differentiator Power of Variables*

To measure the ability of the three variables in differentiate the formed clusters simultaneously (multivariate) then statistic testing that can be used is T^2Hotelling analysis for 2 populations (cluster) and MANOVA for advanced cases more than 2 populations (clusters).

2.4.1 *T^2Hotelling*

In univariate case (one variable), the analysis to test the mean difference among 2 populations used is t-test analysis. The statistical testing that used as the formula below [14]:

$$t = \frac{\overline{y_i - \bar{y}_2}}{Sgab\sqrt{\frac{1}{n_1}+\frac{1}{n_2}}} \quad (5)$$

From the formula above, it notes that $y_1$and y_2 is the means value of variables of population 1 and 2. For case using some variables in 2 populations, t-test analysis can not be performed, so that it will need multivariate analysis to test the difference of 2 mean vectors using T^2Hotelling which is can perform an analysis to some variables of 2 populations simultaneously [14].

$$T^2 = (\bar{y}_1 - \bar{y}_2)' \left[\left(\frac{1}{n_1} + \frac{1}{n_2}\right) S_{gab}\right]^{-1} (\bar{y}_1 - \bar{y}_2) \quad (6)$$

From the equation above, $\left(\frac{1}{n_1} + \frac{1}{n_2}\right) S_{gab}$ as covarians to $(\bar{y}_1 - \bar{y}_2)$.

2.4.2 *MANOVA*

Multivariate analysis of variance is an advanced method of univariate technique ANOVA to predict the difference between some groups (>2 populations). MANOVA is capable to performs the testing to determine the difference between >2 populations by using some variables simultaneously. If there are some multiple variate then this analysis will give the dimension of differences that will differentiate between some groups better.

To perform the testing for differences means vector between k populations note in a form null hypotheses $H_0: \mu_1 = \mu_2 = \cdots = \mu_k$ using statistics Wilks Lambda testing with statistics F approach.

$$\Lambda = \frac{|E|}{|E| + |H|} \quad (7)$$

H is sum squares matrix between groups and E is sum squares matrix in the groups.

Table 1. MANOVA Difference of Mean Vectors of Populations

Source of Variation	Sum Squares Matrix	Degree of Freedom
Effect	$H = \sum_{i=1}^{k} n_i (\bar{y}_{i.} - \bar{y}_{..})(\bar{y}_{i.} - \bar{y}_{..})'$	k-1
Residual	$E = \sum_{i=1}^{k} \sum_{j=1}^{n_i} (\bar{y}_{ij} - \bar{y}_{i.})(\bar{y}_{ij} - \bar{y}_{i.})'$	$\sum_{i=1}^{k} n_i - k$
Total	$H + E = \sum_{i=1}^{k} \sum_{j=1}^{n_i} (\bar{y}_{ij} - \bar{y})(\bar{y}_{ij} - \bar{y})'$	$\sum_{i=1}^{k} n_i - 1$

2.4.3 *Evaluation of Grouping Result*

After performing the grouping process with cluster analysis, then we need to evaluate the result of the grouping of one cluster. This occurs because there is only one cluster that will be taken as the answer of the problem in this study that is the population of countries which is have similarities to Indonesian's. The objective of this evaluation is that population to be right on target as the objectives of this study. There are things that should be noted as criteria of success of this grouping, the population of countries which is has similarities to Indonesian's [7]:

- There is Indonesia inside of that group.
  The existence of Indonesia inside of the cluster or group is very important as a criteria of success, because we will look for the countries which is have the similarities to Indonesian's so Indonesia should be there in the cluster.
- Inside of the cluster or group, there are both countries which is experience the rice crisis in 2008 and countries which did not experience the rice crisis in 2008.
- The cluster or group is the group that have the most number of cluster's members. The more countries inside the cluster or group which is have the similarities to Indonesian's the better to perform the study which is aim for testing of indicator called *Unprecedented Restlessness* (UR) as indicator of rice crisis.

3. Methodology

3.1. Data and Variables

The data sources for this study is a secondary data of 38 countries which is obtained by taking the countries that consumes rice as their staple or main food, and also rice cannot be substitute with another food [7]. Variables in this study are below:
X1: Democratic Atmosphere
X2: Percent Rice Consumption
X3: Restlessness

Table 2. Framework of Data

No	Countries	*Polity*	Rice Consumption (%)	Restlessness
1	Bangladesh	6	76.9031	0.0399934482524908
2	Brazil	8	21.7572	0.00173088466516339
3	Cambodia	2	79.6210	0.0572555778572883
...	...	...	...	...
38	Vietnam	-7	82.0934	0.0361490469875234

3.2. The Steps of Study

3.2.1 Standardized Variables

The data standardizing is performed to overcome the variables that have the difference measurement scaling. In this study, there are 3 variables which is have the difference of unit so it it needs to perform the standardizing of variables that is using normal standard (*Z* score).

The most common form that used to perform data standardizing is to convert each variables to standard score (*Z* score) by subtracting the mean of variables from the observation value then divide it to standard deviation of each variables. The most common way of this method is function normalized distance [2]:
Note:
Z =Normalize Standard Value; X= the valueof observations; μ=Means of population; σ=standard deviation of population

3.2.2 Cluster Analysis: k-Means Method

There are some steps that will be performed:

- Partitioning each observations (countries) to k clusters randomly
- Calculate the centroid value of each clusters
- Move the observations (countries) to the nearest cluster
- Perform the step c repeatedly until there is no possibility of change of cluster's members.

3.2.3 Differentiator Power of Variables

To look for the differentiator power of 3 variables, then it uses the test of T^2Hotelling (2 populations) and MANOVA (>2 populations).

- The formulation of hypotheses

  H_0 : $\mu_1 = \cdots = \mu_k$
  There is no difference of means of k- populations
  (3 variables are not capable to differentiate the formed populations)
  H_1 : There is difference of means of k- populations

(3 variables are not capable to differentiate the formed populations)

- Determine the α value
- Statistics Testing
  T^2Hotelling and Wilks Lambda (Statistics F approach)
- Testing Criteria
  T^2Hotelling : Reject H_0 if $F > F_{\alpha,p,n_1+n_2-2}$, acceptance for others.
  MANOVA : Reject H_0 if $F > F_{\alpha,p,n_1+n_2-2}$, acceptance for others.

3.2.4 Evaluation of Grouping Result

As described in Chapter 2, there are 3 criteria of success of population groupings which is have similar characteristics with Indonesian's, it is becoming a benchmark for the success of getting the appropriate population and meet the goals as the solution for the problem in this study.

4. Result

4.1. Standardized Variables

Table 3 Score of Standardized Variables

No	Countries	Polity	% Rice Consumption (%)	Restlessness
1	Bangladesh	0.44493	0.031265	1.975335
2	Brazil	0.7475	-0.60989	-1.1743
3	Cambodia	-0.1643	0.320521	2.130567
...	...	...	...	...
38	Vietnam	-1.5352	-0.03315	2.271773

From the result of standardized scores as the Table 4.1, then the next is perform the grouping for countries using the K-Means clustering analysis.

4.2. k-Means Clustering Analysis & Differentiator Power of Variables

In choosing the number of clusters that will be performed, then it begin by form the smallest number of cluster that is 2 clusters.

Table 4 Cluster's Members 2 Klaster

2 Cluster	1	3.000
	2	35.000

Table 5 Differentiator Power 2 Klaster

Effect		Value
2 Cluster	**Hotelling's Trace**	**4.703**

From the Table 4 we can see that it formed 2 clusters with the number of members in each clusters are 3 and 35, and Table 5 shows the T2 Hotelling value is 4.703that transformed into statistics $F = 1.4805$. Compared to the value of $F_{0.1:1;36} = 2.8503$ it turns out that the three variables can not yet prove the difference of 2 formed clusters. Therefore, we go back with *clustering* process with form the higher number of clusters that is 3 clusters.

Table 6 Cluster's Members 3 Cluster

3 Cluster	1	6.000
	2	29.000
	3	3.000

Table 7 Differentiator Power 3 Cluster

Source	Dependent Variable	F
3 Cluster	Z_Polity	.458
	Z_Restlessness	85.017
	Z_Rice	29.282

From the Table 6we can see that it formed 3 clusters with the number of members in each clusters are 6, 29 and 3, and Table 7 we can conclude that with performing 3 clusters clustering process, there is only restlessness and percent rice consumption that have statistics F $>F_{0.1;2;35} = 2.46093$ that means there is only 2 variables have the differentiator power of 3 clusters clustering process.Therefore, we go back with *clustering* process with form the higher number of clusters, that is 4 clusters.

Table 8 Cluster's Members 4 Cluster

4 Cluster	1	2.000
	2	29.000
	3	1.000
	4	6.000

Table 9 Differentiator Power 4 Cluster

Source	Dependent Variable	F
4 Cluster	Z_Polity	.996
	Z_Restlessness	56.503
	Z_Rice	31.924

From the Table 8 we can see that it formed 4 clusters with the number of members in each clusters are 2, 29, 1 and 6, and Table 9 we can conclude that with performing 4 clusters clustering process, there is only restlessness and percent rice consumption that have statistics $F>F_{0.1;3;34} = 2.2523$ that means there is only 2 variables have the differentiator power of 4 clusters clustering process. Therefore, we go back with *clustering* process with form the higher number of clusters, that is 5 clusters.

Table 10 Cluster's Members 5 Cluster

5 Cluster	1	1.000
	2	2.000
	3	6.000
	4	28.000
	5	1.000

Table 11 Differentiator Power 5 Cluster

Source	Dependent Variable	F
5 Cluster	Z_Polity	1.480
	Z_Restlessness	49.613
	Z_Rice	21.937

From the Table 10 we can see that it formed 5 clusters with the number of members in each clusters are 1, 2, 6, 28, and 1, Table 11 we can conclude that with performing 5 clusters clustering process, there is only restlessness and percent rice consumption that have statistics $F>F_{0.1;4;33} = 2.1234$ that means there is only 2 variables have the differentiator power of 5 clusters clustering process. Therefore, we go back with *clustering* process with form the higher number of clusters, that is 6 clusters.

Table 12 Cluster's Members 6 Cluster

Cluster	1	6.000
	2	22.000
	3	1.000
	4	7.000
	5	1.000
	6	1.000

Table 13 Differentiator Power 6 Cluster

Source	Dependent Variable	F
6 Cluster	Z_Polity	24.187
	Z_Restlessness	38.164
	Z_Rice	19.696

From the Table 12 we can see that it formed 6 clusters with number of members in each clusters are 6, 22, 1, 7, 1, and 1, Table 13we can conclude that with performing 6 clusters clustering process, all variables that are restlessness, Polity, and percent rice consumption that have statistics $F>F_{0.1;5;32} = 2.0360$ that means the three variables have the differentiator power of 6 clusters clustering process. Therefore, in this step clustering process stopped.

Further, evaluation of the criteria of success for population which is has similarities to Indonesian's. From the result of clustering process with 6 clusters, it shows all criteria is fulfilled by cluster 2, that is there is Indonesia in the cluster, there are both crisis and non crisis countries in 2008, and it is shows the most number of cluster's members compared to another clusters.

5. Conclusion and Suggestion

5.1. Conclusion
The conclusion of this study is there are 21 countries which is have similarities to Indonesian's based on three characteristics (democratic atmosphere, rice consumption, and rice purchasing ability (restlessness)), that are Brazil, Costa Rica, Djibouti, Dominican Republic, Ecuador, Fiji, Guinea-Bissau, Haiti, India, Jamaica, Japan, South Korea, Madagascar, Malaysia, Mali, Nicaragua, Panama, Peru, Senegal, Sierra Leonne and Suriname.

5.2. Suggestion
The suggestion based on this study result is so that this study can become preference for the next study which is perform the analysis to test the strength and predictability of indicator *Unprecedented Restlessness* (UR) as indicator of rice crisis.

References
[1] AFP. (2008, April 13). Haiti PM Ousted Over Soaring Food Prices.
[2] F.Hair, J., JR, A. R., Tatham, R. L., & Black, W. C. (1998). *Multivariate Data Analysis.* New Jersey: Prentice Hall International Inc.
[3] FAO. (2015). 1 Oktober 2015. *www.faostat3.fao.org.*
[4] FAO. (2010). *The Rice Crisis: Markets, Policies, and Food Security.* London: Earthscan.
[5] FAO. (2008). *The State of Food Insecurityin the World 2008.* Roma: FAO FIAT PANIS.
[6] Hajarisman, N. (2008). *Statistika Multivariat.* Bandung.
[7] Hidayat, Y. (2015). Developing Unprecendented Restlessness as the New Strong Indicator of Rice Crisis at National Level. *Global Journal of Pureand Applied Mathematics.*
[8] Hidayat, Y. (2013). Forecasting Framework of Rice Crises Time in Bandung-Indonesia.
[9] Johnson, R. A. (2002). *Applied Multivariate Statistical Analysis.* New Jersey: Prentice Hall.
[10] Marshall, M. G. (2014). Polity IV Project: Data User's Manual. 16-17.
[11] Mattjik, A. A. (2011). *Sidik Peubah Ganda.* Bogor: Institut Pertanian Bogor Press.
[12] Mohanty, S. (2013). Trends in Global Rice Consumption.
[13] Mukhopadhyay, P. (2009). *Multivariate Statistical Analysis.* Singapore: World Scientific.
[14] Rencher, A. C. (2002). *Methods of Multivariate Analysis.* Canada: WILEYINTERSCIENCE.
[15] Rokach, O. M. (2005). *Data Mining and Knowledge Discovery Handbook. Secaucus, NJ,* USA: Springer-Verlag New York, Inc.
[16] Srihari, S. (2008). Principal Component Analysis & Multidimensional Scaling.

IOP Conf. Series: Materials Science and Engineering **166** (2017) 012024 doi:10.1088/1757-899X/166/1/012024

The nurse scheduling problem: a goal programming and nonlinear optimization approaches

L. Hakim, T. Bakhtiar, and Jaharuddin

Department of Mathematics, Faculty of Mathematics and Natural Sciences, Bogor Agricultural University, Kampus IPB Dramaga, Bogor 16680, Indonesia

E-mail: tbakhtiar@ipb.ac.id

Abstract. Nurses scheduling is an activity of allocating nurses to conduct a set of tasks at certain room at a hospital or health centre within a certain period. One of obstacles in the nurse scheduling is the lack of resources in order to fulfil the needs of the hospital. Nurse scheduling which is undertaken manually will be at risk of not fulfilling some nursing rules set by the hospital. Therefore, this study aimed to perform scheduling models that satisfy all the specific rules set by the management of Bogor State Hospital. We have developed three models to overcome the scheduling needs. Model 1 is designed to schedule nurses who are solely assigned to a certain inpatient unit and Model 2 is constructed to manage nurses who are assigned to an inpatient room as well as at Polyclinic room as conjunct nurses. As the assignment of nurses on each shift is uneven, then we propose Model 3 to minimize the variance of the workload in order to achieve equitable assignment on every shift. The first two models are formulated in goal programming framework, while the last model is in nonlinear optimization form.

1. Introduction

Staff scheduling is a universal problem that can be encountered in many organizations, such as call centers, educational institution, industry, hospital, and any other public services. It is one of the most important aspects of workforce management strategy and the one that is most prone to errors or issues as there are many entities should be considered, such as the staff turnover, employee availability, time between rotations, unusual periods of activity, and even the last minute shift changes.

The nurse scheduling problem (NSP) is a variant of staff scheduling problems which appoints nurses to shifts as well as rooms per day taking both hard constraints, i.e., hospital requirements, and soft constraints, i.e., nurse's preferences, into account. Thus, the timetable is commonly designed to maximize the preferences of the nurses and to minimize the total penalty cost from violations of the soft constraints [3]. Various models and techniques have been proposed to address the NSP, ranging from simple to advanced approaches.

It is proposed in [4] a simple heuristic approach, flexible and easy enough to be implemented on spreadsheets, and requiring almost no investment. In [5], it is considered an exact branch-and-price algorithm for solving the nurse scheduling problem incorporating multiple objectives and discuss different branching and pruning strategies. While, a search technique for nurse scheduling, which deals with it as a multi-objective problem, based on Pareto search methodology is presented by [2].

In this present work we formulate a specific NSP in a hospital within the framework of goal programming. We aimed to minimize the total penalty cost of violations of the number of days off as well as that of assigned shifts. As an alternate we also proposed a nonlinear optimization problem whose objective is to evenly distribute the workload among nurses by minimizing the variance.

2. Problem description

We considered an NSP at Bogor State Hospital which has 76 nurses, including several chief nurses, to be allocated into 4 inpatient rooms, namely Flamboyant, Vanda, Dahlia, and Pafio. The number of nurses assigned to each room is provided by Table 1. It can be seen from the number of chiefs that nurses in Dahlia room is divided into three teams, while those of Flamboyant and Vanda rooms into two teams, and there is only one team for Pafio room. The number of nurses required in each room and each shift is given by Table 2.

Table 1. The number of nurses available in each room

	Flamboyant	Dahlia	Vanda	Pafio
Nurse	23	20	13	12
Chief nurse	2	3	2	1
Total	25	23	15	13

Table 2. The number of nurses required in each room

Shift	Flamboyant	Dahlia	Vanda	Pafio
Morning	8	7	4	3
Afternoon	6	6	4	3
Night	6	6	4	3

The following regulations are imposed by hospital management and considered as a number of constraints, either hard or soft, in the scheduling models:

1. A day consists of three shifts of seven to ten hours, namely morning shift (7 am to 2 pm), afternoon shift (2 pm to 9 pm), and night shift (9 pm to 7 am).
2. A nurse has a total of 22 to 24 shifts within one scheduling period. In this case study, one scheduling period is equivalent to 28 days.
3. A nurse may have at most 3 consecutive night shifts.
4. A nurse should not be assigned morning shift immediately after night shift.
5. A nurse should have one day-off immediately after 2 consecutive night shifts and 2 consecutive days-off immediately after 3 consecutive night shifts.
6. There is a possibility that a nurse may be on duty for at most 6 consecutive days.
7. Chief nurse should always be assigned in morning shift and have day-off in Sunday.
8. Each nurse should have about 6 days-off within one scheduling period.
9. Each nurse should have about 6 morning shifts, 8 afternoon shifts, and 8 night shifts within one scheduling period.
10. All nurses are assigned in the same room as indicated in Table 1.
11. Beyond the typical constraints commonly imposed by a scheduling problem, we also consider a specific requirement, namely the nurses assigned in Pafio room, excluding the chief nurse, are also designated to have a limited number of shifts in Polyclinic room. This additional work is considered as an overtime.
12. Overtime shifts in Polyclinic room are conducted in Friday and Saturday during the morning and afternoon shifts, involving 4 nurses in each shifts. Each nurse should have 4 to 6 overtime shifts within period.

3. Mathematical models

Since all nurses are assigned in the same room, then the NSP for Flamboyant, Dahlia, and Vanda rooms can be formulated in a single model, i.e., Model 1. While that of Pafio room we introduced

Model 2. Both Models 1 and 2 are expressed in goal programming setting. A nonlinear optimization model is formulated in Model 3, aiming to evenly distribute the workload among nurses. In this development we mainly follow [1].

3.1. Model 1

To facilitate our models, we define the following sets: $\mathbb{D} = \{1,2,\dots,n\}$ the set of all days within one scheduling period, $\mathbb{L} \subset \mathbb{D}$ is the set of all Sundays, $\mathbb{R} = \{1,2,\dots,m\}$ is the set of all nurses, including the chief nurses, and $\mathbb{T} \subset \mathbb{N}$ is the set of all chief nurses. For Flamboyant room we have $m = 25$ and $\mathbb{T} = \{1,14\}$, for Dahlia room we have $m = 23$ and $\mathbb{T} = \{1,9,17\}$, and for Vanda room we have $m = 15$ and $\mathbb{T} = \{1,9\}$. We also define the following decision variables:

$$P_{ij} = \begin{cases} 1, & \text{if nurse } i \text{ is assigned a morning shift at day } j \\ 0, & \text{otherwise,} \end{cases}$$
$$S_{ij} = \begin{cases} 1, & \text{if nurse } i \text{ is assigned an afternoon shift at day } j \\ 0, & \text{otherwise,} \end{cases}$$
$$M_{ij} = \begin{cases} 1, & \text{if nurse } i \text{ is assigned a night shift at day } j \\ 0, & \text{otherwise,} \end{cases}$$
$$H_{ij} = \begin{cases} 1, & \text{if nurse } i \text{ has a day-off at day } j \\ 0, & \text{otherwise.} \end{cases}$$

The abovementioned regulations asked by the hospital management are then expressed in a set of hard and soft constraints alongside with other constraints commonly required in scheduling formulations. The following are the hard constraints:

1. The minimum number of nurses required in each shift every day should be fulfilled:
   $\sum_{i\in\mathbb{R}} P_{ij} \ge b_p, \sum_{i\in\mathbb{R}} S_{ij} \ge b_s, \sum_{i\in\mathbb{R}} M_{ij} \ge b_m, \forall j \in \mathbb{D}$.
   Here we set b_p, b_s, b_m according to Table 2.
2. In a day, a nurse should be assigned only one shift. Otherwise, he/she gets a day-off:
   $P_{ij} + S_{ij} + M_{ij} + H_{ij} = 1, \forall i \in \mathbb{R}, \forall j \in \mathbb{D}$.
3. A nurse should not be assigned morning shift immediately after night shift:
   $M_{ij} + P_{i(j+1)} \le 1, \forall i \in \mathbb{R}, \forall j \in \mathbb{D}$.
4. A nurse is not assigned more than α consecutive night shifts:
   $M_{ij} + M_{i(j+1)} + \dots + M_{i(j+\alpha-1)} \le \alpha, \forall i \in \mathbb{R} - \mathbb{T}, \forall j = 1,2, \dots n - \alpha + 1$.
   In this case we set $\alpha = 3$.
5. There is no possibility that a nurse is on duty more than β consecutive days:
   $H_{ij} + H_{i(j+1)} + \dots + H_{i(j+\beta)} \ge 1, \forall i \in \mathbb{R} - \mathbb{T}, \forall j = 1,2,\dots,n-\beta$,
   where $\beta = 6$.
6. Each nurse should have one day-off immediately after 2 consecutive night shifts and 2 consecutive days-off immediately after 3 consecutive night shifts:
   $M_{ij} + M_{i(j+1)} + M_{i(j+2)} - \left(H_{i(j+1)} + H_{i(j+2)} + H_{i(j+3)} + H_{i(j+4)}\right) \le 1, \forall i \in \mathbb{R} - \mathbb{T}, \forall j = 1,2,\dots,n-4$,
   $M_{ij} + M_{i(j+1)} + S_{i(j+2)} \le 2, \forall i \in \mathbb{R} - \mathbb{T}, \forall j = 1,2,\dots,n-2$.
7. Chief nurse should always be assigned in morning shift and should have day-off in Sunday:
   $\sum_{j\in\mathbb{D}} P_{ij} = n - \gamma, \sum_{j\in\mathbb{L}} H_{i,j} = \gamma, \forall i \in \mathbb{T}$,
   where γ is the number of Sundays within period of scheduling.

The remaining requirements regarding the ideal level for the number of morning, afternoon, and night shifts as well as the number of days-off are then formulated as soft constraints, i.e., goals. Thus, we attempt to achieve the ideal level as close as possible. We define deviation variables d_{ki}^- and d_{ki}^+, which are nonnegative integers, to measure the shortage and excess from the goal, respectively.

1. Each nurse should have about δ morning shifts, ε afternoon shifts, and θ night shifts within one scheduling period:

$\sum_{j\in\mathbb{D}} P_{ij} + d_{1i}^- - d_{1i}^+ = \delta, \sum_{j\in\mathbb{D}} S_{ij} + d_{2i}^- - d_{2i}^+ = \varepsilon, \sum_{j\in\mathbb{D}} M_{ij} + d_{3i}^- - d_{3i}^+ = \theta, \forall i \in \mathbb{R} - \mathbb{T}$.
For our case we have $\delta = 6$, $\varepsilon = 8$, and $\theta = 8$.

2. Each nurse should have about γ days-off within one scheduling period:
$\sum_{j\in\mathbb{D}} H_{ij} + d_{4i}^- - d_{4i}^+ = \gamma,$
where $\gamma = 6$.

By this goal programming framework, we then aim to minimize the shortage of the number of morning shifts and days-off from their ideal levels, as well as the excess of the number of afternoon and night shifts from their ideal levels, measured for all nurses, not including chief nurses. Thus, we consider the following objective function:

$$\min Z := \sum_{i\in\mathbb{R}-\mathbb{T}} (d_{1i}^- + d_{2i}^+ + d_{3i}^+ + d_{4i}^-).$$

3.2. Model 2

Model 2 is devoted to undertake the NSP of Pafio room, where nurses of this room are required to have a number of additional shifts at Polyclinic room, counting as overtime shifts. This problem is obviously an extension of that formulated previously. Thus, it is only needed to add several new constraints to Model 1. For this, we define by $\mathbb{K}$ the set of all days for overtime shift, i.e., Friday and Saturday in the scheduling period. We also introduce the following new decision variables.

$$LP_{ij} = \begin{cases} 1, & \text{if nurse } i \text{ is assigned a morning overtime shift at day } j \\ 0, & \text{otherwise,} \end{cases}$$
$$LS_{ij} = \begin{cases} 1, & \text{if nurse } i \text{ is assigned an afternoon overtime shift at day } j \\ 0, & \text{otherwise.} \end{cases}$$

With the same objective function, we impose the following constraints in addition to those of Model 1.

1. In Friday or Saturday, a nurse should have either morning overtime shift, afternoon overtime shift, night shift, or day-off:
$LP_{ij} + LS_{ij} + M_{ij} + H_{ij} \leq 1, \forall i \in \mathbb{R} - \mathbb{T}, \forall j \in \mathbb{K}$.
2. In Friday or Saturday, morning shift and morning overtime shift cannot be assigned in the same time. So do afternoon shift and afternoon overtime shift:
$P_{ij} + LP_{ij} \leq 1, S_{ij} + LS_{ij} \leq 1, \forall i \in \mathbb{R} - \mathbb{T}, \forall j \in \mathbb{K}$.
3. The number of nurses required for overtime should be fulfilled:
$\sum_{i\in\mathbb{R}-\mathbb{T}} LP_{ij} = b_l, \sum_{i\in\mathbb{R}-\mathbb{T}} LS_{ij} = b_l, \forall j \in \mathbb{K},$
where b_l is the required number of nurses for overtime. Here we have $b_l = 4$.
4. As before, a nurse should not be assigned morning overtime shift immediately after night shift:
$M_{i(j-1)} + LP_{ij} \leq 1, \forall i \in \mathbb{R} - \mathbb{T}, \forall j \in \mathbb{K}$.
5. Each nurse will only have a limited number of overtimes:
$\tau \leq \sum_{j\in\mathbb{K}} (LP_{ij} + LS_{ij}) \leq \varphi, \forall i \in \mathbb{R} - \mathbb{T},$
where $\tau = 4$ and $\varphi = 6$.

3.3. Model 3

Model 3 is intended to evenly distribute the workload and day-off among nurses by minimizing the variance. As the formula of variance is nonlinear, this leads to a nonlinear optimization problem. In fact, we only need to design a new model by keeping all hard constraints prescribed in Model 1 and change its objective function into the following variance formulae:

$$\min Z := \sum_{k=1}^{4} V_k,$$

where

$$V_1 = \frac{\sum_{i\in\mathbb{R}-\mathbb{T}} G_i^2}{m-|\mathbb{T}|} - \left(\frac{\sum_{i\in\mathbb{R}-\mathbb{T}}\sum_{j\in\mathbb{D}} P_{ij}}{m-|\mathbb{T}|}\right)^2, V_2 = \frac{\sum_{i\in\mathbb{R}-\mathbb{T}} N_i^2}{m-|\mathbb{T}|} - \left(\frac{\sum_{i\in\mathbb{R}-\mathbb{T}}\sum_{j\in\mathbb{D}} S_{ij}}{m-|\mathbb{T}|}\right)^2,$$

$$V_3 = \frac{\sum_{i\in\mathbb{R}-\mathbb{T}} B_i^2}{m-|\mathbb{T}|} - \left(\frac{\sum_{i\in\mathbb{R}-\mathbb{T}}\sum_{j\in\mathbb{D}} M_{ij}}{m-|\mathbb{T}|}\right)^2, V_4 = \frac{\sum_{i\in\mathbb{R}-\mathbb{T}} Y_i^2}{m-|\mathbb{T}|} - \left(\frac{\sum_{i\in\mathbb{R}-\mathbb{T}}\sum_{j\in\mathbb{D}} H_{ij}}{m-|\mathbb{T}|}\right)^2.$$

In this formulation, we define by G_i, N_i, B_i, and Y_i represent the number of morning shifts, afternoon shifts, night shifts, and days-off assigned to nurse i during the period, respectively. In this model, those of chief nurses are excluded from the variance calculation as $|\mathbb{T}|$ denotes the cardinality of the set $\mathbb{T}$. Thus, V_i denotes the variance of the workload among nurses within shift i, where i can be morning, afternoon, night, or day-off.

4. Result and discussion

Table 3 provides the difference between the value of objective function Z obtained by the OR models, i.e., Models 1 and 2, and that discovered in manual rooster applied by the hospital.

Table 3. The value of objective function by Models 1 and 2

Method	Inpatient Room			
	Flamboyant	Dahlia	Vanda	Pafio
Manual	29	42	41	20
OR Model	6	38	32	0

Since the objective function Z accounts the shortage of the number of morning shifts and days-off from their ideal levels, as well as the excess of the number of afternoon and night shifts from their ideal levels, it can be shown that in general by using OR modeling we are able to minimize the shortage of the number of morning shifts and days-off as it is preferable to have more and reduce the excess of the number of afternoon and night shifts it is preferable to have less.

Table 4. The percentage level of constraints accomplishment

No	Component	Flamboyant		Dahlia		Vanda	
		Manual	Model 1	Manual	Model 1	Manual	Model 1
1	Fulfillment of the required nurses in						
	morning shift	82.14	100	89.29	100	100	100
	afternoon shift	82.14	100	25	100	82.14	100
	night shift	100	100	100	100	100	100
2	Total of 22-24 shifts in one period	96	100	73.91	100	93.33	100
3	Day-off in Sunday for chief nurses	100	100	100	100	100	100
4	Overtime	0	0	100	0	38.46	0
5	Duties more than 6 consecutive days	13.04	0	20	0	7.69	0
6	Duties more than 3 consecutive night shifts	0	0	5	0	0	0
7	One day-off after 2 consecutive night shifts	100	100	100	100	100	100
8	Two consecutive days-off after 3 consecutive night shifts	91.67	100	90	100	53.85	100
9	No morning shift immediately after night shift	100	100	100	100	100	100

Table 4 provides the percentage level of constraints accomplishment by using manual rooster and by performing OR model. Obviously utilizing OR models give a better achievement. By OR model we can assure that there will be sufficient number of nurses in each shift, while manual rooster it is shown that it was an inadequate number of nurses in Flamboyant, Dahlia, and Vanda rooms. It is also guaranteed by the model that all nurses will have 22-24 shifts within one scheduling period, while manual scheduling of Dahlia room shows a 74 percent of nurses with the preferred number of shifts. Table 5 presents the shift, overtime, and day-off distribution among nurses in Pafio room by running Model 2.

Table 5. The shift, overtime, and day-off distribution in Pafio room

Nurse	Shift			Overtime		Day-off
	Morning	Afternoon	Night	Morning	Afternoon	
1*	24	0	0	0	0	4
2	6	8	7	2	3	7
3	6	7	8	2	2	7
4	6	8	8	2	2	6
5	7	8	7	3	3	6
6	10	8	4	4	2	6
7	6	8	8	3	3	6
8	6	8	7	4	2	7
9	8	8	6	2	4	6
10	6	8	8	3	1	6
11	8	8	6	2	4	6
12	6	8	8	2	3	6
13	6	7	8	3	3	7

*: chief nurse, not assigned to have overtime.

Table 6. The variance of workload and day-off

Inpatient Room	Method	Shift			
		Morning	Afternoon	Night	Day-off
Vanda	Manual	3.08	3.69	0.59	0.17
	Model 1	0.76	0.42	0.59	0.59
	Model 3	0	0	0	0
Pafio	Manual	4.16	5.25	0	0.47
	Model 2	1.66	0.15	1.5	0.24
	Model 3	0	0	0	0

Even though nurse scheduling by using OR model provides better solutions than manual approach and satisfies all requirements addressed by the hospital, there still exists some less preferable facts regarding the distribution of shifts and days-off among nurses. For example, nurse scheduling in Vanda room by using Model 1 provides variance of 0.59 on the number of night shifts and days-off as depicted by Table 6. It means that nurses responsible in this room possess different number of night shifts and days-off, where the former is ranging from 8 to 10 times within period and the later is ranging from 4 to 6 times. Of course, having more night shifts but less days-off is less preferable for most nurses.

Model 3 is developed to reduce the difference in shift workload and day-off among nurses by minimizing the variance. Table 6 shows that, in the case of Vanda and Pafio rooms, the variance can be minimized to zero, indicating that all nurses, not including the chief nurses, have the same number of morning, afternoon, night, and day-off shifts. However, this fair distribution of workload and day-off is compensated by the increase of computation time. Regarding this drawback, Model 3 can still be considered as an alternate model in preparing nurses timetable.

IOP Conf. Series: Materials Science and Engineering **166** (2017) 012024 doi:10.1088/1757-899X/166/1/012024

5. Conclusion

In this paper we have developed three mathematical models to address the nurse scheduling problem arising at Bogor State Hospital. The first two models are formulated in term of goal programming and the last model is a nonlinear programming of variance minimization. It is shown that all conditions required by the hospital can be fulfilled by Models 1 and 2 and the variance due to shift workload and day-off among nurses can be minimized by Model 3.

References

[1] M.N. Azaiez and S.S. Al-Sharif, Comp. Oper. Res. **32**, 491-507 (2005).
[2] E.K. Burke, J. Li , and R. Qu, Ann. Oper. Res. **196**(1), 91-109 (2012).
[3] A.A. Constantino, D. Landa-Silva, E.L. de Melo, C.F.X. de Mendonça, D.B. Rizzato, and W. Romão, Ann. Oper. Res. **218**(1), 165-183 (2013).
[4] A. Legrain, H. Bouarab, and N. Lahrichi, J. Med. Syst. **39**, 160 (2015).
[5] B. Maenhout and M. Vanhoucke, J. Sched. **13**, 77-93 (2010).

Courses timetabling problem by minimizing the number of less preferable time slots

M. Oktavia, A. Aman and T. Bakhtiar

Division of Operations Research, Department of Mathematics, Faculty of Mathematics and Natural Sciences, Bogor Agricultural University, Indonesia

Abstract. In an organization with large number of resources, timetabling is one of the most important factors of management strategy and the one that is most prone to errors or issues. Timetabling the perfect organization plan is quite a task, thus the aid of operations research or management strategy approaches is obligation. Timetabling in educational institutions can roughly be categorized into school timetabling, course timetabling, and examination timetabling, which differ from each other by their entities involved such as the type of events, the kind of institution, and the type and the relative influence of constraints. Education timetabling problem is generally a kind of complex combinatorial problem consisting of NP-complete sub-problems. It is required that the requested timetable fulfills a set of hard and soft constraints of various types. In this paper we consider a courses timetabling problem at university whose objective is to minimize the number of less preferable time slots. We mean by less preferable time slots are those devoted in early morning (07.00 – 07.50 AM) or those in the late afternoon (17.00 – 17.50 AM) that in fact beyond the working hour, those scheduled during the lunch break (12.00 – 12.50 AM), those scheduled in Wednesday 10.00 – 11.50 AM that coincides with Department Meeting, and those in Saturday which should be in fact devoted for day-off. In some cases, timetable with a number of activities scheduled in abovementioned time slots are commonly encountered. The courses timetabling for the Educational Program of General Competence (PPKU) students at odd semester at Bogor Agricultural University (IPB) has been modelled in the framework of the integer linear programming. We solved the optimization problem heuristically by categorizing all the groups into seven clusters.

1. Introduction

The timetabling is one of the problem that often encountered in arranging the assignments at an organisation includes educational institutions. When we want to solved problem used a design of mathematical model, then we need to consider in depth the various kinds of constraints related to their policies or regulations. Thus, being able to create a timetable that is as ideal as possible.

Even though not so complicated as in the university, timetabling issues at the school level are still interesting to be studied because it can provide the solution with a better quality. Such as the research conducted by Birbas *et al.* [1] which solved the problem assignments of optimal working shift from each teacher. The problem is modelled using an integer programming framework. Cangaiovic and Schreuder [9] proposed a special case of a teacher-class timetabling problem. This case is considering a partial request between the topics in the curriculum and specific requirements based on their daily lectures. The problem is modelled as a discrete lexicography model with heuristic procedure. This method is a combination of two different approaches, specifically: on the level of public used a specific heuristic approach and at the level of daily used graph colouring method. Another timetabling problem at the high school level covered by Saviniec *et al.* [9] that applied the three algorithms of

Iterated Local Search (ILS) includes two operators of new neighbourhood which be proposed to solve the problem from literature heuristically. This study examined seven cases. The results showed that this method is effective and efficient to solve the problem, because these methods are able to find the optimal solution for all cases.

The 0-1 integer programming model of timetabling problems at the university presented by Daskalaki *et al.* [2]. The objective function of the problem formulation is to minimize the cost function. This model is used to solve timetabling cases at the Department of Engineering for five years with a large number of lectures and lecturers. Dimopoulou and Miliotis [3] reported the results of the design and implementation of PC-based computer systems to help the construction of a combined schedule of lectures and exams at the university. The difficulties to be found is the limited availability of classroom space and increase the flexibility of the courses that can be elected by students, which makes the problem becomes very tight. The system used integer programming (IP) model that assigned the course in a specific both the time slot and room. This study has been successfully applied at the Athens University of Economics and Business. In other cases, Schimmelpfeng and Helber [7] solved timetabling problems using integer programming that has been implemented at the Economics and Management Faculty of Hannover University. A problem of determining the course time slot in a faculty studied by Ismayilova *et al.* [5] also developed timetabling of courses as integer programming problem with the multi-objective functions. The model is designed considering both the administration's and instructor's preferences. Both modelling and solving such problem are difficult task due to the size, the varied nature, and conflicting objectives of the problem. Analytic Hierarchy Process (AHP) and Analytic Network Process (ANP) is used to consider the purpose of different and contradictory.

The courses timetabling problem at university with large numbers of students is a difficult task, particularly to ensure that there are no courses which overlap. Most timetabling prepared manually, or at least with the aid of a spreadsheet program. However, the manual process requires verification number of experts (e.g. lecturers and supervisors) before it can be approved. Thus, several researchers solved the problem with heuristically analysis. In this research, proposed a solution of the courses timetabling problem at university. Various soft and hard constraints of timetabling parameters considered, such as the number of subjects, tutorials, classrooms, teachers, students, and workloads. Simulated annealing method is used to obtain optimal solutions and sub-optimal. Illustration was applied to the timetabling problem of Tamhidi Program in Universiti Sains Islam Malaysia as a case study [12]. Hertz [8] defined the daily quantum as a number of total period that scheduled into consecutive period a day for each chapter. Daily quantum has a minimum and maximum number of courses each day, so there are no arbitrary chapter on every single day. The completion of this problem using Tabu Search Method for obtaining a schedule of feasible lectures. Studenovsky [11] arranged the University Course timetabling Problem (UCTP). In particular, this study presented UCTP into two issues, as the timetabling of time slot and the timetabling of room. Researchers proved that in this case UCTP is a polynomial form that be reduced.

Hanum *et al.* [4] considered a capacitated timetabling problem of the invigilator exam. By using the framework of Goal Programming in formulating the problem, where a number of constraints associated with this type of exam, exam controller type, time availability, and several other preferences classified into primary and secondary constraints. The model was applied into a simple case of the exam invigilator in the Department of Mathematics, Bogor Agricultural University. A study at the University Malaysia Pahang done by Kahar and Kendall [6] modelled the capacitated examination problems with calculating the distance between the rooms and divided the test in several rooms. Construction heuristically is built to produce solution with good quality for real world problems that be encountered. Fahrion and Dollansky [10] proposed both the room and lecturer based on the faculty educational plan and the selection of fixed assignment. To speed up the process of finding a solution, then is made the of priority scheme assumption of simple heuristic.

In this paper, we consider the courses timetabling of Educational Program of General Competence (PPKU) that be carried out by the Directorate of PPKU used manual timetabling system. The constraint that often hamper the timetable process is there are no warnings at software Microsoft Excel if courses timetable is overlap. So that, the schedule-makers must made the timetabling carefully and checking manually in order to obtain a schedule that can be applied without any courses those collide. Because of the currently timetabling method is still using manual method, it will hinder the directorate to produce the timetable of courses would be desirable.

Therefore, the authors tried to design the courses timetabling of PPKU at first semester with framework of integer linear programming to minimize the number of courses at less preferred time slots. The design of this model, hoping to eliminate some types of schedules those less effective, such as the tutorial day that be scheduled before its lecture day for each of the corresponding subjects, and subjects which has two classes of tutorial those be held on different both the day or time slot.

The outline of the paper is as follows: The problem formulation and the corresponding mathematical model are presented in Section 2. Section 3 provides an illustrative example. Section 4 develops the solution approach and numerical results. Some conclusions drawn from the study are presented in Section 5.

2. Problem formulation

The timetabling model involves about 3600 students of PPKU that has been organized into 34 groups, when each group will take a different set of courses. Every group is then classified into seven classes, each class will take the same set of courses. Among 34 groups, we have 9 groups of agriculture class, 9 groups of science and technology class, 6 groups of economy and management class, 2 groups of social and humanity class, 5 groups of flora, fauna, and human class, 2 groups of chemical and biochemical class, and 1 group of international class. Overall, it is offered 35 different courses that will be organized in 26 large rooms of capacity 130, 7 small rooms of capacity 80, 1 Laboratory of Biology, 1 Laboratory of Physic, 1 Laboratory of Chemistry, 1 mosque for Islamic Religious Education tutorial, and 1 gymnasium for sport. The courses will be scheduled at six work days (Monday-Saturday) with eleven time slots a day. Duration of each time slot is fifty minutes.

Each group is scheduled to follow a number of courses. Each group those are in the same class will take same of courses. Each course has different time of face to face. A few courses must be scheduled only once time of slot, but should be scheduled more than once time at respectively time slot for anyone else. There are courses those be accompanied by tutorial or practicum. For the subjects those be accompanied by tutorial, implementation of the lecture day should come before its tutorial day for each of the corresponding subjects. Each tutorial those has two classes are scheduled on the same both of the day and time slot for the same subjects, but at the different of rooms. Timetabling is designed certainly that each course which be taken by each group are timetabled once a week. The learning activities are not recommended on Friday (11.00 AM - 13.00 PM). The courses on the less preferred time slots are expected to be avoided or be minimized. The number of courses at every day should be limited, so that students can be absorbing the materials of learning optimally.

Modelling of the problem is encountered by defined a few of indices, parameters, and variables. Further described the constraints in the form of mathematical equality or inequality which gave a demonstration of rules and regulations which applicable at that institution. Recently presented the objective function which stated the purpose of optimization that will be established.

2.1 Parameters

Let

$I = \{i : i = 1, 2, \ldots, p\}$ be the set of days,

$J = \{j : j = 1, 2, \ldots, q\}$ be the set of time slots,

$U_b = \{(i, j) : (i, j) = 1, 2, \ldots, u_b\}$ be the set days and time slots which have been used by previously cluster at Laboratory of Biology,

$U_p = \{(i, j) : (i, j) = 1, 2, \ldots, u_p\}$ be the set days and time slots which have been used by previously cluster at Laboratory of Physics,

$U_k = \{(i, j) : (i, j) = 1, 2, \ldots, u_k\}$ be the set days and time slots which have been used by previously cluster at Laboratory of Chemistry,

$K = \{k : k = 1, 2, \ldots, r\}$ be the set of groups,

$L = \{l : l = 1, 2, \ldots, s\}$ be the set of courses,

$\bar{L}_k = \{\bar{l}_k : \bar{l}_k = 1, 2, \ldots, \bar{s}\}$ be the set of courses which should be taken by group k,

$\tilde{L}_k$={$\tilde{l}_k$: $\tilde{l}_k$ = 1,2, . . . ,$\tilde{s}$ } be the set of courses which should not be taken by group *k*,
$\dot{L}_k$={$\dot{l}_k$: $\dot{l}_k$ = 1,2, . . . ,$\dot{s}$ } be the set of lectures which should be taken by group *k*,
$\ddot{L}_k$={$\ddot{l}_k$: $\ddot{l}_k$ = 1,2, . . . ,$\ddot{s}$ } be the set of tutorial one which should be taken by group *k*,
$\dddot{L}_k$={$\dddot{l}_k$: $\dddot{l}_k$ = 1,2, . . . ,$\dddot{s}$ } be the set of tutorial two which should be taken by group *k*,
$\hat{L}_k$={$\hat{l}_k$: $\hat{l}_k$ = 1,2, . . . ,$\hat{s}$ } be the set of practicums which should be taken by group *k*,
L_o= {l_o : l_o = 1,2, . . .,s_o} be the set of tutorial one that have one class,
L_t= {l_t : l_t = 1,2, . . .,s_t} be the set of tutorial one that have two class,
N = {n : n = 1, 2, . . . , t} be the set of rooms,
N_b= {n_b: n_b= 1,2, . . . ,u} be the set of large rooms, $n_b \in n$
N_s= {n_s: n_s= 1,2, . . . ,v} be the set of small rooms, $n_s \in n$
N_p= {n_p: n_p= 1,2, . . . ,w} be the set of practicum rooms, $n_p \in n$
t = {t : t = 1,2,3} be the time of face to face,
M be the big number,
y_{min} be the minimum number of courses that might be assigned to each group during the day,
y_{max} be the maximum number of courses that might be assigned to each group during the day,
$sesi1_{min}$ be the minimum number of courses at first time slot that might be assigned to each group during the week,
$sesi1_{max}$ be the maximum number of courses at first time slot that might be assigned to each group during the week,
$sesi11_{min}$ be the minimum number of courses at last time slot that might be assigned to each group during the week,
$sesi11_{max}$ be the maximum number of courses at last time slot that might be assigned to each group during the week,
d_l be the time of face to face for courses l.

2.2 Decision variables

x_{lkijn}: 1 if the course l is scheduled for group k on day i time slots j in the room n and 0 if others.
y_{lkin}: 1 if the course l is scheduled for group k on day i in the room n and 0 if others.

2.3 Objective functions

- Minimize S1, where S1 is the courses of early morning (07.00 AM - 07.50 AM)

$$\sum_l \sum_k \sum_i \sum_n x_{lkijn} = S_1, \qquad \forall j = 1, j \in J \tag{1}$$

- Minimize S6, where S6 is the courses of lunch break time (12.00 AM - 12.50 AM)

$$\sum_l \sum_k \sum_i \sum_n x_{lkijn} = S_6, \qquad \forall j = 6, j \in J \tag{2}$$

- Minimize S11, where S11 is the courses of late afternoon (17.00 AM - 17.50 AM)

$$\sum_l \sum_k \sum_i \sum_n x_{lkijn} = S_{11}, \qquad \forall j = 11, j \in J \tag{3}$$

- Minimize R, where R is the courses of Wednesday meeting (10.00 AM - 11.50 AM)

$$\sum_l \sum_k \sum_n x_{lkijn} = R, \qquad \forall i = 3, j = 4 - 5, i \in I, j \in J \tag{4}$$

- Minimizing S, where S is the courses of Saturday.

$$\sum_{l}\sum_{k}\sum_{n} x_{lkijn} = S, \qquad \forall\, i = 6, j \in J \tag{5}$$

Formulation of the integer linear programming model of the problem under consideration is given below.
Minimize

$$S_1 + S_6 + S_{11} + R + S \tag{6}$$

2.4 Constraints
Based on the terms and conditions of the timetabling, then there are several constraints as follow.

$$\sum_{\bar{l}_k}\sum_{n} x_{\bar{l}_k kijn} \le 1, \qquad \forall i, j, k \tag{7}$$

$$\sum_{l}\sum_{k} x_{lkijn} \le 1, \qquad \forall i, j, n \tag{8}$$

$$\sum_{i}\sum_{j}\sum_{n} x_{lkijn} = d_l, \qquad \forall l, k \tag{9}$$

$$\sum_{j} x_{lkijn} = y_{lkin} d_l, \qquad \forall l, k, i, n \tag{10}$$

$$x_{lki1n} \le x_{lkitn}, \qquad \forall l, k, i, 1 \le t \le d_l, n \tag{11.1}$$

$$x_{lki(j+1)n} \le x_{lkijn} + x_{lki(j+t)n}, \qquad \forall l, k, i, j + t \le 10, 1 \le t \le d_l, n \tag{11.2}$$

$$x_{lki10n} \le x_{lki(10-t)n}, \qquad \forall l, k, i, 1 \le t \le d_l - 1, n \tag{11.3}$$

$$\sum_{i}\sum_{n} y_{\bar{l}_k kin} = 1, \qquad \forall \bar{l}_k, k \tag{12.1}$$

$$\sum_{n_b} y_{\bar{l}_k kin_b} = 1, \qquad \forall \bar{l}_k = 2, k = 1, i = 1 \tag{12.2}$$

$$\sum_{\ddot{l}_k}\sum_{n_l = n_{\dot{l}} + n_{\ddot{l}}} x_{\ddot{l}_k kijn_l} - M\left(\sum_{\ddot{l}_k}\sum_{n_l = n_{\dot{l}} + n_{\ddot{l}}} x_{\ddot{l}_k kijn_l}\right) \le 0, \quad \forall\, i, j, k \tag{13}$$

$$x_{lkijn} = 0, \qquad \forall l, k, i = 5, j = 5\&6, n \tag{14}$$

$$x_{\tilde{l}_k kijn} = 0, \qquad \forall \tilde{l}_k, k, i, j, n \tag{15}$$

$$x_{\dot{l}_k kijn} = 0, \qquad \forall \dot{l}_k, k, i, j, n = n_s, n_p \tag{16.1}$$

$$x_{l_o kijn} = 0, \qquad \forall l_o, k, i, j, n = n_s, n_p \tag{16.2}$$

$$x_{l_t kijn} = 0, \qquad \forall \dot{l}_t, k, i, j, n = n_p \tag{16.3}$$

$$x_{\hat{l}_k kijn} = 0, \qquad \forall \hat{l}_k, k, i, j, n = n_b, n_s \tag{16.4}$$

$$2 \le \sum_{l}\sum_{n} y_{lkin} \le 4 \qquad \forall k, i \tag{17}$$

$$2 \le \sum_{l}\sum_{i}\sum_{n} x_{lkijn} \le 4 \qquad \forall k, j = 1 \tag{18}$$

$$2 \le \sum_{l}\sum_{i}\sum_{n} x_{lkijn} \le 4 \qquad \forall k, j = 11 \tag{19}$$

$$\left(\sum_{k} x_{3,k,i,j,6} + \sum_{k} x_{9,k,i,j,7} + \sum_{k} x_{23,k,i,j,8} + \sum_{k} x_{33,k,i,j,8}\right) = 1, \qquad \forall\, i \le 5, j = 1 \tag{20}$$

$x_{\hat{l}_k kijn_p} = 0,$ $\forall (i,j) \in U_b$ (21.1)
$x_{\hat{l}_k kijn_p} = 0,$ $\forall (i,j) \in U_p$ (21.2)
$x_{\hat{l}_k kijn_p} = 0,$ $\forall (i,j) \in U_k$ (21.3)
$\sum_k \sum_n x_{\dot{l}_k kijn} \leq 4,$ $\forall \dot{l}_k, i, j$ (22)
$x_{lkijn} = \{0,1\},$ $\forall l, k, i, j, n$ (23.1)
$y_{lkin} = \{0,1\},$ $\forall l, k, i, n$ (23.2)

The constraint sets (7) ensure that each group is scheduled to follow exactly one course on its time slot for every day. The constraint set (8) guarantees that each room used by exactly one group on its time slot for every day. Based on the capacity and utility, we can categorized the rooms into several types, as follow.

- Large rooms: useable for lectures (the courses which be delivered by lecturers, not by assistants. Where those being assistants are students from second, third, or fourth levels. We can called as mainly courses), tutorials (the courses which be delivered by assistants. But, a few tutorial be delivered by lecturers. We can called as additional courses). Each tutorial has one class or two classes.
- Small rooms: useable for a tutorials that has two classes.
- A Laboratory of Biology: useable for practicums of Biology.
- A Laboratory of Physics: useable for practicums of Physics.
- A Laboratory of Chemistry: useable for practicums of Chemistry and Elementary Chemistry.
- An Al-Hurriyyah Mosque: useable for tutorial of Islamic Educational.
- A Gymnasium: useable for practicums of sport.

The constraint set (9) and (10) guarantees that each course should be scheduled based on the time of face to face for each group and each course are scheduled exactly once a day respectively. The constraint set (11.1), (11.2), and (11.3) respectively guarantees that each course with time of face to face just over one hour are scheduled at respectively time slots. There are three possibilities conditions for scheduled each course with time of face to face more than one hour as follow.

- The courses might be scheduled in the first slot,
- The courses might be scheduled not both in the first and the last slot, or
- The courses might be scheduled in the last slot.

The constraint sets (12.1) ensure that each course which be taken are scheduled exactly once a week, the constraint sets (12.2) guarantees each course those be accompanied by tutorial, then the lecture day is scheduled before tutorial day for each of the corresponding subjects. For example, lecture Biology would be taken by group P1 be scheduled at Monday in one of large rooms, we defined as constraint (12.2). For another courses be scheduled at the fixed time like we listed at Table 5. The constraint sets (13) ensure that the tutorial one and the tutorial two held on the same both the day and time slot. There are no courses on Friday at 11.00 AM - 13.00 PM which is stated on the constraint sets (14). The constraint sets (15) ensure that if there are subjects those should not be taken for each group, so they will not be scheduled. The constraint set (16.1), (16.2), (16.3), and (16.4) respectively guarantees that each mainly courses are not scheduled on the small rooms and laboratories, each tutorial those has one class is not scheduled on the small rooms and the laboratories, each tutorial those has two classes are not scheduled on the laboratories, and each practicum are not scheduled on the large and small rooms. The constraint set (17), (18), and (19) guarantees that each student attend minimum two courses and a maximum four courses for every day, each student attend minimum two courses and a maximum four courses at the first time slots for every day, and each student attend minimum two courses and a maximum four courses at the last time slot every day respectively. The constraint sets (20) ensure that each laboratory starts at 07.00 AM in Monday - Friday, this constraint is applied to Cluster One through Cluster Three. The constraint set (21.1), (21.2) and (21.3) respectively guarantees that each group not used Laboratory of Biology, Laboratory of Physics, and Laboratory of Chemistry both on the days and time slots those had been scheduled for the previously clusters. The constraint sets (22) ensure that each mainly course with the same subject are scheduled maximum four times on same of the day and time slot, it is associated with the availability of the lecturers. The constraint sets (23.1) and (23.2) ensure that $x_{lkijn}$and y_{lkin}decision variable integer zero or one.

3. An application—a demonstrative example

We consider a particular problem with 35 courses, 34 groups, 6 days, 11 time slots, and 6 lecturers. The courses that must be taken by each group is shown in Table 1.

Table 1. The courses that must be taken by each group.

Index Number	Course	Type of Course	d_l	Group						
				P1 - P9	Q1 - Q9	R1 - R6	S1 - S2	T1 - T5	U1 - U2	Int
1	Pengantar Kewirausahaan	K	2		√	√	√		√	
2	Biologi	K	2	√				√		√
3		P	3	√				√		√
4	Biologi Umum	K	2			√	√			
5	Ekonomi Umum	K	2	√			√			
6		R1	2	√			√			
7		R2	2	√			√			
8	Fisika	K	2		√				√	
9		P	3		√				√	
10	Pendidikan Agama Islam	K	2	√		√		√		√
11		R	2	√		√		√		
12	Bahasa Indonesia	K	1	√			√	√		√
13		R1	2	√			√	√		√
14		R2	2	√			√	√		√
15	Pengantar Ilmu Pertanian	K	2	√	√	√	√	√	√	√
16	Bahasa Inggris	K	2	√			√	√	√	√
17		R1	2	√			√	√	√	√
18		R2	2	√			√	√	√	
19	Pendidikan Pancasila	K	1		√	√				
20		R1	2		√	√				
21		R2	2		√	√				
22	Kimia	K	2		√			√		√
23		P	3		√			√		√
24	Sosiologi Umum	K	2		√	√	√		√	
25		R1	2		√	√	√		√	
26		R2	2		√	√	√		√	
27	Pengantar Matematika	K	2		√	√			√	
28		R1	2		√	√			√	
29	Landasan Matematika	K	2	√						
30		R1	2	√						
31	Fisika Umum	K	2			√	√			
32	Kimia Dasar	K	2						√	
33		P	3						√	
34	Olah Raga	P	2		√			√	√	√
35	Olah Raga	P	2	√	√	√	√	√	√	√

Note: d_l = time of face to face, K = Lecture, R1 = The tutorial 1, R2 = The tutorial 2, and P = Practicum.

4. The solution

Because of the desired result is not obtained by once modeling, thus we break down the problem into seven clusters where each cluster consists of several groups and rooms so there are no intersection between one room and the others. Each program of the clusters will be ran by gradually process. The partition of the members and the rooms which can be used by each cluster is shown in Table 2 and Table 3 respectively. However, every laboratory can be used by all of cluster at the different both of day and time slot.

Table 2. Members of each cluster.

Index Number	Cluster 1	Cluster 2	Cluster 3	Cluster 4	Cluster 5	Cluster 6	Cluster 7
1	P1	P2	P3	P4	P5	P6	P8
2	Q1	Q2	Q3	Q4	Q5	P7	P9
3	R1	R2	R3	R4	R5	Q6	Q8
4	S1	S2	U1	U2	Int	Q7	Q9
5	T1	T2	T3	T4	T5	R6	

Table 3. Room for each cluster.

Index Number	Cluster 1	Cluster 2	Cluster 3	Cluster 4	Cluster 5	Cluster 6	Cluster 7
1	ICCR 101	ICCR 105	ICCR 109	ICCR 203	ICCR 207	ICCR 211	ICCR 215
2	ICCR 102	ICCR 106	ICCR 110	ICCR 204	ICCR 208	ICCR 212	ICCR 216
3	ICCR 103	ICCR 107	ICCR 201	ICCR 205	ICCR 209	ICCR 213	RTL 2.05
4	ICCR 104	ICCR 108	ICCR 202	ICCR 206	ICCR 210	ICCR 214	LAB BIO
5	RTL 1.01	RTL 2 FIS	RTL 2.01	RTL 2.02	RTL 2.03	RTL 2.04	LAB FIS
6	LAB BIO	LAB BIO	LAB BIO	LAB BIO	LAB BIO	LAB BIO	LAB KIM
7	LAB FIS	LAB FIS	LAB FIS	LAB FIS	LAB FIS	LAB FIS	AlHur
8	LAB KIM	LAB KIM	LAB KIM	LAB KIM	LAB KIM	LAB KIM	Gym
9	AlHur	AlHur	AlHur	AlHur	AlHur	AlHur	
10	Gym	Gym	Gym	Gym	Gym	Gym	

In reality there are still exist the tutorial's day that become before its lecture's day for each of the corresponding subjects as listed in Table 4.

Table 4. Class which tutorial before its lecture.

Course	Group
Pengantar Matematika	Q8
Pengantar Matematika	R1
Bahasa Indonesia	T4
Pengantar Matematika	U1

With operations research modeling is expected there are no such that conditions. Therefore, the authors conducted a lecture and tutorial time scheduling so that making lectures held on the day before the day of the tutorial for each of the corresponding subjects as attached in Table 5. At modeling of mathematic be presented in constraint twelve.

Table 5. The schedule time of the lecture and tutorial day for each of the corresponding subjects to each cluster.

Index Number	Course	Type of Course	Cluster 1 2	Cluster 3 4	Cluster 5 6 7
1	Pengantar Kewirausahaan	K	Monday	Tuesday	Wednesday
2	Biologi	K	Monday	Tuesday	Wednesday
4	Biologi Umum	K	Monday	Tuesday	Wednesday
5		K	Monday	Tuesday	Wednesday
6	Ekonomi Umum	R1	After Monday	After Tuesday	After Wednesday
7		R2	After Monday	After Tuesday	After Wednesday
8	Fisika	K	Monday	Tuesday	Wednesday
10	Pendidikan Agama Islam	K	Tuesday	Wednesday	Monday
11		R	After Tuesday	After Wednesday	After Monday
12	Bahasa Indonesia	K	Tuesday	Wednesday	Monday
13		R1	After Tuesday	After Wednesday	After Monday
14		R2	After Tuesday	After Wednesday	After Monday
15	Pengantar Ilmu Pertanian	K	Tuesday	Wednesday	Monday
16		K	Tuesday	Wednesday	Monday
17	Bahasa Inggris	R1	After Tuesday	After Wednesday	After Monday
18		R2	After Tuesday	After Wednesday	After Monday
19	Pendidikan Pancasila		Tuesday	Wednesday	Monday
20			After Tuesday	After Wednesday	After Monday
21			After Tuesday	After Wednesday	After Monday
22	Kimia	K	Wednesday	Monday	Tuesday
24		K	Wednesday	Monday	Tuesday
25	Sosiologi Umum	R1	After Wednesday	After Monday	After Tuesday
26		R2	After Wednesday	After Monday	After Tuesday
27	Pengantar Matematika	K	Wednesday	Monday	Tuesday
28		R1	After Wednesday	After Monday	After Tuesday
29	Landasan Matematika	K	Wednesday	Monday	Tuesday
30		R1	After Wednesday	After Monday	After Tuesday
31	Fisika Umum	K	Wednesday	Monday	Tuesday
32	Kimia Dasar	K	Wednesday	Monday	Tuesday

5. Conclusions

Our result, whose summary is provided in Table 6, reveals that timetabling based on operations research may reduce the number of course sessions which scheduled in less preferable time slots. For instance, now there are only 19 course sessions scheduled in the early morning slot, which is equivalent to a 68 percent reduction compared to that produced manually. Moreover, a hundred percent reduction attained for managing late afternoon sessions.

Table 6. The number of less preferable time slots by manual and operation research methods.

Less Preferable Time Slot	Time	Number of sessions		Reduction
		Manual Method	OR Method	
Early morning	07.00 – 07.50	59	19	68%
Lunch break time	12.00 – 12.50	22	3	86%
Late afternoon	17.00 – 17.50	10	0	100%
Wednesday meeting	10.00 – 11.50	30	10	67%
Saturday	all time	9	7	22%

References

[1] T. Birbas, S. Daskalaki, and E. Housos, School timetabling for quality student and teacher schedules, *Journal of Scheduling*, 12 (2009), 177-197, doi: 10.1007/s10951-008-0088-2
[2] S. Daskalaki, T. Birbas, E. Housos, An integer programming formulation for a case study in university timetabling, *European Journal of Operational Research*, 153 (2004), 117-135.
[3] M. Dimopoulou, P. Miliotis, Implementation of a university course and examination timetabling system, *European Journal of Operational Research*, 130 (2001), 202-213.
[4] F. Hanum, M.A. Romliyah, T. Bakhtiar, exam invigilators assignment problem: a goal programming approach, *Applied Mathematical Sciences*, 9(58) (2015), 2871-2880, doi: 10.12988/ams.2015.52134.
[5] N.A. Ismayilova, M. Sagir, R.N. Gasimov, A multiobjective faculty-course-time slot assignment problem with preferences, *Journal of Mathematical and Computer Modelling*, 46 (2007), 1017-1029.
[6] M.N.M. Kahar and G. Kendall, The examination timetabling problem at Universiti Malaysia Pahang: comparison of a constructive heuristic with an existing software solution, *European Journal of Operational Research*, 207(2) (2010), 557-565, doi: 10.1016/j.ejor.2010.04.011.
[7] K. Schimmelpfeng, S. Helber, Application of a real-world university-course timetabling model solved by integer programming, *OR Spectrum*, 29 (2007), 783-803.
[8] A. Hertz, Finding a feasible course schedule using Tabu search, *Discrete Applied Mathematics*, 35 (1992), 255-270.
[9] M. Cangaiovic, J.A.M. Schreuder, Modelling and solving an acyclic multi-period timetabling problem, *Discrete Applied Mathematics*, 35 (1992), 177- 195.
[10] R. Fahrion, G. Dollansky, Construction of university faculty timetables using logic programming techniques, *Discrete Applied Maahematics*, 35 (1992), 221-236.
[11] J. Studenovský, Polynomial reduction of time_space scheduling to time scheduling, *Discrete Applied Mathematics*, 157 (2009), 1364-1378.
[12] N. Basir, W. Ismail, N.M. Norwawi, A Simulated Annealing for Tahmidi Course Timetabling, *Elsevier B.V*, 2013.
[13] L. Saviniec, A.A. Constantino, W. Romao, Solving the high school timetabling problem to optimality by using ils algorithms, Simpósio Brasileiro de Pesquisa Operacional, 2013.

Green supply chain management strategy selection using analytic network process: case study at PT XYZ

W. Adelina and R.D. Kusumastuti*

Department of Management, Faculty of Economics and Business, Universitas Indonesia, Kampus UI Depok 16424, Indonesia

*Email: ratih.dyah@ui.ac.id

Abstract. This study is about business strategy selection for green supply chain management (GSCM) for PT XYZ by using Analytic Network Process (ANP). GSCM is initiated as a response to reduce environmental impacts from industrial activities. The purposes of this study are identifying criteria and sub criteria in selecting GSCM Strategy, and analysing a suitable GSCM strategy for PT XYZ. This study proposes ANP network with 6 criteria and 29 sub criteria, which are obtained from the literature and experts' judgements. One of the six criteria contains GSCM strategy options, namely *risk-based strategy, efficiency-based strategy, innovation-based strategy,* and *closed loop strategy*. ANP solves complex GSCM strategy-selection by using a more structured process and considering green perspectives from experts. The result indicates that innovation-based strategy is the most suitable green supply chain management strategy for PT XYZ.

1. Introduction

Significant development of the manufacturing industry in the last three decades has led to the trend of mass production that affects the environment. Some of the environmental impacts of mass production are the depletion of natural resources and increased pollution of water, air, and soil that affect human health. According to data from the U.S Department of Energy [1], carbon emission has increased significantly from 1990. Emissions from fossil fuels and industrial processes contribute to about 78%, while agricultural activities, deforestation, and land conversion are the second largest contributors [2].

Reports from United Nations Environment Program (UNEP) indicate that there is a gap between efforts to reduce environmental impacts and earth's level of damage [3]. The gap occurs because only few companies consider social and environmental factors in their business decisions.

Every activity in the company's supply chain affects the environment. Therefore, the companies must find ways to minimize the environment impacts from their supply chain activities. Protection activities toward environment performed by the companies are not just a form of responsible action toward the environment, but also a way to respond to public concerns over environmental issues. In addition, increased national and international legislation have also made companies to be more responsible towards the environment issues [4].

One of the popular programs that reflects environmental friendly business activities is green supply chain management [5], which is the integration of environmental principles into supply chain management including material procurement, production process, distribution to end customers, and end-of-life product management [6]. The green supply chain management (GSCM) can lower the costs leading to significant advantages and profit [4].

PT XYZ (the name of the company is not revealed for confidentiality reasons) is a multinational company producing personal and skin care products with thirteen different brands. In the Asia-Pacific region, the company produces two brands of skin care and protection products. The supply chain activities of PT XYZ are from procurement, manufacturing, until distribution to the end customers. Due to limited resources and increasing customer demand, the company is very concerned about environmental issues related to manufacturing process, from the beginning up to the end of it. The company has an objective to produce high quality products and at the same time minimizing the environmental impacts. Thus the company requires an approach to select the most suitable green supply chain management strategy to be implemented.

This paper aims to develop a decision network model to select the suitable green supply chain management strategy for PT XYZ by using analytic network process (ANP) [7]. ANP is used as it provides a general framework to deal with decisions without making assumptions about the independence of higher-level elements from lower-level elements, and about the independence of the elements within a level as in a hierarchy [8]. It focuses on dependent and feedback relations among factors in the network, imitating the decision-making environment [5], and thus it is suitable for GSCM strategy selection.

The remainder of the paper is presented as follows. Relevant literature review is explained in Section 2. Research methodology and ANP model are presented in Section 3, followed by analysis of findings in Section 4, and conclusions and recommendations in Section 5.

2. Literature review

According to Zsidisin and Siferd [9], GSCM is "the set of supply chain management policies held, actions taken and relationships formed in response to concerns related to the natural environment with regard to the design, acquisition, production, distribution, use, re-use and disposal of the firm's goods and services". The scope of GSCM practices implementation are from green purchasing to integrated life cycle management supply chains flowing from supplier, through to manufacturer, customer and closing the loop with reverse logistics [10].

According to Nikbaksh [4], GSCM is originated from two concepts, namely supply chain management and environmental management. The council of Supply Chain Management Professionals (CSCMP) defines supply chain management as the planning and management of all activities involved in sourcing and procurement, conversion, and all logistics management activities, it also includes coordination and collaboration with channel partners, which can be suppliers, intermediaries, third party service providers, and customers [11]. Environmental management, on the other hand, can be defined as the management of human's interaction with the environment and their impacts on the environment [4].

Diabat and Govindan [12] discuss the eleven drivers of GSCM obtained from the literature, namely certification of suppliers' environmental management system, environmental collaboration with suppliers, collaboration between product designers and suppliers to reduce and eliminate product environmental impacts, government regulation and legislation, green design, ISO 14001 certification, integrating quality environmental management into planning and operation process, reducing energy consumption, reusing and recycling materials and packaging, environmental collaboration with customers, and reverse logistics.

The advantages of GSCM are reducing cost and risk, improving productivity, increasing property value, improving public image, and creating healthier environment, while the main barriers of GSCM are environmental requirement costs, lack of green awareness, technological barriers, and lack of environmental information, knowledge and trainings [4].

Companies require a strategy to implement GSCM. Simpson and Samson [13] propose a typology of GSCM strategies, namely: risk-based, efficiency-based, innovation-based, and closed-loop strategies. Risk-based strategies are based on minimal inter-organizational engagement, and the ultimate goal is risk-minimization by passively accomplishing environmental programs. Efficiency-based strategies are more complex than risk-based strategies. They are primarily directed at firms to simultaneously reduce cost, meet operational optimization, and decrease environmental degradation. Innovation-based strategies guide companies to develop products from product life-cycles, enact stricter environmental requirements upon their suppliers, and even trains them to adjust operational processes to follow the newest environmental regulations. Lastly, closed-loop strategies are the most complicated strategies. They aim at seamlessly integrate issues of economic, operational, and environmental performance, therefore require high levels of control over the capture and return of used materials.

Chen *et al.* [5] also propose the company to estimate their situations to determine the amount of resources to invest in green management, allowing the strategic green attitude to develop GSCM strategies. The perspectives of green management and their characteristics are presented in Table 1.

Table 1. Green management perspectives [5]

Green Management Perspectives	Characteristics
Proactive innovation	1. Improving green management capability and make it as a part of corporate strategy 2. strengthening green management performance through innovation
Active integration	1. External: developing cooperative relationship with supply chain partners 2. Internal: incorporate environmental program with other business functions
Receptive-learning	1. Introducing environment-related initiatives non spontaneously 2. Take a benchmark (organization) as model
Reactive response	1. Behave in conformity with environmental regulations 2. Obey customers' environmental instructions

Previous studies have been conducted to discuss the GSCM strategy selection. For instance, Zhao *et al.* [14] describe the strategy selection to minimize carbon emission and environmental risks in the green supply chain by using game theory. Chen *et al.* [5] propose an approach to select GSCM using ANP. The decision network consists of five clusters (green design, green manufacturing, green purchasing, green marketing and service, and GSCM strategies) with a total of eighteen sub-criteria. The GSCM strategies alternatives are adopted from Simpson and Samson [13]. Muralidhar *et al.* [15], on the other hand, use fuzzy-analytic hierarchy process (fuzzy-AHP) and Topsis to evaluate GSCM strategies. The first-level criteria are green procurement, green manufacturing, customer service, and environment management, and the total number of sub-criteria is sixteen. Luthra *et al.* [16] use AHP to rank strategies to implement GSCM in Indian manufacturing industry. They identify more than thirty strategies from literature and discussions with experts, and classify the strategies into four groups based on direct or indirect role in GSCM, namely nonmembers of supply chain, downward stream supply chain members, organizational perspective and upward stream supply chain members. Odeyale *et al.* [17] use fuzzy-AHP to select GSCM strategy, and the criteria are the same as in Chen *et al.* [5], and the strategies are adopted from Simpson and Samson [13].

Various methods with different criteria have been used to determine/rank the GSCM strategies. However different company may have different circumstances and thus may require different criteria to choose the most-suitable GSCM for them.

3. Research methodology and ANP model

3.1. Research methodology

The research objective is to develop decision network model to select the best-suitable GSCM strategy for PT XYZ by using ANP. The ANP is a generalization of AHP. The basic structure is an influence network of clusters and nodes contained within the clusters, and the priorities are determined the same way as in AHP by using pairwise comparisons and judgement [7].

In particular, a literature review is conducted to identify GSCM strategy alternatives and the selection criteria. A questionnaire is then distributed to three SCM experts (from the Indonesian Supply Chain organization, PT XYZ, and academician) to revise/complement the identified criteria. Based on the experts' evaluations, a second questionnaire is developed to determine the relationships (inner and outer dependence) among all criteria and sub-criteria. The same experts fill out the second questionnaire. Based on the second questionnaire result, a decision network model is developed, and thereafter, a third questionnaire is developed to confirm the GSCM perspective of the company and for pairwise comparisons between related criteria and sub-criteria. Nine-scale pairwise comparisons are performed by two experts representing PT XYZ, and the results are processed to determine the limiting (global) weights using Superdecision software.

3.2. ANP model

This study proposes ANP network with 6 criteria and 29 sub-criteria, which are obtained from the literature and through structured interviews with experts from the company. One of the six criteria contains GSCM strategy options, namely risk-based strategy, efficiency-based strategy, innovation-based strategy, and closed loop strategy. The criteria in the network model are basically adopted from GSCM definition [9] and practices scope [10]. Most of the criteria are the same as in Chen *et al.* [5] and Muralidhar *et al.* [15]. The GSCM strategy alternatives are adopted from Simpson and Samson [13] as well. The sub-criteria, however, are taken from several literature as can be seen in Table 2.

Results of the first questionnaire, the experts propose another sub-criterion in the "green design" criterion, which is "product and packaging designs that take into account green logistics". The second questionnaire is then distributed to the same three experts to determine the relationships (inner and outer dependencies) between criteria and sub-criteria, and then based on the questionnaire results, a decision network model is developed (see Figure 1).

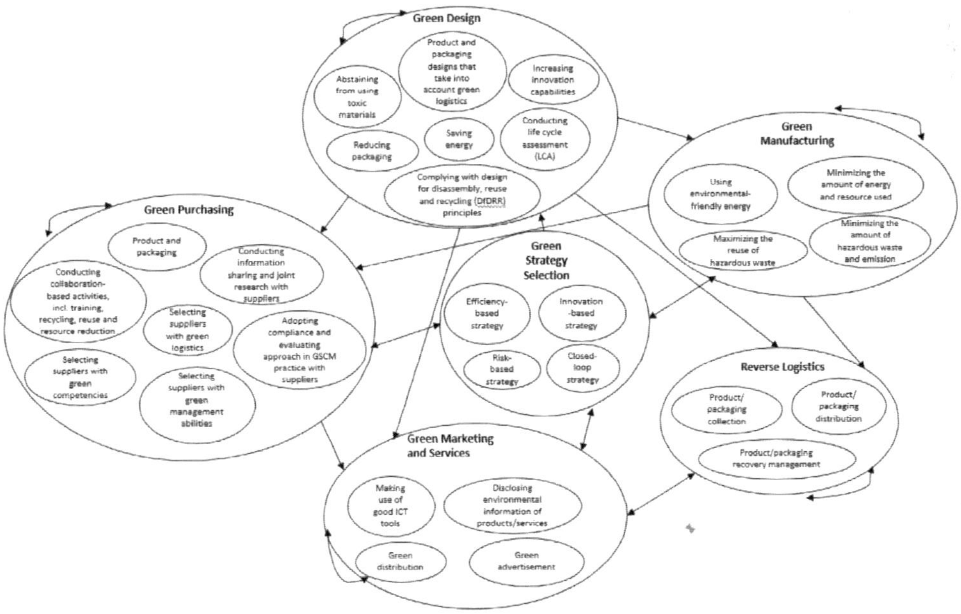

Figure 1. ANP model for GSCM strategy selection at PT. XYZ

Table 2. GSCM strategy selection criteria and sub-criteria

Criteria	Sub-criteria	Sources
a. Green Design: Optimizing product design by considering its impacts on the environment, resources, and product life cycle without decreasing the product function, quality, efficiency, and competitiveness in the market	(1) Abstaining from using toxic materials	[5], [18]
	(2) Complying with design for disassembly, reuse, and recycling (DfDRR) principles	[18]
	(3) Saving energy	[5]. [19]
	(4) Increasing innovation capability	[5]
	(5) Reducing packaging	[18]
	(6) Conducting life-cycle assessment/analysis (LCA) on procducts	[18]. [19]
b. **Green Purchasing:** Adopting GSCM compliance and evaluative approach practices towards suppliers by evaluating the suppliers based on environmental criteria and requiring them to develop and monitor some forms of environment management system	(1) Conducting collaboration activities, including training, recycling facility, reuse and resources reduction	[18]
	(2) Adopting compliance approach in GSCM practices towards suppliers	
	(3) Conducting environment information sharing and joint research	
	(4) Selecting suppliers with green logistics	[17]
	(5) selecting suppliers with green competencies	[5], [17]
	(6) Selecting suppliers with green image	
	(7) Selecting suppliers with green management abilities	
c. Green Manufacturing: The production process uses input with relatively low **environmental** impacts and high efficiency, and producing output with relatively low environmental impact	(1) Minimizing the amount of energy and resources used	[5], [19]
	(2) Using environmental-friendly energy	[5]
	(3) Minimizing the amount of hazardous emission and waste	[5], [19]
	(4) Maximizing the reuse of hazardous waste	[5]
d. Green Marketing and Services: Green marketing should evolve into a strategic activity, including manipulating STP (segmenting, targeting, and positioning) and 4P (product, price, place, and promotion) activities, greening logistics, and developing green alliances	(1) Making good use of ICT tools	[5], [17]
	(2) Disclosing environmental information of products/services	[5]
	(3) Green distribution	[19]
	(5) Green advertisement	[19]
e. Reverse Logistics: The process of planning, implementing, and controlling the efficient and effective flow of cost, raw materials, in-process inventory, final product and information from the point of consumption to the point of origin for the purpose of value recovery or proper disposal	(1) Product/packaging collection	[18]
	(2) Product/packaging recovery management	
	(3) Product/packaging re-distribution	
f. Green Strategy Alternatives: Classification of environmental-friendly supply chain strategies based on company behaviour towards environmental management, level of commitment to resources, and level of complexity in the implementation	(1) Risk-based strategy	[13]
	(2) Efficiency-based strategy	
	(3) Innovation-based strategy	
	(4) Closed-loop strategy	

4. Results and discussions

4.1. GSCM strategy selection

A third questionnaire is distributed to two experts representing the SCM division in PT XYZ to determine the pairwise comparisons between all related criteria and sub-criteria. The questionnaire results are then processed using Superdecision software, and the resulting limiting weights of all criteria and sub-criteria can be seen in Table 3.

Table 3. GSCM strategy selection criteria and sub-criteria weights

Criteria	Weights of Criteria	Sub-criteria	Weights of Sub-criteria	
			Normalized By Cluster	Limiting
Green Design	0.183	(1) Product and packaging designs that take into account green logistics	0.223	0.041
		(2) Conducting life-cycle assessment/analysis (LCA) on products	0.136	0.025
		(3) Complying with design for disassembly, reuse, and recycling (DfDRR) principles	0.112	0.020
		(4) Saving energy	0.065	0.012
		(5) Reducing packaging	0.080	0.015
		(6) Increasing innovation capability	0.339	0.062
		(7) Abstaining from using toxic materials	0.046	0.008
Green Manufacturing	0.247	(1) Maximizing the reuse of hazardous waste	0.100	0.025
		(2) Minimizing the amount of energy and resources used	0.231	0.057
		(3) Minimizing the amount of hazardous emission and waste	0.495	0.122
		(4) Using environmental-friendly energy	0.175	0.043
Green Marketing & Services	0.073	(1) Green distribution	0.339	0.025
		(2) Green advertisement	0.491	0.036
		(3) Disclosing environmental information of products/services	0.061	0.004
		(4) Making good use of ICT tools	0.108	0.008
Green Purchasing	0.166	(1) Conducting environment information sharing and joint research	0.080	0.013
		(2) selecting suppliers with green competencies	0.145	0.024
		(3) Selecting suppliers with green image	0.020	0.003
		(4) Selecting suppliers with green logistics	0.059	0.010
		(5) Selecting suppliers with green management abilities	0.376	0.063
		(6) Conducting collaboration activities, including training, recycling facility, reuse and resources reduction	0.255	0.042
		(7) Adopting compliance approach in GSCM practices towards suppliers	0.065	0.011
Green Strategy Alternatives	0.273	(1) Risk-based strategy	0.045	0.012
		(2) Efficiency-based strategy	0.096	0.026
		(3) Innovation-based strategy	0.455	0.124
		(4) Closed-loop strategy	0.405	0.111
Reverse Logistics	0.057	(1) Product/packaging collection	0.088	0.005
		(2) Product/packaging re-distribution	0.336	0.019
		(3) Product/packaging recovery management	0.576	0.033

It can be seen from Table 3, that among the selection criteria, “green manufacturing” has the highest priority (0.247), followed by “green design” (0.183), “green purchasing” (0.166), “green marketing & services” (0.073), and “reverse logistics” (0.057). The results indicate that the proper GSCM strategy for PT XYZ is highly determined by the implementation of green manufacturing in the company, especially in minimizing the amount of hazardous emission and waste (0.495) and in minimizing the amount energy and resources used (0.231). In terms of green design, increasing innovation capability, and also having product and packaging designs that take into account green logistics have the highest importance, while concerning purchasing, selecting suppliers with green management abilities and conducting collaboration activities, including training, recycling facility, reuse and resources reduction have the highest importance in the respective clusters.

The most suitable GSCM strategy for PT XYZ is indicated by the highest sub-criterion weight in the “Green strategy alternatives” cluster. It can be seen from Table 4 that sub-criterion with the highest local weight (normalized) is “innovation-based strategy” with the weight of 0.4548. Therefore, innovation-based strategy is the most suitable GSCM strategy for PT XYZ.

Based on the definition, the strategy requires high investment on resources and strong commitment to improve the environmental-based innovation capability at PT XYZ. High investment on resources are used to support innovations performed by the company, while strong commitment is needed because in implementing the strategy, environmental-based supply chain activities must be conducted continuously. Therefore, the strong commitment is needed to ensure that the company is consistent in conducting those environmental-based supply chain activities.

Table 4. GSCM strategy alternatives sub-criteria weights

GSCM Strategy Alternatives	Limiting Weights	Normalized Weights	Idealized weights	Rank
Risk-based strategy	0.012	0.045	0.099	4
Efficiency-based strategy	0.026	0.096	0.211	3
Innovation-based strategy	0.124	0.455	1	1
Closed-loop strategy	0.111	0.405	0.890	2

Based on the third questionnaire’s results, it is confirmed that that the company has an active integration GSCM perspective, whereas the company has put an effort to build cooperative relationship with the supply chain partners (suppliers, distributors, and retailers) by establishing environmental standards towards the third parties that support the business processes. Furthermore, PT XYZ has already had environment program incorporated in the business process, for instance, the company has an effort to reduce the waste and also uses environmental friendly packaging design. However, compared to XYZ global, especially XYZ German, there is an inconsistency of perspective towards green management. In terms of the characteristics, XYZ global has a proactive innovation, whereby environmental friendly business activities have become a strategy and green management process has involved innovation. This indicates that the environment management of PT XYZ has not been fully adjusted to XYZ global. Therefore, with innovation-based strategy, PT XYZ can strengthen the innovation and invest more resources to support GSCM implementation.

4.2. Sensitivity analysis

Sensitivity analysis is conducted to understand how the optimal solution is responding to changes in network parameter inputs. If the chosen strategy still has the highest weight when parameter inputs are changed, the chosen strategy is stable and thus can be implemented by PT XYZ.

Based on the results, “improving innovation capability” sub-criterion has high influence on “innovation-based strategy” sub-criterion (the chosen strategy), and thus we want to know whether changes in the weight of “improving innovation capability” can change the weight of the optimal strategy. The sensitivity analysis results are presented in Figure 2.

It can be seen that there is no significant change on the weights of the GSCM strategy alternatives due to changes in the weight of "improving innovation capability" sub-criterion. "Innovation-based strategy" sub-criterion still has the highest weight. When the weight of "improving innovation capability" sub-criterion is reduced to 80%, the closed-loop strategy and risk-based strategy weights increase by 3% and 2% respectively, while the other two sub-criteria slowly decrease from the initial values.

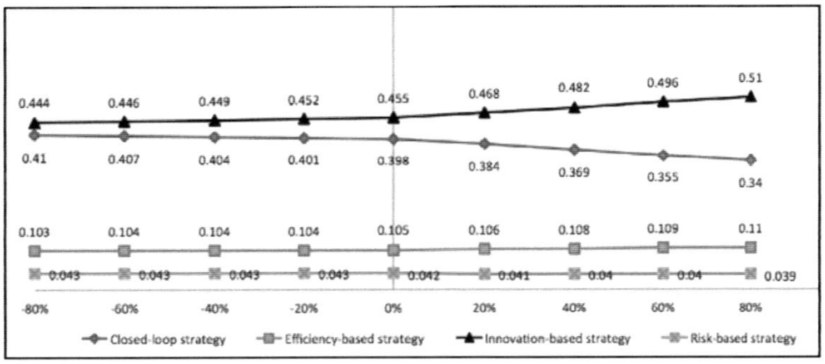

Figure 2. Sensitivity analysis results

Therefore, it can be concluded that "innovation-based strategy" is not sensitive to changes in the weight of "improving innovation capability" (its most influential sub-criterion). It implies that PT XYZ has high preference toward the strategy because as a manufacturer of highly competitive products, it requires strong innovation to meet the market requirements. By involving innovation in environment management, PT XYZ can support the corporate vision sustainably.

5. Conclusions and recommendations

This research aims at helping PT XYZ to select the most-suitable GSCM strategy, which is a multiple criteria decision-making problem. We use ANP to solve the problem as the method models the relationships between criteria and sub-criteria in a network that can represent real world problem. By modeling the problem as a network, the decision-maker can understand the complex problem more clearly and in a systematic way.

The criteria and sub-criteria are obtained from literature review with an addition of one sub-criterion that is obtained from the experts. The criteria are namely green design, green purchasing, green manufacturing, green marketing and services, reverse logistics and GSCM strategy alternatives (risk-based, efficiency-based, innovation-based, and closed-loop strategies). The total number of sub-criteria is 29. Results of pairwise comparisons indicate that innovation-based strategy has the highest weight among the GSCM strategy alternatives sub-criteria, and thus is selected as the most-suitable GCSM strategy for PT XYZ. The results also show that the strategy selection process is highly determined by green manufacturing, followed by green design, green purchasing, green marketing and services, and reverse logistics. Sensitivity analysis results show that the chosen strategy is not sensitive to changes in the weight of its most influential sub-criterion ("improving innovation capability").

The chosen strategy enacts stricter environment requirement upon suppliers and thus, forces companies to invest more resources and cultivate innovative capabilities for green management. Therefore, it requires strong commitment from the top management of PT XYZ.

In order to implement the strategy, PT XYZ can start by conducting activities that can stimulate innovation capability improvement as well communicating their environment vision and mission internally and externally. The company may also form a cross-functional development team that is encouraged to create environment-based innovations that can be integrated into the supply chain through environment-related research. PT XYZ may also involve external parties to enrich the innovative ideas by conducting joint research or by collaborating with experts or other stakeholders, such as the suppliers.

Therefore, there is knowledge sharing activities between PT XYZ and the supply chain partners that may lead to the creation of outstanding green products that can open a new market for the company.

References

[1] U.S. Department of Energy. (2015). *Global, regional, and national fossil-fuel CO2 emissions.* Carbon Dioxide Information Analysis Center. Oak Ridge National Laboratory. Retrieved from http://cdiac.ornl.gov/trends/emis/tre_glob_2011.html.

[2] Intergovermental Panel on Climate Change. (2014). *Climate change 2014: Mitigation of climate change.* Cambridge: Cambridge University Press.

[3] United Nations Environment Program. (2002). *The state of the planet is getting worse but for many it's still "business as usual".* Retrieved from United Nations Environment Program (UNEP): http://www.unep.org/Documents.Multilingual/Default.asp?DocumentID=248&ArticleID=3049

[4] Nikbakhsh, E. (2009). Green supply chain management. Book chapter in *Contributions in Management Science*, pp. 195-220, Springer.

[5] Chen, C., Shih, H., Shyur, H., & Wu, K. (2012). A business strategy selection of green supply chain management via an analytic network process. *Computers and Mathematics with Applications*, 64, 2544-2557.

[6] Srivastava, S. (2007). Green supply-chain management: A state-of-the-art literature review. *International Journal of Management Reviews*, 9(1), 53-80.

[7] Saaty, T. (1996). *Decision making with dependence and feedback: The analytic network process.* Pittsburg: RWS Publications.

[8] Saaty, T. (2009). *Theory and applications of the analytic network process: Decision making with benefits, opportunities, costs, and risks.* Pittsburg: RWS Publications.

[9] Zsidin, G.A., & Siferd, S.A. (2001). Environmental purchasing: a framework for theory development. *European Journal of Purchasing & Supply Management*, 7(1), pp. 61-73.

[10] Zhu, Q., Sarkis, J., & Lai, K. (2008). Confirmation of a measurement model for green supply chain management practices implementation. *International Journal of Production Economics*, 111(2), pp. 261–273.

[11] CSCMP. (2016). CSCMP supply chain management definitions and glossary. Retrieved from https://cscmp.org/supply-chain-management-definitions.

[12] Diabat, A., & Govindan, K. (2011). An analysis of the drivers affecting the implementation of green supply chain Management. *Resources, Conservation and Recycling*, 55, pp. 659–667.

[13] Simpson, D., & Samson, D. (2008). Developing strategies for green supply chain management. *Decision Line*, 39(4), pp. 12–15.

[14] Zhao, R., Neighbour, G., Hanj, J., McGuire, M., & Deutz, P. (2012). Using game theory to describe strategy selection for environmental risk and carbon emissions reduction in the green supply chain. *Journal of Loss Prevention in the Process Industries*, 25, pp. 927-936.

[15] Muralidhar, P., Ravindranath, K., & Srihari, V. (2012). Evaluation of green supply chain management strategies using fuzzy AHP and TOPSIS. *IOSR Journal of Engineering*, 2(4), pp. 824-830.

[16] Luthra, S., Garg, D., & Haleem, A. (2013). Identifying and ranking of strategies to implement green supply chain management in Indian manufacturing industry using Analytical Hierarchy Process. *Journal of Industrial Engineering and Management*, 6(4), pp. 930-962.

[17] Odeyale, S.O., Oguntola, A.J., & Odeyale, E.O. (2014). Evaluation and selection of an effective green supply chain management strategy: A case study. *International Journal of Research Studies in Management*, 3(1), pp. 27-39.

[18] Laosirihongthong, T., Adebanjo, D., & Tan, K. C. (2013). Green supply chain management practices and performance. *Industrial Management & Data Systems*, 113(8), pp. 1088 - 1109.

[19] Ghobakhloo, M., Tang, S.H, Zulkifli, N., & Ariffin, M.K.A. (2013). An Integrated Framework of Green Supply Chain Management Implementation. *International Journal of Innovation, Management and Technology*, 4(1), pp. 86-89.

A DRM for steady water absorption from different types of periodic channels

I. Solekhudin

Department of Mathematics, Faculty of Mathematics and Natural Sciences, Universitas Gadjah Mada, INDONESIA

E-mail: imams@ugm.ac.id

Abstract. In this paper, problems involving steady infiltration from a homogeneous soil with water absorption by plant roots are studied. Different geometries of periodic channels are considered. The governing equation of the problems is a Richards equation. This equation may not be solved analytically. Hence, a numerical method is needed. To do so, the equation is transformed into a modified Helmholtz equation using a set of transformations involving Kirchhoff transformation, dimensionless variables and an exponential transformation. The modified Helmholtz equation is then solved numerically using a Dual Reciprocity Method (DRM) with a predictor-corrector scheme. Results obtained from channels with different geometries are presented.

1. Introduction

Problems involving water infiltration from channels into a homogeneous soil have been studied by numerous researchers, such as Solekhudin and Ang [6, 7, 8], and Clements *et al.* [2]. While Clements *et al.* studied infiltration from channels with impermeable layers [2], Solekhudin and Ang examined infiltration from periodic channels with different geometries [7]. Time-dependent infiltration from a circular channels has been considered by Clements and Lobo [3]. In these studies, water absorption by plant roots has not been taken into account. Hence, Solekhudin and Ang studied problems involving infiltration from periodic trapezoidal channels with water absorption by plant roots [6, 8]. However, in their studies Solekhudin and Ang did not consider different types of channels geometry. In this paper, we consider different types of channels geometry with water absorption by plant roots. The root distribution in this study is different from that in [6, 8].

2. Problem Formulation

Referred to a Cartesian coordinates $OXYZ$ consider an isotropic soil, Pima Clay Loam (PCL), in the region $Z \geq 0$ with OZ positively downward. Periodic irrigation channels are created on the surface of soil. The channels has surface area $2L$ per unit length the OY direction. The distance between the centre of two adjacent channels is $2(L+D)$. A row of crops are planted and equidistant between two adjacent channels. Water flows from the channels into the soil. An illustration of the description is shown in Figure 1. The root-water uptake used in this study is adopted from one of those reported in [10], which is Root A.

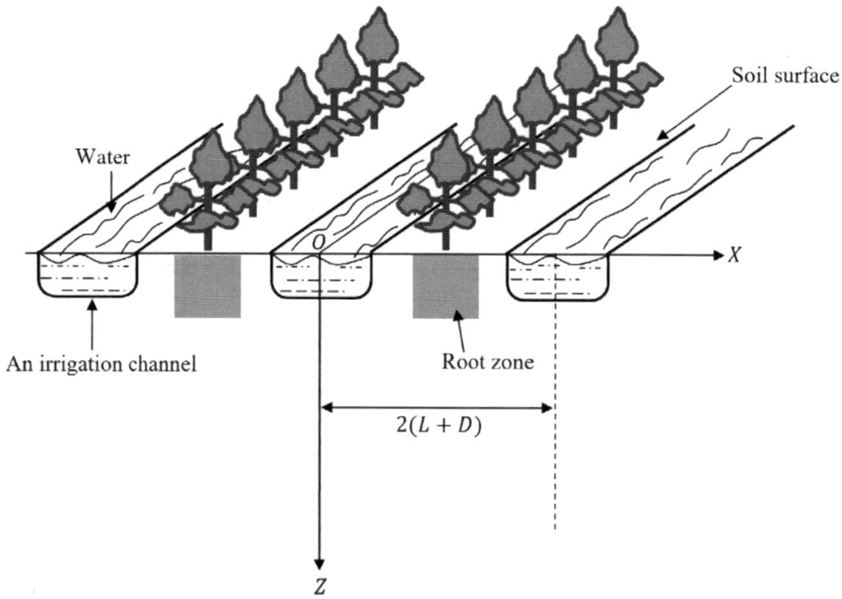

Figure 1. Periodic irrigation channels with rows of crops.

It is assumed that geometries of the channels and root zones do not alter in the OY direction. Hence, the problem is symmetry about the planes $X = \pm k(L+D)$, $k = 0, 1, 2, \cdots$. Due to the symmetry of the problem it is sufficient to consider a semi infinite region defined by $0 \leq X \leq L+D$ and $Z \geq 0$. This region is denoted by R bounded by C. The fluxes over the surface of channels are uniform, that is v_0, while over the surface of soil outside the channels are 0. There are no fluxes across $X = 0$ and $X = L+D$ as the problem symmetrical about them. Following Batu [1], we have $\partial\Theta/\partial X \rightarrow 0$ and $\partial\Theta/\partial Z \rightarrow 0$ as $X^2+Z^2 \rightarrow \infty$. Given this situation, we wish to determine suction potential and the water uptake from the three different geometries of periodic channels.

3. Basic Equations

In this study, equations and method of solution used are as discussed in Solekhudin and Ang [6]. However, for completeness, the equations and the method are briefly presented in this section. The governing equation of water infiltration with root-water uptake that may be used is

$$\frac{\partial}{\partial X}\left(K\frac{\partial \psi}{\partial X}\right)+\frac{\partial}{\partial Z}\left(K\frac{\partial \psi}{\partial Z}\right)-\frac{\partial K}{\partial X}=S(X,Z,\psi), \tag{1}$$

where K is the hydraulic conductivity, ψ is the suction potential and S is the root-water uptake function formulated as

$$S(X,Z,\psi)=\gamma(\psi)S_m(X,Z), \tag{2}$$

where

$$S_m(X,Z)=\frac{L_t\beta(X,Z)T_{pot}}{\int_0^{Z_m}\int_{L+D-X_m}^{L+D}\beta(X,Z)\,dX\,dZ}. \tag{3}$$

Here γ is the dimensionless water stress response function, L_t is the width of the soil surface associated with the transpiration rate, T_{pot} is the potential transpiration, and β is the spatial root-water uptake distribution. The function γ used in this study is the same as that used in [8]. The spatial root-water uptake distribution is formulated as

$$\beta(X,Z)=\left(1-\frac{L+D-X}{X_m}\right)\left(1-\frac{Z}{Z_m}\right)e^{-\left(\frac{p_Z}{Z_m}|Z^*-Z|+\frac{p_X}{X_m}|X^*-(L+D-X)|\right)}, \tag{4}$$

for $L + D - X_m \leq X \leq L + D,\ 0 \leq Z \leq Z_m$.

where X_m is half of the width of the root zone, Z_m is the depth of the root zone, and p_Z, p_X, X^*, and Z^* are empirical parameters. The flux normal to the surface with outward pointing normal $\mathbf{n} = (n_1,n_2)$ is given by

$$F = U\,n_1 + V\,n_2, \tag{5}$$

where

$$U = -\frac{\partial\Theta}{\partial X} \text{ and } V = \alpha\Theta - \frac{\partial\Theta}{\partial Z}$$

Using the Kirchhoff transformation

$$\Theta = \int_{-\infty}^{\psi} K ds. \tag{6}$$

where Θ is the Matric Flux Potential (MFP), an exponential relationship between K and ψ,

$$K = K_s e^{\alpha\psi},\ \alpha > 0 \tag{7}$$

where K_s is the saturated hydraulic conductivity, the suction potential can be formulated as

$$\psi = \frac{1}{\alpha}\ln\left(\frac{\alpha\Theta}{K_s}\right). \tag{8}$$

Using Equations (6), (7) and substituting dimensionless variables

$$x=\frac{\alpha}{2}X;\ z=\frac{\alpha}{2}Z;\ \Phi=\frac{\pi\Theta}{v_0 L};\ u=\frac{2\pi}{v_0\alpha L}U;\ v=\frac{2\pi}{v_0\alpha L}V;\ f=\frac{2\pi}{v_0\alpha L}F; \tag{9}$$

into Equation (1) yield

$$\frac{\partial^2\Phi}{\partial x^2}+\frac{\partial^2\Phi}{\partial z^2}-2\frac{\partial\Phi}{\partial z}=\gamma^*(\Phi)s^*(x,z), \tag{10}$$

where

$$s^*(x,z)=\frac{2\pi}{\alpha L}\frac{l_t\beta^*(x,z)}{\int_0^{z_m}\int_{b-x_m}^{b}\beta^*(x,z)dxdz}\frac{T_{pot}}{v_0} \tag{11}$$

and

$$\beta^*(x,z)=\left(1-\frac{b-x}{x_m}\right)\left(1-\frac{z}{z_m}\right)e^{-\left(p_Z\left|Z^*-\frac{2z}{\alpha}\right|+p_X\left|X^*-\frac{2(b-x)}{\alpha}\right|\right)}, \tag{12}$$

Here $l_t = \alpha L_t/2$, $x_m = \alpha X_m/2$, $z_m = \alpha Z_m/2$ and $b = \alpha(L + D)/2$. Applying transformation

$$\Phi = \phi e^{z}, \tag{13}$$

into Equation (10) yields

$$\frac{\partial^2\phi}{\partial x^2}+\frac{\partial^2\phi}{\partial z^2}-\phi=\gamma^*(\phi)s^*(x,z)e^{-z}. \tag{14}$$

Equation (14) subject to boundary conditions

$$\frac{\partial\phi}{\partial n}=\frac{2\pi}{\alpha L}e^{-z}-n_2\phi \text{ on the surface of the channel,} \tag{15}$$

$$\frac{\partial\phi}{\partial n}=-\phi \text{ on the surface of soil outside the channel,} \tag{16}$$

$$\frac{\partial \phi}{\partial n} = 0, \quad x = 0 \text{ and } z \geq 0, \tag{17}$$

$$\frac{\partial \phi}{\partial n} = 0, \quad x = b \text{ and } z \geq 0, \tag{18}$$

$$\frac{\partial \phi}{\partial n} = -\phi, \quad 0 \leq x \leq b \text{ and } z = \infty. \tag{19}$$

An integral equation for solving Equation (14) is

$$\lambda(\xi,\eta) = \iint_R \varphi(x,z;\xi,\eta)\left[\phi(x,z) + \gamma^*(\phi)s^*(x,z)e^{-z}\right]dxdz$$
$$+ \int_C \left[\phi(x,z)\frac{\partial}{\partial n}(\varphi(x,z;\xi,\eta)) - \varphi(x,z;\xi,\eta)\frac{\partial}{\partial n}(\phi(x,z))\right]ds(x,z), \tag{20}$$

where $\varphi(x,z;\xi,\eta) = (1/4\pi)\ln[(x-\xi)^2 + (y-\eta)^2]$ is the fundamental solution of Laplace equation, and

$$\lambda(\xi,\eta) = \begin{cases} 1, & \text{for } (\xi,\eta) \in R \\ 1/2, & \text{for } (\xi,\eta) \text{ on smooth part of } C \end{cases}$$

Equation (20) may be solved using a DRBEM and a predictor-corrector simultaneously. Readers may refer to [6] for the detail of the method.

4. Results and Discussion

The method described in the preceding section is tested on problems involving infiltration from three different geometries periodic channels into Pima Clay Loam (PCL) with root-water uptake process. The channels considered in this study are trapezoidal, semi-circular and triangular channels. We set $L = D =$ 50 cm. The width of the three geometries of channels is the same, which is $4L/\pi$. For trapezoidal channels, the depth of the channels is $3L/2\pi$. It is assumed that the width and depth of the root zone are $2X_m$ = 100 cm and Z_m = 100 cm, respectively. The potential transpiration rate, T_{pot}, is 4 cm/day, which was also used by Li et al [4], and Šimunek and Hopmans [5] in their studies. The values of experimental parameters α and K_s of the soil are 0.014 cm^{-1} and 9.9 cm/day, respectively [7]. The root-water stress response function γ used here is identical to that reported by Utset *et al.* [9], which can be seen graphically in Figure 2. The value of h_3 for T_{pot} = 0.4 cm/day is interpolated from $h_{3,a}$ and $h_{3,b}$, and we have h_3 = 470. Values of parameters appear in the root-water uptake function S are summarized in Table 1. The Parameter values in Table 1 are the same as those reported in [10].

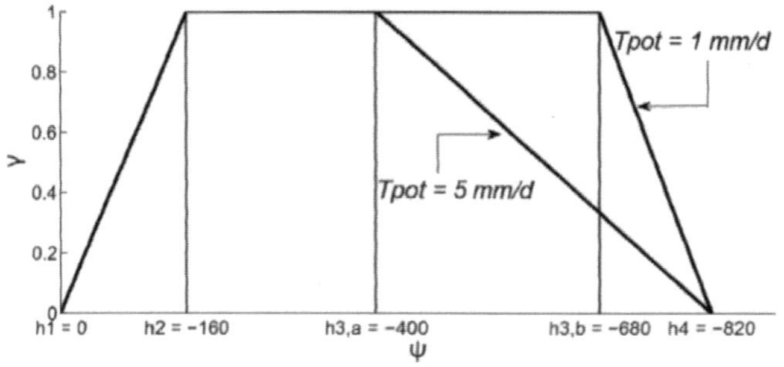

Figure 2. Graph of Root-water stress response function reported by Utset *et al.* [9]

Table 1. Parameters appear in the root-water uptake function.

P_Z	P_X	Z^*	X^*
1.00	1.00	0.00 cm	0.00 cm

The DRM with the predictor-corrector scheme is employed to obtain numerical solutions to equation (14). Using the numerical solutions, Transformation (13) and (9), numerical values of suction potential, ψ, can be obtained using Equation (8). Substituting ψ to equation (2) yield values of root uptake function, S. To employ the DRM, the domain must be bounded by a simple closed curve. The domain is set to be between $z = 0$ and $z = 4$, sufficient depth for boundary conditions to be applied without significant impact to values of Φ in the domain. To employ the method, the boundary must be discretized into a number of line segments, and a number of interior points are chosen as collocation points. The numbers of line segments and interior collocation points for three different geometries of channels are summarized in Table 2.

Table 2. Numbers of line segments and interior points chosen for three different geometries of channels.

	Triangular	**Semicircular**	**Trapezoidal**
Line Segments	200	201	202
Interior points	622	620	619

These numbers of line segments and interior points are chosen in such a way, such that an optimum computational time and the convergence of the values of Φ are achieved after several computational experiments. Some of the results are presented graphically in Figures 4 and 5. Figure 4 shows values of dimensionless MFP, Φ, for the three channels, at $x = 0.385$, $x = 0.525$ and $x = 0.665$. The three values of x are outside the channels. Specifically, Figure 3(a) shows the values of Φ for semicircular channels, Figures 3(b) and (c) show the values of Φ for trapezoidal and triangular channels, respectively. Dashed lines are the values of Φ for infiltration without root-water uptake. Solid lines are the values of Φ for the correspondence infiltration with root-water uptake. It can be seen that the values of Φ for infiltration without root-water uptake is higher than those with root-water uptake. This means that water content in the soil for infiltration without root-water uptake is higher than that in the soil for infiltration with root-water uptake. This result is expected as some amount of water in the soil absorbed by the plant roots for infiltration with root-water uptake. Thus, there is a decrease in the water content in the soil. The values of Φ increases as z increases, and convergent to a certain value of Φ. This means that water contents at shallower level of soil are lower than those at deeper. This is expected due to the assumption that there are no water fluxes across the surface of soil outside the channels. It can also be seen that Φ at $x = 0.385$ is the highest. This result indicate that water content at the location near the channels are higher than those further at the same level of soil depth.

The values of Φ for infiltration from semicircular and trapezoidal channels are about the same. However, the values of Φ for infiltration from triangular channels are lower than those from the others. This indicate that water contents in the soil for infiltration from triangular channels are the lowest.

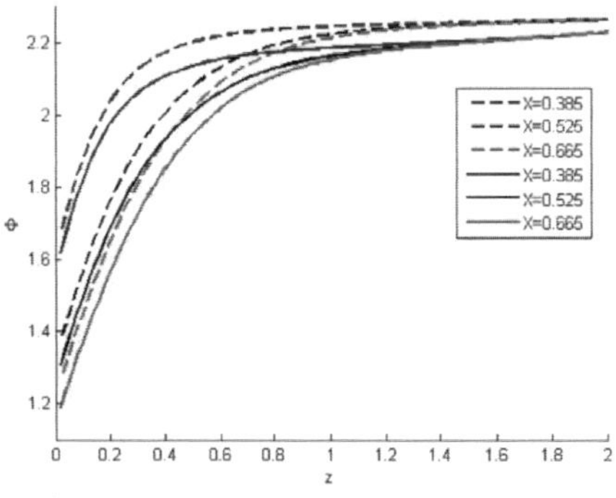

(a) Semicircular Channels

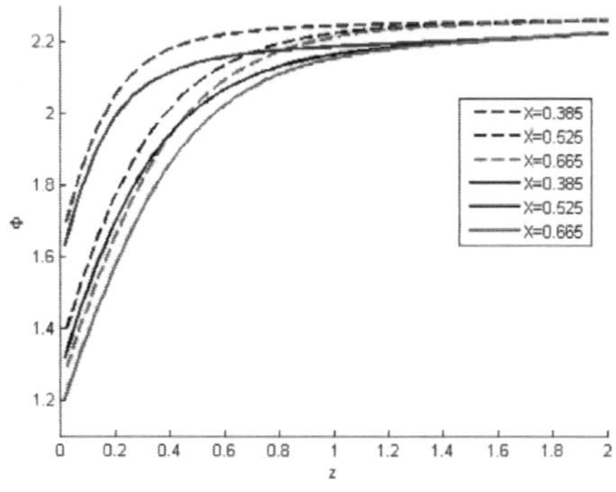

(b) Trapezoidal Channels

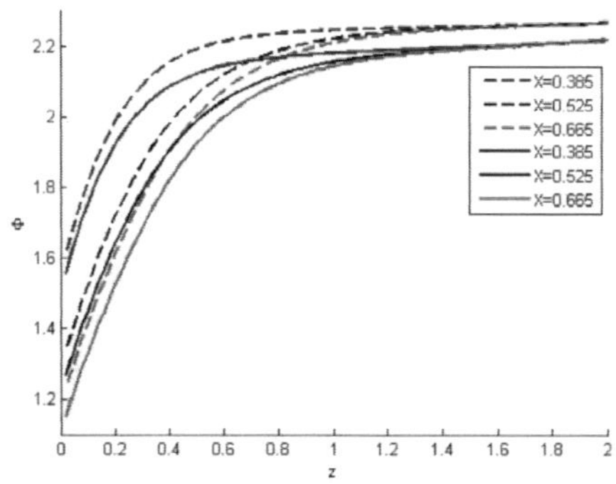

(c) Triangular Channels

Figure 3. Dimensionless MFP at selected values of x along z-axis.

Figure 4 shows values of suction potential, ψ, and corresponding values of root-water uptake, S, at three different values of X along the root zone. The three values of X are 55 cm, 75 cm, and 95 cm. The two values, $X = 55$ cm and $X = 95$ cm, are near the boundary of the root zone, and the other value, $X = 75$ cm, is in the middle of the root zone. It can be seen that the graph of ψ and S for infiltration from trapezoidal and semicircular channels have the same fashion. This means that water content in the soil as well as amount of water absorbed by plant roots for infiltration from trapezoidal and semicircular channels are about the same. For infiltration from triangular channels, the value of ψ are the smallest among those from the other types of channels, at any value of Z. Since the values of ψ are between −118 and −75, from Figure 3 higher ψ implies smaller value of response function, and hence smaller values of S as shown in Figure 5. This indicates that the amount of water absorbed by the roots from the soil with triangular channels is higher than those from the soil with other two channels. This result can clearly be observed from Figure 5. The highest uptake or water absorption for any values of X occurs at the surface of the soil. This is expected, as the values of X^* and Z^* are both 0 cm. This result indicates that the highest amount of water absorbed by the plant roots is at the distance of 0 cm from the plant.

IOP Conf. Series: Materials Science and Engineering **166** (2017) 012027 doi:10.1088/1757-899X/166/1/012027

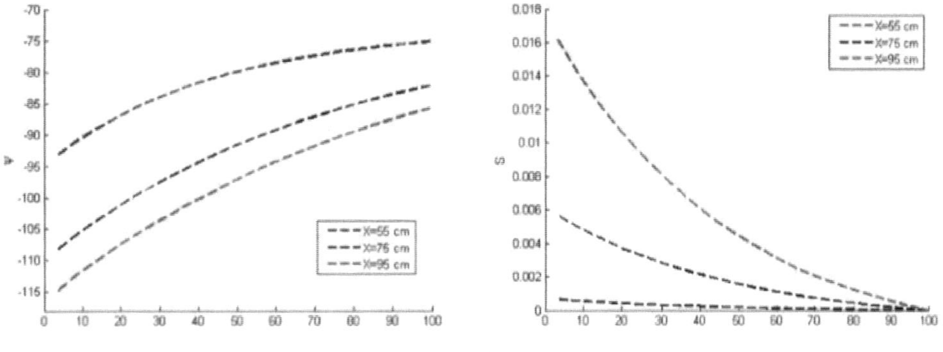

(a) Semicircular Channels

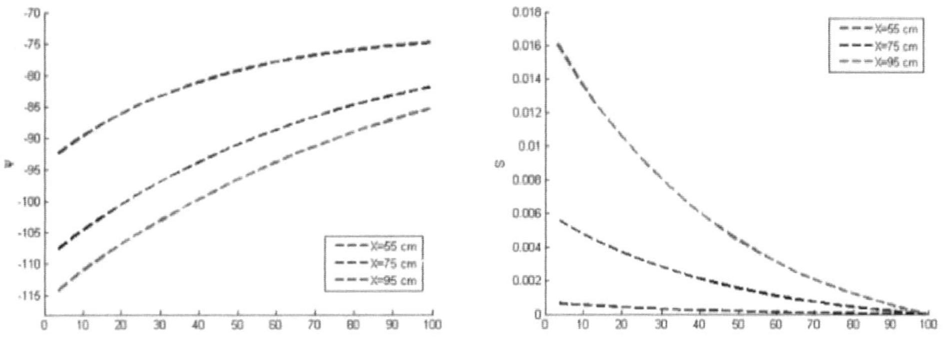

(b) Trapezoidal Channels

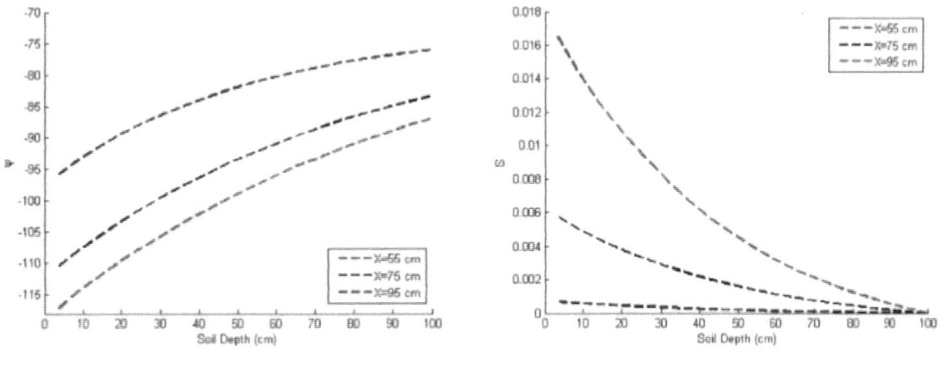

(c) Triangular Channels

Figure 4. Suction potential and corresponding root-water uptake at selected values of X.

5. Concluding Remarks

Problems involving steady infiltration from periodic channels with root-water uptake in a homogeneous soil have been solved numerically. Three different geometries of periodic channels with the same width and surface area were considered. The problems are solved numerically by employing a DRM with a predictor-corrector scheme. Using the method, numerical solutions of the dimensionless MFP are obtained. From the dimensionless MFP obtained and empirical parameters, suction potential can be calculated. Moreover, water uptake can also be calculated.

In this study, the results indicate that triangular channels provide lower suction potential than other two channels geometries. In contrast, triangular channels results in higher uptake of moisture than other two channels geometries.

References

[1] V Batu 1978 *Soil Sci. Soc. Am. J.* **42** 544
[2] DL Clements, M Lobo and N Widana 2007 *El. J. of Bound. Elem.* **5** 1
[3] DL Clements and M Lobo 2010 *Engineering Analysis with Boundary Elements* **34** 1100
[4] KY Li, R De Jong and JB Boisvert 2001 *J. of Hydrol.* **252** 189
[5] J Šimunek and JW Hopmans 2009 *Ecol. Model* **220** 505
[6] I Solekhudin and KC Ang 2012 *Engineering Analysis with Boundary Elements* **36** 1199
[7] I Solekhudin and KC Ang 2013 *ANZIAM J.* **54** 171
[8] I Solekhudin and KC Ang 2015 *Engineering Analysis with Boundary Elements* **50** 141
[9] A Utset, ME Ruiz, J Garcia and RA Feddes 2000 *Potato Res.* **43** 19
[10] JA Vrugt, JW Hopmans and J Šimunek 2001 *Soil Sci. Soc. Am. J.* **65** 1027

Dynamic optimization approach for integrated supplier selection and tracking control of single product inventory system with product discount

Sutrisno[1], Widowati[2] and R. Heru Tjahjana[3]

Department of Mathematics, Diponegoro University, Jl. Prof. Soedarto, SH, Tembalang, Semarang, Indonesia, 50275

Emails: [1]tresno.math@undip.ac.id, [2]widowati_math@undip.ac.id, [3]heru_tjahjana@undip.ac.id

Abstract. In this paper, we propose a mathematical model in the form of dynamic/multi-stage optimization to solve an integrated supplier selection problem and tracking control problem of single product inventory system with product discount. The product discount will be stated as a piece-wise linear function. We use dynamic programming to solve this proposed optimization to determine the optimal supplier and the optimal product volume that will be purchased from the optimal supplier for each time period so that the inventory level tracks a reference trajectory given by decision maker with minimal total cost. We give a numerical experiment to evaluate the proposed model. From the result, the optimal supplier was determined for each time period and the inventory level follows the given reference well.

1. Introduction

Logistics & Supply Chain Management (LSCM) involves the management of many components that are procurement, movement and storage of materials, parts and finished product [1]. Supplier selection and inventory control system are two important components in LSCM. Supplier selection problem is appeared when there are several supplier alternatives that have to be determined who the optimal supplier(s) is. In other hand, manufacturer has to decide when and how many he has to store the product in his inventory in order to satisfy the demand with minimal cost. Some manufacturer decides to control his inventory level so that the stock level of the product is located at some point as close as possible to a reference point or trajectory given by the decision maker.

To find the optimal strategy so that the total cost is minimal, several mathematical models were developed by many researchers. We can classify the model into single product or multi product model. For single product model, [2, 3, 4] were determined the optimal strategy for inventory controlling by without solving the supplier selection problem. In the other hand, [5, 6, 7] were solved the dynamic supplier selection problem, but the inventory of the product was not controlled. By applying some assumption, several researchers were developed the model for integrated supplier selection and inventory control problem. Reference [8] was developed a model for integrated supplier selection problem and inventory system by assuming that the decision maker was not decided to control the inventory level to follows some reference point. In the other hand, [9] was integrated the production control with multi supplier without inventory level controlling and [10] was integrated the supplier selection problem and optimal control problem of a single product by assuming that there is no product discount on the purchasing cost.

In this paper, we formulate a mathematical model in the dynamic/multi-stage optimization form to determine the optimal strategy for an integrated supplier selection problem and trajectory tracking control problem of an inventory system with product discount. The dynamic programming will be used to solve this optimization. A numerical experiment will be performed to evaluate and simulate the model and to illustrate how the problem will be solved.

2. Mathematical Model

Suppose that a manufacturer will purchase a product from S suppliers with different purchase cost per unit. Let $X_{t,s}$ denotes the product volume purchased from supplier s at time period t and $U_{t,s}$ denotes the purchasing cost per unit from supplier $s \in S$ at time period $t \in T$ where S denotes the set of the suppliers and T denotes the set of time period. Assume that there is product discount from supplier s with the scheme as follows

$$U_{t,s} = \begin{cases} U_{t,s}^{(1)}, & \text{if } 0 = d_s^{(0)} < X_{t,s} \le d_s^{(1)} \\ U_{t,s}^{(2)}, & \text{if } \quad d_s^{(1)} < X_{t,s} \le d_s^{(1)} \\ \vdots & \\ U_{t,s}^{(J)}, & \text{if } \quad d_s^{(J-1)} < X_{t,s} \le d_s^{(J)} \end{cases} , \forall t \in T, \forall s \in S \tag{1}$$

or

$$U_{t,s} = U_{t,s}^{(j)}, \quad \text{if } d_s^{(j-1)} < X_{t,s} \le d_s^{(j)}, \forall t \in T, \forall s \in S \tag{2}$$

where $d_s^{(j)}, j = 0,1,2,...,J$ is a constant number decided by the supplier s. This product discount scheme can also be called as price level. Let H_t denotes the holding cost per unit at time period t, D_t denotes the demand of the product at time period t and I_t denotes the inventory/stock level of the product at time period t. The decision maker decides that the inventory level will be controlled so that it will be located at some point as close as possible to a reference point or trajectory given by the decision maker. Let r_t denotes the reference point at time period t, we define $(I_t - r_t)^2$ as the track reference objective function. Assume that the cost components in this problem are purchasing cost and holding cost, then we define the following optimization which is minimizing the total cost i.e.

$$\min \begin{cases} \sum_{t=1}^{T}\sum_{s=1}^{S} U_{t,s}^{(1)} X_{t,s} + \sum_{t=1}^{T} H_t I_t + \sum_{t=1}^{T} (I_t - r_t)^2, & \text{if } 0 = d_s^{(0)} < X_{t,s} \le d_s^{(1)} \\ \sum_{t=1}^{T}\sum_{s=1}^{S} U_{t,s}^{(2)} X_{t,s} + \sum_{t=1}^{T} H_t I_t + \sum_{t=1}^{T} (I_t - r_t)^2, & \text{if } \quad d_s^{(1)} < X_{t,s} \le d_s^{(1)} \\ \vdots & \\ \sum_{t=1}^{T}\sum_{s=1}^{S} U_{t,s}^{(J)} X_{t,s} + \sum_{t=1}^{T} H_t I_t + \sum_{t=1}^{T} (I_t - r_t)^2, & \text{if } \quad d_s^{(J-1)} < X_{t,s} \le d_s^{(J)} \end{cases} \tag{3}$$

or

$$\min \left\{ \sum_{t=1}^{T}\sum_{s=1}^{S} U_{t,s}^{(j)} X_{t,s} + \sum_{t=1}^{T} H_t I_t + \sum_{t=1}^{T} (I_t - r_t)^2, \quad \text{if } d_s^{(j-1)} < X_{t,s} \le d_s^{(j)}, j = 1,2,...,J. \right. \tag{4}$$

Next we formulate the constraints for this optimization model. To ensure that the demand of the product will be satisfied at each time period, we give the first constraint as follows

$$I_{t-1}+\sum_{s=1}^{S}X_{t,s}-I_t \geq D_t, \forall t\in T. \tag{5}$$

If the supplier s has maximum capacity C_s to supply the product for any time period t, then we have the second constraint as follows

$$X_{t,s} \leq C_s, \forall t\in T, \forall s\in S. \tag{6}$$

For any time period, if the inventory has maximum capacity M, then we have the third constraint as follows

$$I_t \leq M, \forall t\in T. \tag{7}$$

Integer constraint for the purchased product volume can be written as follows

$$X_{t,s} \in \{0,1,2,\ldots\}, \forall t\in T, \forall s\in S. \tag{8}$$

3. Dynamic Programming

Dynamic programming or multi-stage programming is a technique that can be used to solve an optimization problem by working backward (or forward) iteration from the end of a problem toward the beginning of the problem [11]. Dynamic programming divides the problem into stages where each stage has several associate states and uses the recursion function to do backward or forward iteration. There are two types of the dynamic programming which are deterministic when all of parameters are known with certainty or stochastic where at least one of the parameters uncertain or random. Dynamic programming can be used to solve many optimization problem classes such as dynamic linear programming, dynamic quadratic programming, dynamic nonlinear programming, dynamic integer programming and mixed integer programming.

Several optimization tools were developed that can be used to perform a dynamic programming. In this paper, we use LINGO 15.0 given in [12] to solve the optimization problem (3) where the model class is pure integer quadratic programming (PIQP).

4. Numerical Experiment

Suppose that a manufacture has four suppliers which are s_1, s_2, s_3 and s_4. For any time period, the purchasing cost from s_1 is \$12/unit if the manufacturer purchasing less or equal to 100 units and it will be \$9/unit if the manufacturer purchasing the product for more than 100 units but no more than 120 units since the maximum capacity of supplier s_1 is 120 units. The purchasing cost from supplier s_2 is \$13/unit if the manufacturer purchasing less or equal to 150 units and it will be \$9/unit if the manufacturer purchasing more than 150 units but no more than 200 units since the maximum capacity of supplier s_2 is 200 units. The purchasing cost from supplier s_3 is \$13/unit if the manufacturer purchasing less or equal to 100 units and it will be \$9/unit if the manufacturer purchasing more than 100 units but no more than 200 units since the maximum capacity of supplier s_3 is 200 units. Finally, the purchasing cost from supplier s_4 is \$14/unit if the manufacturer purchasing less or equal to 120 units and it will be \$8/unit if the manufacturer purchasing more than 120 units but no more than 150 units since the maximum capacity of supplier s_4 is 150 units. The purchasing cost per product unit from all suppliers is summarized by Table (1) and the demand of the product for time period 1 to 12 is given by Table (2).

Suppose that the initial inventory level is 0 item and the holding cost is \$1/unit/period. Finally, the warehouse's maximum capacity is 200 units/period. The decision maker desires that the inventory/stock level must be located at some point as close as possible to the reference level 100 units with minimal cost.

IOP Conf. Series: Materials Science and Engineering **166** (2017) 012028 doi:10.1088/1757-899X/166/1/012028

Table 1. Purchasing cost & supplier capacity

Supplier	Purchasing cost	Supplier capacity (units/period)
s_1	$\begin{cases} \$12 \text{ if } X_{t,1} \le 100 \\ \$9 \text{ if } X_{t,1} > 100 \end{cases}$	120
s_2	$\begin{cases} \$13 \text{ if } X_{t,2} \le 150 \\ \$10 \text{ if } X_{t,2} > 150 \end{cases}$	200
s_3	$\begin{cases} \$13 \text{ if } X_{t,2} \le 100 \\ \$9 \text{ if } X_{t,2} > 100 \end{cases}$	200
s_4	$\begin{cases} \$14 \text{ if } X_{t,2} \le 120 \\ \$8 \text{ if } X_{t,2} > 120 \end{cases}$	150

Table 2. Demand of the product

Time period (t)	Demand (unit)
1	150
2	70
3	80
4	120
5	140
6	100
7	320
8	250
9	270
10	180
11	310
12	270

We solve optimization problem (4) for 12 time periods with 3-by-3 of time periods by using LINGO 15.0 in Windows 8 AMD A6 2.7GHz of processor and 4 GB of Memory. The optimal strategy for this problem is summarized in Table 3.

Table 3. Solution of the problem

Time period (t)	$X_{t,s}$				Demand (unit)	Inventory (unit)
	s_1	s_2	s_3	s_4		
1	101	0	0	150	150	101
2	68	0	0	0	70	99
3	75	0	0	0	80	94
4	0	0	0	125	120	99
5	0	0	0	138	140	97
6	0	0	101	0	100	98
7	0	0	171	150	320	99
8	0	0	101	150	250	100
9	115	0	0	150	270	95
10	33	0	0	150	180	98
11	0	0	162	150	310	100
12	115	0	0	150	270	95

Table 3 shows the optimal strategy for time period 1 to 12. It can be seen that at time period 1, the manufacturer has to purchase 101 units from supplier s_1 and 150 units from supplier s_4 while s_2 and s_3 was not selected to supply the product. At time period 2, the manufacturer has to purchase only 68 units from supplier s_1. The evolution of the inventory level of the product and its reference point can be seen in Fig. 1.

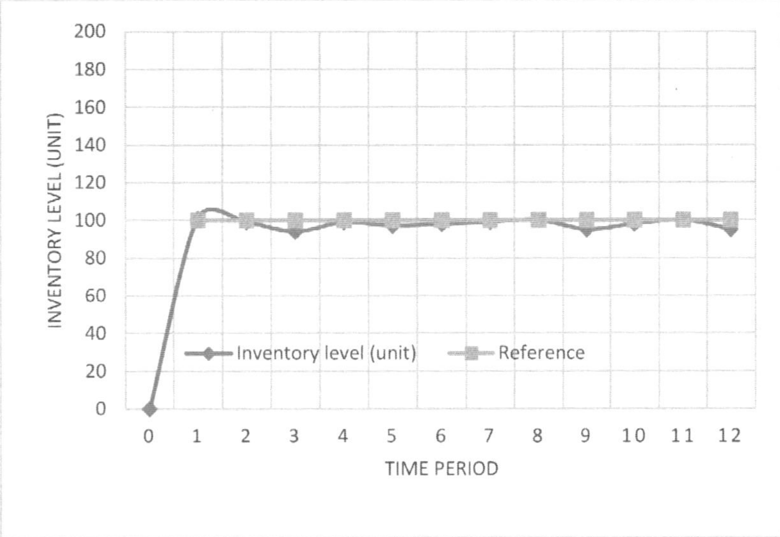

Figure 1. The evolution of the inventory level of the product and its reference trajectory

From Fig. 1, it can be seen that the initial inventory level is 0 unit and the reference inventory is 100 unit for all time period. At time period 1 to 12, the inventory level is sufficiently closed to the reference point. The largest deviation between the actual inventory level and the reference inventory is 6 units at time period 3 and the smallest deviation is 0 unit at time periods 8 and 11. It can be concluded that the inventory level follows the reference point well.

5. Conclusion and Future Works

In this paper, a mathematical model in the form of dynamic/multi-stage optimization was considered to solve an integrated supplier selection problem and tracking control problem of single product inventory system with product discount. The optimal strategy was determined by using dynamic programming. Numerical experiment was considered and from the result, it can be conclude that the supplier selection problem was solved and the inventory/stock level followed the reference point well.

In other work, we propose the mathematical model in the form of stochastic dynamic optimization to solve the integrated supplier selection problem and trajectory tracking control problem for inventory system with product discount in probabilistic environment. In the future works, we will develop the model for multi-product inventory system including order cost, service level parameter, late delivery parameter, etc. We will also develop the mathematical model with carrier selection problem solving.

Acknowledgment
Authors would like to thank Universitas Diponegoro for funding support under Professorship Research Program, PNBP DIPA Universitas Diponegoro 2016.

References

[1] M. Christopher, Logistics & Supply Chain Management, Great Britain: Pearson Education Limited, 2011.

[2] Y. Arda and J.-C. Hennet, "Inventory control in a multi-supplier system," *Int. J. Production Economics,* vol. 104, pp. 249-259, 2006.

[3] W. Guerrero, T. Yeung and C. Guéret, "Joint-optimization of inventory policies on a multi-product multi-echelon pharmaceutical system with batching and ordering constraints," *European Journal of Operational Research,* vol. 231, pp. 98-108, 2013.

[4] S. Minner, "Multiple-supplier inventory models in supply chain management: A review," *Int. J. Production Economics,* Vols. 81-82, pp. 265-279, 2003.

[5] N. R. Ware, S. P. Singh and D. K. Banwet, "A mixed-integer non-linear program to model dynamic supplier Selection Problem," *Expert Systems with Applications,* vol. 41, pp. 671-678, 2014.

[6] C. Haksever and J. Moussourakis, "A model for optimizing multi-product inventory systems with multiple constraints," *International Journal of Production Economics,* vol. 97, pp. 18-30, 2005.

[7] S. S. Kara, "Supplier selection with an integrated methodology in unknown environment," *Expert Systems with Applications,* vol. 38, pp. 2133-2139, 2011.

[8] C. Guo and L. Xueping, "A multi-echelon inventory system with supplier selection and order allocation under stochastic demand," *Int. J. Production Economics,* vol. 151, pp. 37-47, 2014.

[9] A. Hajji, A. Gharbi, J.-P. Kenne and R. Pellerin, "Production control and replenishment strategy with multiple suppliers," *European Journal of Operational Research,* vol. 208, pp. 67-74, 2011.

[10] Sutrisno and P. A. Wicaksono, "Optimal Strategy for Multi-product Inventory System with Supplier Selection by Using Model Predictive Control," in *Procedia Manufacturing of Industrial Engineering and Service Science (IESS),* Yogyakarta, 2015.

[11] W. L. Winston, Operations Research: Applications and Algorithms, 2004.

[12] L. S. Team, LINGO: The Modeling Language and Optimizer, Illinois: Lindo Systems Inc., 2014.

[13] S. M. Mousavi, S. Niaki, A. Bahreininejad and S. N. Musa, "Multi-Item Multiperiodic Inventory Control Problem with Variable Demand and Discounts: A Particle Swarm Optimization Algorithm," *The Scientific World Journal,* vol. 2014, pp. 1-16, 2014.

Combination of nearest neighbor and heuristics algorithms for sequential two dimensional loading capacitated vehicle routing problem

A.A. Sarwono, T.J. Ai and S.S. Wigati

Department of Industrial Engineering, Universitas Atma Jaya Yogyakarta, Jl. Babarsari No. 44 Yogyakarta 55281, Indonesia

E-mails: asa.adiaji@yahoo.com, jinai@mail.uajy.ac.id, yayan@mail.uajy.ac.id

Abstract. Vehicle Routing Problem (VRP) is a method for determining the optimal route of vehicles in order to serve customers starting from depot. Combination of the two most important problems in distribution logistics, which is called the two dimensional loading vehicle routing problem, is considered in this paper. This problem combines the loading of the freight into the vehicles and the successive routing of the vehicles along the route. Moreover, an additional feature of last-in-first-out loading sequencesis also considered. In the sequential two dimensional loading capacitated vehicle routing problem (sequential 2L-CVRP), the loading must be compatible with the trip sequence: when the vehicle arrives at a customer *i*, there must be no obstacle (items for other customers) between the item of *i* and the loading door (rear part) of the vehicle. In other words, it is not necessary to move *non-i*'s items whenever the unloading process of the items of *i*. According with aforementioned conditions, a program to solve sequential 2L-CVRP is required. A nearest neighbor algorithm for solving the routing problem is presented, in which the loading component of the problem is solved through a collection of 5 packing heuristics.

1. Introduction

Distribution is one of the logistic system components which has the responsibility to handle material movement among facilities. Even though, it is believe that the distribution process has a broader scope than just transportation of goods. Distribution is the factor that determine total profit of company because it affects the supply chain cost and direct customers satisfaction, simultaneously [1]. Survey said that distribution cost of American company is equal to 20% of cost of goods sold of a product [2]. So, distribution planning is one of the most important activities to make supply chain process more efficient. Scope of distribution problem consists of several aspects, such as route determination, facility location, and delivery capacity [3].

The vehicle routing problem (VRP) is one of the most frequently studied problems in distribution. It consists of finding an optimal set of trips for a fleet of vehicles which must serve a given set of customers. Several versions of the vehicle routing problem exist in the literature, in particular the capacitated VRP (CVRP). In this variant, customers with known demands are spread over an undirected network and must be visited using a fleet of identical vehicles with limited capacity [4]. CVRP has been solved with many methods, heuristics or metaheuristics in [5], [6], [7], [8], and [9]. Gendreau *et al.* [10] has developed Tabu search algorithm and Fuellerer *et al.* [11] used Ant Colony System (ACO) to solve CVRP problem. Exact method also used in [12], [13], and [14] which is able to solve this CVRP properly. Beside the problem settings defined in the CVRP, we should know that loading and unloading problem to-from container is a common setting in real distribution cases.

Two-dimensional bin packing problem (2BPP) is the term for defining a loading problem which seen the problem in two dimensions only. Exact algorithms and lower bounds was used by [15], [16], [17], [18], [19], [20], and [21] to solve the 2BPP. A lot of heuristics methods have also been used to solve this problem [22]. Beside that, metaheuristics method such as GRASP algorithm and tabusearch [23, 24], and also iterated local search method [25] have been used to solve the 2BPP.

This paper addresses the two-dimensional loading capacitated vehicle routing problem, which is denoted as 2L-CVRP [26, 27]. The 2L-CVRP is a variant of one of the most frequently studied combinatorial optimization problems, the capacitated vehicle routing problem (CVRP). If it is viewed from the loading type, there are two version of 2L-CVRP, unrestricted and sequential loading.

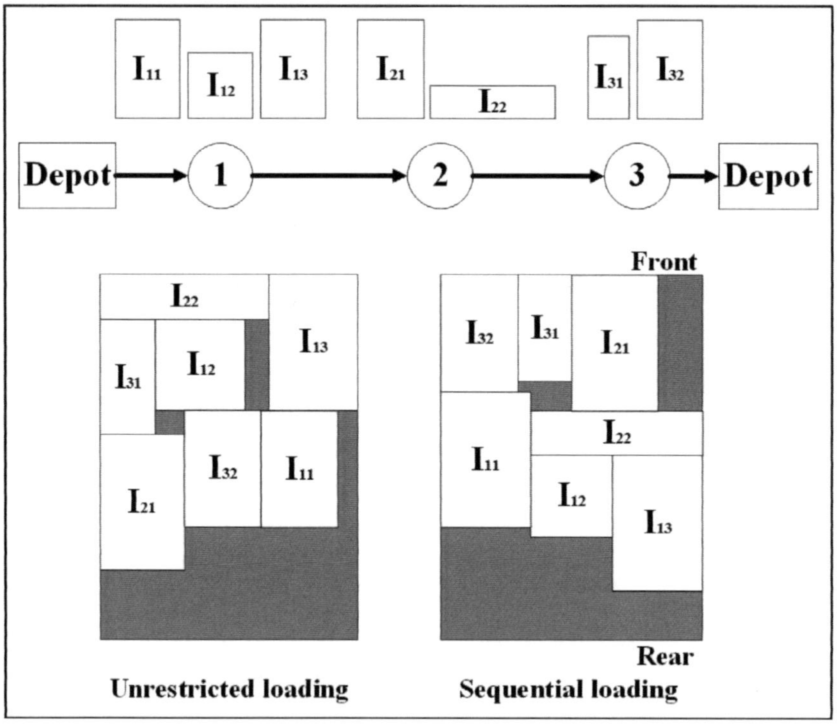

Figure 1. Example of a 2L-CVRP route with unrestricted and sequential loadings

Figure1 depicts that unrestricted version does not pay attention between items location in container and the order of customers that should be visited. This version is used for the vehicle which the container can be unloaded vertically using crane. Beside, sequential version have additional feature named last-in-first-out, in other words there is no need to move *non-i*'s items when unload the items of customer *i*. In this version, we need to pay attention in items arrangement inside the container, because it should be compatible with the order of customers that should be visited. Specifically, thie paper adresses the 2L-CVRP with sequential version of loading, named sequential 2L-CVRP.

The 2L-CVRP was first solved by an exact algorithm which used the branch-and-cut technique [28]. In the test dataset, their approach can deal with the instances with no more than 30 customers and 91 items. As for the larger scale problems, a metaheuristic approach was proposed by Gendreau et al. [29]. Precisely, the Tabu search was employed for routing aspects of the problem. Usually, the means of lower bounds, heuristics, local search and a truncated branch-and-bound were used to check the loading feasibility. 180 problem instances were tested in their work. The number of customers went up to 255 and the items up to 786. Recently, a new method, the guided Tabu search [30], which combines the Tabu search with guided local search, was presented. For checking the feasibility of loading, a collection of packing heuristics was used. To accelerate the algorithm, two strategies that reduced the neighborhoods explored, and record of the loading feasibility information, were employed. A nature inspired metaheuristic algorithm, an effective heuristic based on ant colony optimization, has been proposed by Fuellerer *et al.* [11]. The costs of four different loading configurations were compared in this work.

The importance of the 2L-CVRP is mainly reflected in two aspects. Theoretically, composed of two NP-hard optimization problems (CVRP and 2BPP), it is also a high complexity NP-hard problem. For practical applications, this problem may exist at many companies. An efficient method to solve this problem can significantly reduce costs for the companies.

2. Problem description

This paper considers the sequential version of the 2L-CVRP defined as follows:

- Like the VRP, the problem is based on a complete undirected graph with a set of n + 1 nodes. Node 0 is a depot while node 1 to n are customers. The edge $[i, j]$ between any two distinct nodes i and j models a shortest path of length $d_{ij} = d_{ji}$.
- The depot contains a virtually unlimited fleet of homogeneous vehicles with a capacity Q (maximum weight) and a rectangular loading surface of length L and width W. The loading surface is denoted as $A = WL$.
- A set D of m items must be delivered to customers. Item $t(t = 1, 2, \dots, m)$ has a length $l_t \leq L$ and a width $w_t \leq W$. When loaded in a vehicle, its l-edge and w-edge must be respectively parallel to the vehicle L and W-edges, i.e., rotations are not allowed.
- D is partitioned into n subsets $D_1 \cup D_2 \cup \dots \cup D_n$, where D is the subset of items requested by customer i, with a known total weight q_i.
- Each set D_i must be loaded into a single vehicle.
- Each customer must be visited by one, and only one vehicle, once.
- Every vehicle starts from, and ends at, the central depot.
- The weight of the items loaded in a vehicle must not exceed the capacity of the vehicle Q.
- All the items in a vehicle must be loaded in area A. Overlapping loading is not permitted.
- When a vehicle is visiting customer i, all of the items in the set of D_i can be unloaded from the vehicle by means of forklift trucks parallel to the length dimension of the vehicle surface, without moving other items required by other customers

The sequential 2L-CVRP objective is to determine a set of routes of minimal total length that satisfy the following constraints: (a) every route starts and ends at the central depot, (b) the demand of every customer is totally covered, (c) each customer is visited once, (d) the total weight of all items demanded by the set of customers covered by a route must not exceed the capacity of the vehicle Q, (e) there must be a non-overlapping loading of all items demanded by the set of customers covered by a route into the $L \times W$ loading surface of the vehicles and (f) the loading of the items must ensure that whenever a customer i is visited, all items in the set D_i can be unloaded by employing a sequence of straight movements (one per item) parallel to the length dimension of the vehicle surface. In other words, no item of customer j, visited after customer i, can be placed between items of customer i and the rear part (loading door) of the vehicle.

3. Proposed algorithm

The proposed algorithm for the solution of the 2L-CVRP employs nearest neighbor (NN) to solve route determination problem. Regarding loading constraints of the problem, the bundle of packing heuristics designed in Zachariadis *et al.* [7] are applied.

3.1. Route determination

Route-first-cluster-second concept is used in this research. The meaning of route-first concept is that the determination of the route made at the beginning, in which nearest neighbor method is applied here. After that, routes are cut to form clusters as the cluster-second concept whenever one of these following two constraints are violated: (1) if the weight of customers items exceed the container weight capacity (weight capacity constraint) and if (2) the item cannot be loaded into the same container (area capacity constraint). If one of these two constraints violation happened, a new container are needed and consequently a new route cluster is created.

We are proposing two technique to form route clusters, which are called regular cutting NN and skip cutting NN. In regular cutting NN, routes are cutting off if the next customer cannot fulfill one of the two constraints that has been explained above, which is the weight of customers items exceed the container weight capacity or the item that should be delivered to the customer is not enough anymore if loaded on the same container. In skip cutting NN, the route should be cutting off if two constraints feasibility from all customers have been checked. If in that checking process the customer can fulfill two constraints, so that customer's item can be loaded into the same container.

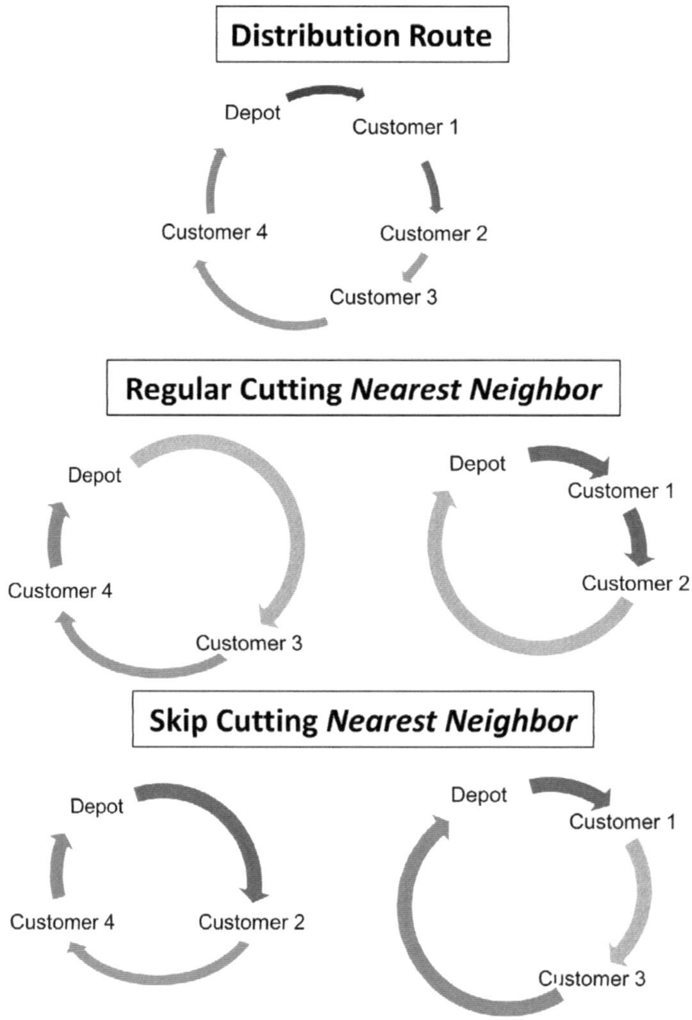

Figure 2. The difference between regular and skip cutting NN

The bottom part of figure 2 (skip cutting NN) illustrates that customer 2 cannot fulfill one of two existing loading constraints, but customer 3 can fulfill those two constraints and then the condition of customer 4 are same with customer 2. So, this condition affect the routes cutting process, customer 1 and 3 are loaded on the same container, while customer 2 and 4 are loaded on the different container. From figure 2 we can see that in skip cutting NN, all of customer are checked for its loading feasibility.

3.2. The bundle of packing heuristics

For loading items in the container, there are five heuristic methods used in this research which are combined with the two proposed route determination methods. These five heuristics are bottom-left fill (W axis), bottom-left fill (L axis), maximum touching perimeter, maximum touching perimeter (no walls), and minimum area. Every heuristic method have its own characteristics for determining the location of an item to be loaded into the container. Each customer have at least 1 item and not more than 5 item to be loaded. Order of each customers item that should be loaded are determined from area of the item which sorted from smallest area.

When new item loaded into the container, there will be a new *posList*. Let *posList* denote a list of available loading positions for the items. In the beginning, the only available loading position lies in the front left corner (0, 0) of the vehicle, so *posList* = {(0, 0)}. Whenever an item is inserted, its loading position is erased from the *posList*, while new loading positions are generated and added into the *posList*. In this way, the positions of holes that may be created between the placed items are stored into the *posList* and may be later filled by the subsequent items.

The position for the placement of an item is selected from the list of available positions *posList* and must not lead to any loading constraint violation (overlapping or sequential constraint). As later explained, it is determined by the packing heuristic currently employed. If all items are packed onto the loading surface, the route is considered to be feasible in terms of the loading constraints of the problem. If, on the other hand, the insertion of an item into any available position leads to loading constraint violations, the method empties the loading surface, *posList* is set equal to {(0, 0)}, and the next packing heuristic is employed from the beginning. If none of the five available packing heuristics manages to produce a feasible loading, the heuristic bundle is applied to the second ordering of the items. If again, no feasible loading is obtained, the examined route is considered to be infeasible regarding the loading constraints.

As mentioned earlier, the loading position of an inserted item is determined by the packing heuristic currently in use. This position must be feasible, i.e., it must not lead to any overlaps or sequence constraint violations. Each of the proposed five packing heuristics $Heur_i$(i =1...5) employs a different criterion for selecting the loading position of an item:

$Heur_1$: *Bottom-Left Fill (W-axis)* [31]

From the *feasible* available loading positions of *posList*, the position selected is the one with the minimum W-axis coordinate, breaking ties by minimum L-axis coordinate. Using this heuristic, the packing tends to evolve in the form of strips parallel to the L-axis.

$Heur_2$: *Bottom-Left Fill (L-axis)* [31]

From the *feasible* available loading positions of *posList*, the position selected is the one with the minimum L-axis coordinate, breaking ties by minimum W-axis coordinate. Using this heuristic, the packing tends to evolve in the form of strips parallel to the W-axis.

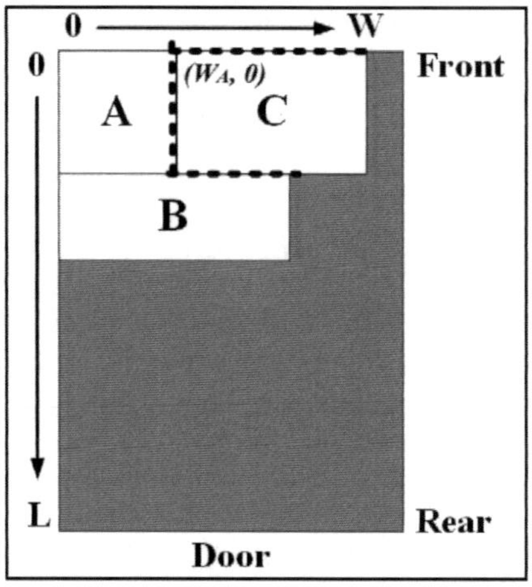

Figure 3. Calculating the perimeter for Maximum Touching Perimeter heuristic.

$Heur_3$: *Maximum Touching Perimeter* [32]

For each of the *feasible* available positions of *posList*, the total touching perimeter of the inserted item is calculated. The total touching perimeter is evaluated as the sum of the common edges of the inserted item with the edges of the already inserted items, and the edges of the loading surface of the vehicle as seen in figure 3: the total touching perimeter of item C placed in position $(w_A, 0)$ is demonstrated by the bold dotted lines, and is equal to $l_C + w_C + (w_B - w_A)$. Term l_C corresponds to the common edges of items C and A, term w_C corresponds to the common edges of item C and the loading surface, and term $(w_B - w_A)$ corresponds to the common edges of items C and B. The item is placed into the loading position that maximises the value of touching perimeter.

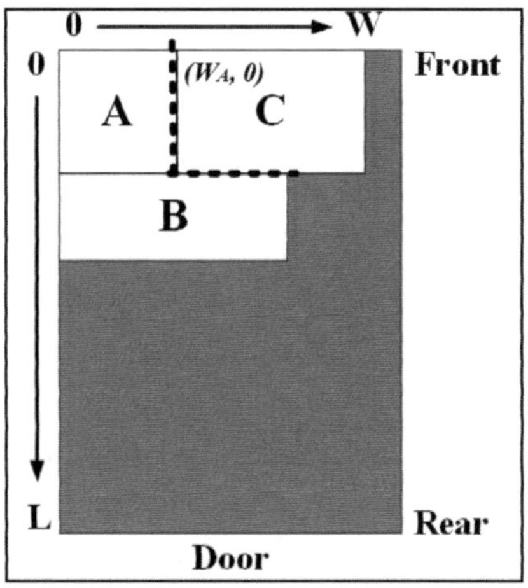

Figure 4. Calculating the perimeter for Maximum Touching Perimeter No Walls heuristic.

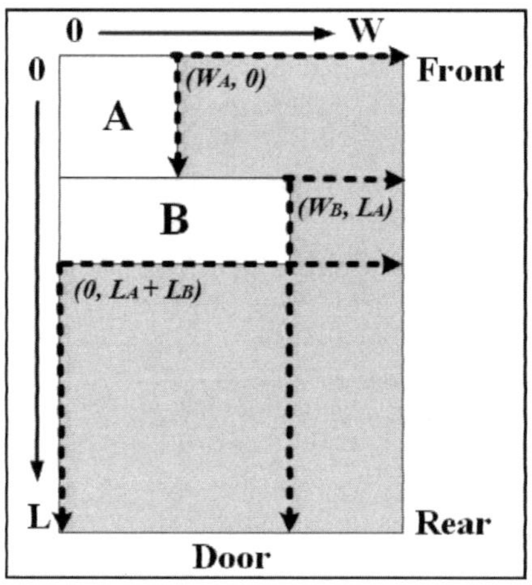

Figure 5. Calculating the rectangular areas of the loading positions.

$Heur_4$: *Maximum Touching Perimeter No Walls* [32]

As in the case of the Max Touching Perimeter heuristic, for each of the *feasible* available positions of *posList*, the total touching perimeter of the inserted item is calculated. In this case, the total touching perimeter is evaluated as the sum of the common edges of the inserted item with the edges of the already inserted items. The common edges of the item and the loading surface are not taken into account. The evaluation of the touching perimeter is presented in figure 4: the total touching perimeter of item C placed in position $(w_A, 0)$ is demonstrated by the bold dotted lines, and is equal to $l_C + (w_B - w_A)$. Term l_C corresponds to the common edges of items C and A and term $(w_B - w_A)$ corresponds to the common edges of items C and B. The item is placed into the loading position that maximises the value of touching perimeter.

$Heur_5$: *Minimum Area* [30]

For each of the *feasible* available positions of *posList*, the area of its corresponding rectangular surface is calculated, as demonstrated in figure 5. The area of the rectangular surface determined by the loading position $(w_A, 0)$ equal to $(W - w_A) \times l_A$, the area corresponding to loading position (w_B, l_A) is equal to $(W - w_B) \times (L - l_A)$, and the area of the surface corresponding to position $(0, l_A + l_B)$ is equal to $W \times (L - l_A - l_B)$. The loading position selected is the one yielding the minimum surface area.

4. Computational results

4.1. Benchmark intances characteristics

All algorithms were implemented and tested on benchmark instances from the literature. In order to verify the performance of the proposed algorithm, the proposed algorithm was tested by 180 sequential 2L-CVRP benchmark problem instances that was used in [26], [28], [29], and [30]. The datasets are available at http://www.or.deis.unibo.it/research.html. These instances were derived from 36 CVRP instances, whose the description can be found in [33], by expressing the customer demand as a set of two-dimensional, weighted and rectangular items. To generate the aforementioned item sets, five classes of the item demand characteristics are introduced

```
Instance:  E016-03m.dat
Class:  3
   15 --- number of customers (no depot)
    3 --- number of vehicles
   31 --- number of items
Capacity - height - width of vehicles
    90       40       20
Node - x - y - demand
    0     30.0     40.0      0.0
    1     37.0     52.0      7.0
    2     49.0     49.0     30.0
    3     52.0     64.0     16.0
    4     20.0     26.0      9.0
    5     40.0     30.0     21.0
    6     21.0     47.0     15.0
    7     17.0     63.0     19.0
    8     31.0     62.0     23.0
    9     52.0     33.0     11.0
   10     51.0     21.0      5.0
   11     42.0     41.0     19.0
   12     31.0     32.0     29.0
   13      5.0     25.0     23.0
   14     12.0     42.0     21.0
   15     36.0     16.0     10.0
Node - number of items - h - w for each item
    0   0
    1   2    4   11   13    6
    2   2    9    5    5   15
    3   2   13    8   26    2
    4   1   10    5
    5   2   13    7   31    2
    6   2   12    8    6   11
    7   1    5   10
    8   3    9    7    7    8   12    3
    9   2   22    2    7   13
   10   3   11    5    6   12   29    2
   11   3    9    6   15    8   10    5
   12   1   13    3
   13   3   17    3   14    5    5    9
   14   1    6    9
   15   3    7    6    8    5   16    7
```

Figure 6. Data set contents

- Class 1: with each customer is associated a single item of width and length equal to nil. The problems of Class 1 are in fact pure CVRP instances, as every customer sequence is feasible in terms of the loading constraints of the problem examined. They are used to test the algorithmic effectiveness in terms of the routing aspects of the problem examined.
- Classes 2–5: with each customer i, a set of m_i is uniformly distributed within a given range (see table 1, column 2). Each item is classified into one of the three shape categories, with equal probability. The three categories are *vertical* (the relative lengths are greater than the relative widths), *homogeneous* (the relative lengths and widths are generated in the same intervals), and *horizontal* (the relative lengths are smaller than the relative widths). The dimensions (width and length) of an item are uniformly distributed into the ranges determined by this item's shape category (see table 1, column 3-8).

Table 1. The characteristics of items of classes 2-5 instances

		Vertical		Homogenous		Horizontal	
Class	m_i	Length	Width	Length	Width	Length	Width
2	[1, 2]	[0.4L, 0.9L]	[0.1W, 0.2W]	[0.2L, 0.5L]	[0.2W, 0.5W]	[0.1L, 0.2L]	[0.4W, 0.9W]
3	[1, 3]	[0.3L, 0.8L]	[0.1W, 0.2W]	[0.2L, 0.4L]	[0.2W, 0.4W]	[0.1L, 0.2L]	[0.3W, 0.8W]
4	[1, 4]	[0.2L, 0.7L]	[0.1W, 0.2W]	[0.1L, 0.4L]	[0.1W, 0.4W]	[0.1L, 0.2L]	[0.2W, 0.7W]
5	[1, 5]	[0.1L, 0.6L]	[0.1W, 0.2W]	[0.1L, 0.3L]	[0.1W, 0.3W]	[0.1L, 0.2L]	[0.1W, 0.6W]

Table 2. The characteristics of classes of 2-5 instances

		Number of items of classes 2-5						Number of items of classes 2-5			
Inst	n	2	3	4	5	Inst	n	2	3	4	5
1	15	24	31	37	45	19	50	82	103	134	157
2	15	25	31	40	48	20	71	104	151	178	226
3	20	29	46	44	49	21	75	114	164	168	202
4	20	32	43	50	62	22	75	112	154	198	236
5	21	31	37	41	57	23	75	112	155	179	225
6	21	33	40	57	56	24	75	124	152	195	215
7	22	32	41	51	55	25	100	157	212	254	311
8	22	29	42	48	52	26	100	147	198	247	310
9	25	40	61	63	91	27	100	152	211	245	320
10	29	43	49	72	86	28	120	183	242	299	384
11	29	43	62	74	91	29	134	197	262	342	422
12	30	50	56	82	101	30	150	225	298	366	433
13	32	44	56	78	102	31	199	307	402	513	602
14	32	47	57	65	87	32	199	299	404	497	589
15	32	48	59	84	114	33	199	301	407	499	577
16	35	56	74	93	114	34	240	370	490	604	720
17	40	60	73	96	127	35	252	367	507	634	762
18	44	66	87	112	122	36	255	387	511	606	786

n number of customer

The values $L = 40$ and $W = 20$ were chosen for the dimensions of the loading area. The numbers of customers and items, in the instances for Classes 2–5, are shown in table 2. For details of the datasets, the reader is referred to Gendreau et al. [7] and Zachariadis et al. [8]. As the aforementioned characteristics of data set, each data set contains number of customers, number of items, vehicle's capacity, length & width of container, depot coordinate, customers coordinate, number of items/customer, and length & width of each items that different with the other data set, as we can see in figure 6.

Figure 6 shows that on that data set, there are 15 customers to be served and total items from those customers are 31 item. To know the location for each vertices (depot and customers), *X* and *Y* coordinate are used on the data set, for example, depot are in (30,40) position, customer 1 are in (37,52) position, et cetera. Customer 1 have 2 items that should be delivered which each item have length and width 4×11 and 13×6, respectively, with total weight for those two item are 7. Total weight capacity for the container are 90 with container area is 40×20.

4.2. Results on benchmark instances

There are ten algorithms combination for this research, first five algorithms combination are the combination between general cutting NN and five loading heuristics, while last five algorithms combination are the combination between skip cutting NN and five loading heuristics. From all combination of algorithms, the best algorithm combination will be found in accordance with the performance measurements that has been stated. There are 3 performance measurement that will be used to find the best algorithm combination: *K* (the number of vehicles needed), *D* (total vehicle's mileage), and *U* (containers utility). Good performance are indicated from those three performance measurements value, the smaller *K* and *D* value, the better performance are achieved, same as the bigger *U* value will achieve the better performance.

4.2.1. Loading heuristic methods comparison. Main program which is the combination between nearest neighbor and five loading heuristic algorithms can be used to solve sequential 2L-CVRP problem properly. Summary of data test results for algorithm combination between regular cutting NN and five loading heuristics can be ssen on table 3. SUM row indicated as the total of all results from performance measurements for each methods from 180 data. While AVERAGE row indicated as the average of all results from performance measurements for each methods from 180 data.

Table 3. Data test results summary for algorithm combination between regular cutting NN and five loading heuristics

		Regular Cutting NN				
		Heuristic 1	**Heuristic 2**	**Heuristic 3**	**Heuristic 4**	**Heuristic 5**
SUM	**K**	4352	3954	**3860**	3995	4316
	D	374742	352806	**347892**	354842	372566
	U	68,2	76,04	**77,84**	75,44	69,75
AVERAGE	**K**	24,18	21,97	**21,44**	22,19	23,98
	D	2082	1960	**1933**	1971	2070
	U	38%	42%	**43%**	42%	39%

Table 3 shows that from five algorithm combination between regular cutting NN and five loading heuristics, heuristic 3 (maximum touching perimeter) have the best solution from the other heuristics. Because of that, to prove that these heuristic methods is the best method, five loading heuristics are also combined with skip cutting NN algorithm that can be seen in table 4 which produce the same result as before.

Table 4. Data test results summary for algorithm combination between skip cutting NN and five loading heuristics

		Skip Cutting NN				
		Heuristic 1	**Heuristic 2**	**Heuristic 3**	**Heuristic 4**	**Heuristic 5**
SUM	**K**	3420	3055	**3030**	3077	3348
	D	434026	416526	**410276**	414222	430344
	U	85,12	95,79	**97,08**	96,03	86,95
AVERAGE	**K**	19	16,97	**16,83**	17,09	18,6
	D	2411	2314	**2279**	2301	2391
	U	47%	53%	**54%**	53%	48%

Table 4 shows that from five algorithm combination between skip cutting NN and five loading heuristics, heuristic 3 (maximum touching perimeter) have the best solution from the other heuristics. Maximum touching perimeter are said as the best loading heuristics because have the smallest value of *K* (the number of vehicles needed) and *D* (total vehicle's mileage), and have the biggest value of *U* (containers utility).

4.2.2. Route determination methods comparison. There are two route determination methods, namely regular and skip cutting NN. The comparison between these two methods are summarized in table 5. For each methods of nearest neighbor, averaging five loading results are done for doing this comparison.

Table 5. Data test results summary for route determination methods

		Regular Cutting NN	Skip Cutting NN
SUM	**K**	4095	**3186**
	D	**360569.6**	421078.8
	U	73.45	**92.19**
AVERAGE	**K**	22.75	**17.7**
	D	**2003**	2339
	U	0.408	**0.512**

If viewed from the value of *K* (the number of vehicles needed) and *U* (containers utility), skip cutting NN has the better performance than regular cutting NN. But, if viewed from the value of *D* (total vehicle's mileage), regular cutting NN has the better performance than skip cutting NN.

4.2.3. Algorithms combination comparison. From comparisons above, loading heuristics have been compared at each NN, regular and skip cutting NN. Because of that, last comparison is needed, which is will compare all algorithms combination that used to solve sequential 2L-CVRP problem in this research. The comparison results between those ten algorithms combination for each performance measurement, *K* (the number of vehicles needed), *D* (total vehicle's mileage), and *U* (containers utility), can be seen in table 6.

Table 6 shows that if viewed from performance measurement *K* (the number of vehicles needed) and *U* (containers utility), then combination of algorithm between skip cutting nearest neighbor and maximum touching perimeter have a better performance from the others. Whereas, if viewed from performance measurement *D* (total vehicle's mileage), then combination of algorithm between regular-cutting nearest neighbor and maximum touching perimeter have a better performance from the others.

Table 6. Data test results summary for sequential 2L-CVRP

		Regular Cutting NN					Skip Cutting NN				
		Heuristic 1	Heuristic 2	Heuristic 3	Heuristic 4	Heuristic 5	**Heuristic 1**	**Heuristic 2**	**Heuristic 3**	**Heuristic 4**	**Heuristic 5**
SUM	**K**	4352	3954	3860	3995	4316	3420	3055	**3030**	3077	3348
	D	374742	352806	**347892**	354842	372566	434026	416526	410276	414222	430344
	U	68.20	76.04	77.84	75.44	69.75	85.12	95.79	**97.08**	96.03	86.95
AVERAGE	**K**	24.18	21.97	21.44	22.19	23.98	19.00	16.97	**16.83**	17.09	18.60
	D	2082	1960	**1933**	1971	2070	2411	2314	2279	2301	2391
	U	38%	42%	43%	42%	39%	47%	53%	**54%**	53%	48%

Comparison results from those combination of algorithms depicts that the less container is needed will affect the bigger container utility. Beside, for performance measurement *K* and *U*, skip cutting NN method have a better performance because all customers are checked, so, the container can be used more optimal than regular cutting NN method which directly cut off the route if one of two loading constraints are not fulfilled. Otherwise, for performance measurement *D*, regular cutting NN method have a better performance because this method directly do the cut off if one of two loading constraints are not fulfilled.

5. Conclusions

In this paper, we study a generalisation of the VRP, in which the demand of customers consists of weighted, two-dimensional, rectangular items. This problem is called sequential two dimensional loading constraints capacitated vehicle routing problem (sequential 2L-CVRP) and have three performance measurements to choose the best combination of algorithms that used to test the instances. The 2L-CVRP is of particular theoretical interest as it combines two frequently studied combinatorial optimisation problems, namely the Vehicle Routing Problem, and the two-dimensional bin packing problem. Although 2L-CVRP has several real-life applications in the field of transportation logistics.

Regarding the packing features of the problem examined, our algorithm makes use of a bundle of packing heuristics, producing diverse packing structures in order to increase the probability of obtaining a feasible loading. Those bundle of packing heuristics consist of Bottom Left Fill, Maximum Touching Perimeter, and Minimum area heuristics. The routing aspects of the problem are handled by a Nearest Neighbor method that divided into two cluster, namely regular and skip cutting nearest neighbor.

The conclusion for this research is that from 10 combination algorithms, if viewed from performance measurement *K* (the number of vehicles needed) and *U* (containers utility), then combination of algorithm between skip cutting nearest neighbor and maximum touching perimeter have a better performance from the others. Whereas, if viewed from performance measurement *D* (total vehicle's mileage), then combination of algorithm between regular-cutting nearest neighbor and maximum touching perimeter have a better performance from the other method.

References

[1] Chopra S and Meindl P 2007 *Supply Chain Management Strategy, Planning, and Operation, 3rd edition* (New Jersey: Prentice Hall)

[2] Miltenburg J 2005 *Manufacturing Strategy : How to Formulate and Implement a Winning Plan, 2nd edition* (Oregon: Productivity Press)

[3] Pujawan I N and Mahendrawati E R 2010 *Supply Chain Management, edisi 2* (Surabaya: Guna Widya)

[4] Miao L, Ruan Q, Woghiren K and Ruo Q 2012 A hybrid genetic algorithm for the vehicle routing problem with three-dimensional loading constraints *RAIRO-Operations Research* **46(01)** 63-82

[5] Leung S C, Zheng J, Zhang D and Zhou X 2010 Simulated annealing for the vehicle routing problem with two-dimensional loading constraints *Flexible services and manufacturing journal* **22(1-2)** 61-82

[6] Laporte G and Semet F 2001 Classical heuristics for the capacitated VRP *The vehicle routing problem - Society for Industrial and Applied Mathematics* 109-128

[7] Gendreau M, Laporte G and Potvin J Y 2001 Metaheuristics for the capacitated VRP *The vehicle routing problem - Society for Industrial and Applied Mathematics* 129-154

[8] Cordeau J F, Gendreau M, Laporte G, Potvin J Y and Semet F 2002 A guide to vehicle routing heuristics *Journal of the Operational Research society* 512-522

[9] Cordeau J F, Gendreau M, Hertz A, Laporte G and Sormany J S 2005 New heuristics for the vehicle routing problem *Logistics systems: design and optimization - Springer US* 279-297

[10] Gendreau M, Iori M, Laporte G and Martello S 2006 A tabu search algorithm for a routing and container loading problem *Transportation Science* **40(3)** 342-350

[11] Fuellerer G, Doerner K F, Hartl R F and Iori M 2009 Ant colony optimization for the two-dimensional loading vehicle routing problem *Computers and Operations Research* **36(3)** 655-673

[12] Toth P and Vigo D 2001 Branch-and-bound algorithms for the capacitated VRP *The vehicle routing problem - Society for Industrial and Applied Mathematics* 29-51

[13] Fukasawa R, Longo H, Lysgaard J, Aragão M P D, Reis M, Uchoa E and Werneck R F 2006 Robust branch-and-cut-and-price for the capacitated vehicle routing problem *Mathematical programming* **106(3)** 491-511

[14] Baldacci R, Christofides N and Mingozzi A 2008 An exact algorithm for the vehicle routing problem based on the set partitioning formulation with additional cuts *Mathematical Programming* **115(2)** 351-385

[15] Martello S and Vigo D 1998 Exact solution of the two-dimensional finite bin packing problem *Management science* **44(3)** 388-399

[16] Fekete S P and Schepers J 2001 New classes of fast lower bounds for bin packing problems *Mathematical programming* **91(1)** 11-31

[17] Fekete S P and Schepers J 2004 A general framework for bounds for higher-dimensional orthogonal packing problems *Mathematical Methods of Operations Research* **60(2)** 311-329

[18] Fekete S P, Schepers J and Van der Veen J C 2007 An exact algorithm for higher-dimensional orthogonal packing *Operations Research* **55(3)** 569-587

[19] Boschetti M A and Mingozzi A 2003a The two-dimensional finite bin packing problem. Part I: New lower bounds for the oriented case *Quarterly Journal of the Belgian, French and Italian Operations Research Societies* **1(1)** 27-42

[20] Boschetti M A and Mingozzi A 2003b The two-dimensional finite bin packing problem. Part II: New lower and upper bounds *Quarterly Journal of the Belgian, French and Italian Operations Research Societies* **1(2)** 135-147

[21] Pisinger D and Sigurd M 2007 Using decomposition techniques and constraint programming for solving the two-dimensional bin-packing problem *INFORMS Journal on Computing* **19(1)** 36-51

[22] Lodi A, Martello S and Vigo D 1999 Heuristic and metaheuristic approaches for a class of two-dimensional bin packing problems *INFORMS Journal on Computing* **11(4)** 345-357
[23] Alvarez-Valdes R, Parreño F and Tamarit J M 2005 A GRASP algorithm for constrained two-dimensional non-guillotine cutting problems *Journal of the Operational Research Society* **56(4)** 414-425
[24] Alvarez-Valdés R, Parreño F and Tamarit J M 2007 A tabu search algorithm for a two-dimensional non-guillotine cutting problem *European Journal of Operational Research* **183(3)** 1167-1182
[25] Khebbache S, Prins C and Yalaoui A 2008 Iterated local search algorithm for the constrained two-dimensional non-guillotine cutting problem *Journal of Industrial and Systems Engineering* **2(3)** 164-179
[26] Iori M 2004 *Metaheuristic algorithms for combinatorial optimization problems* PhD thesis DEIS (University of Bologna)
[27] Iori M 2005 Metaheuristic algorithms for combinatorial optimization problems *4OR* **3(2)** 163-166
[28] Iori M, Salazar-González J J and Vigo D 2007 An exact approach for the vehicle routing problem with two-dimensional loading constraints *Transportation Science* **41(2)** 253-264
[29] Gendreau M, Iori M, Laporte G and Martello S 2008 A Tabu search heuristic for the vehicle routing problem with two-dimensional loading constraints *Networks* **51(1)** 4-18
[30] Zachariadis E E, Tarantilis C D and Kiranoudis C T 2009 A guided tabu search for the vehicle routing problem with two-dimensional loading constraints *European Journal of Operational Research* **195(3)** 729-743
[31] Chazelle B 1983 The bottomn-left bin-packing heuristic: An efficient implementation *Computers, IEEE Transactions on* **100(8)** 697-707
[32] Lodi A, Martello S and Vigo D 1999 Heuristic and metaheuristic approaches for a class of two-dimensional bin packing problems *INFORMS Journal on Computing* **11(4)** 345-357
[33] Toth P and Vigo D 2002 The vehicle routing problem. Monographs on discrete mathematics and applications SIAM (Philadelphia)

Appendix

Table A1
Results obtained for the regular cutting nearest neighbor

Inst		Kontainer 1			Kontainer 2			Kontainer 3			Kontainer 4			Kontainer 5		
ance	Class	K	D	U	K	D	U	K	D	U	K	D	U	K	D	U
	1	4	432	0%	4	432	0%	4	432	0%	4	432	0%	4	432	0%
	2	5	498	47%	5	462	47%	5	480	47%	5	492	47%	5	444	47%
	3	5	468	49%	5	492	49%	5	470	49%	5	464	49%	5	460	49%
	4	6	500	47%	5	446	56%	6	472	47%	5	446	56%	6	520	47%
1	5	0	0	0%	0	0	0%	0	0	0%	0	0	0%	0	0	0%
	1	6	494	0%	6	494	0%	6	494	0%	6	494	0%	6	494	0%
	2	6	500	44%	5	480	52%	6	506	44%	5	480	52%	6	484	44%
	3	7	540	42%	6	496	49%	7	540	42%	6	502	49%	6	490	49%
	4	6	442	44%	5	416	53%	5	416	53%	5	416	53%	5	444	53%
2	5	5	456	39%	4	444	49%	3	374	65%	4	424	49%	3	400	65%
	1	5	584	1%	5	584	1%	5	584	1%	5	584	1%	5	584	1%
	2	8	676	43%	7	650	49%	7	618	49%	7	650	49%	7	636	49%
	3	9	730	43%	7	650	55%	7	650	55%	7	616	55%	8	736	48%
	4	7	584	44%	7	638	44%	7	638	44%	7	622	44%	8	644	39%
3	5	6	570	45%	5	566	54%	5	552	54%	5	566	54%	6	580	45%
	1	7	626	0%	7	626	0%	7	626	0%	7	626	0%	7	626	0%
	2	8	698	43%	6	586	58%	6	590	58%	6	586	58%	7	624	49%
	3	8	690	44%	6	578	58%	6	562	58%	7	620	50%	8	714	44%
	4	10	742	38%	9	702	42%	9	718	42%	9	704	42%	9	714	42%
4	5	8	666	35%	7	664	40%	8	666	35%	7	664	40%	7	664	40%
	1	5	640	1%	5	640	1%	5	640	1%	5	640	1%	5	640	1%
	2	6	620	48%	6	680	48%	5	642	58%	6	650	48%	6	692	48%
	3	6	672	50%	5	698	60%	5	664	60%	5	628	60%	5	652	60%
	4	7	800	43%	5	618	61%	5	670	61%	6	698	50%	5	652	61%
5	5	8	820	33%	5	608	53%	5	614	53%	6	680	44%	6	660	44%
	1	7	678	0%	7	678	0%	7	678	0%	7	678	0%	7	678	0%
	2	8	798	40%	8	810	40%	8	794	40%	8	810	40%	8	798	40%
	3	6	672	63%	8	764	47%	7	712	54%	7	706	54%	10	890	38%
	4	9	832	48%	8	754	54%	8	808	54%	8	754	54%	8	728	54%
6	5	6	670	46%	5	588	56%	5	546	56%	5	590	56%	5	572	56%
	1	4	1030	1%	4	1030	1%	4	1030	1%	4	1030	1%	4	1030	1%
	2	7	1376	51%	7	1418	51%	7	1376	51%	7	1418	51%	7	1376	51%
	3	7	1354	47%	6	1216	55%	6	1264	55%	6	1250	55%	6	1294	55%
	4	8	1418	42%	7	1222	48%	7	1214	48%	9	1340	37%	7	1200	48%
7	5	7	1248	42%	5	1144	58%	5	1132	58%	5	1082	58%	6	1254	49%

Table A1 continued

Instance	Class	Kontainer 1			Kontainer 2			Kontainer 3			Kontainer 4			Kontainer 5		
		K	D	U	K	D	U	K	D	U	K	D	U	K	D	U
	1	4	1030	1%	4	1030	1%	4	1030	1%	4	1030	1%	4	1030	1%
	2	7	1236	45%	6	1298	52%	7	1380	45%	6	1170	52%	6	1164	52%
	3	7	1370	50%	7	1372	50%	7	1372	50%	7	1372	50%	7	1372	50%
	4	6	1128	57%	7	1368	49%	6	1224	57%	7	1334	49%	7	1364	49%
8	5	5	1108	45%	4	1096	56%	4	1086	56%	3	964	75%	4	1046	56%
	1	9	886	0%	9	886	0%	9	886	0%	9	886	0%	9	886	0%
	2	10	874	45%	9	882	50%	10	874	45%	9	882	50%	9	882	50%
	3	10	976	49%	11	966	44%	10	940	49%	10	960	49%	12	1086	41%
	4	9	826	53%	9	888	53%	7	748	69%	9	868	53%	9	926	53%
9	5	9	936	47%	8	880	53%	7	800	61%	8	838	53%	8	856	53%
	1	4	792	1%	4	792	1%	4	792	1%	4	792	1%	4	792	1%
	2	10	1470	44%	8	1316	55%	8	1254	55%	8	1240	55%	10	1464	44%
	3	8	1172	49%	7	1078	56%	7	1166	56%	7	1092	56%	8	1222	49%
	4	10	1438	51%	10	1548	51%	11	1544	46%	10	1466	51%	10	1518	51%
10	5	9	1348	49%	8	1294	55%	7	1168	63%	8	1260	55%	8	1254	55%
	1	4	792	1%	4	792	1%	4	792	1%	4	792	1%	4	792	1%
	2	9	1390	51%	8	1208	57%	7	1168	66%	8	1240	57%	9	1400	51%
	3	11	1418	47%	12	1532	43%	10	1416	52%	11	1520	47%	13	1598	40%
	4	12	1640	48%	12	1580	48%	11	1594	53%	12	1700	48%	12	1570	48%
11	5	9	1296	49%	8	1166	55%	8	1196	55%	8	1156	55%	9	1262	49%
	1	10	874	0%	10	874	0%	10	874	0%	10	874	0%	10	874	0%
	2	11	900	51%	9	840	62%	10	872	56%	9	840	62%	11	900	51%
	3	9	794	52%	8	798	58%	8	784	58%	9	796	52%	9	798	52%
	4	14	1026	42%	12	946	49%	11	948	54%	13	980	45%	11	920	54%
12	5	12	966	44%	9	836	58%	9	850	58%	10	914	52%	12	944	44%
	1	3	3166	1%	3	3166	1%	3	3166	1%	3	3166	1%	3	3166	1%
	2	10	4832	48%	9	4546	54%	10	4646	48%	9	4526	54%	10	4488	48%
	3	9	4566	53%	9	4560	53%	9	4700	53%	9	4620	53%	9	4732	53%
	4	12	5134	45%	10	5398	54%	9	4726	60%	9	4740	60%	9	4680	60%
13	5	9	4940	52%	7	4390	67%	7	4224	67%	8	4548	59%	9	4678	52%
	1	5	1214	1%	5	1214	1%	5	1214	1%	5	1214	1%	5	1214	1%
	2	9	2028	51%	9	2074	51%	9	1998	51%	9	1978	51%	9	2086	51%
	3	9	2052	51%	9	1974	51%	9	1944	51%	11	2262	42%	11	2286	42%
	4	10	2140	43%	8	1874	53%	7	1700	61%	7	1746	61%	9	2110	47%
14	5	7	1594	56%	6	1480	65%	6	1488	65%	7	1660	56%	7	1668	56%

Table A1 continued

Instance	Class	Kontainer 1			Kontainer 2			Kontainer 3			Kontainer 4			Kontainer 5		
		K	D	U	K	D	U	K	D	U	K	D	U	K	D	U
	1	5	1214	1%	5	1214	1%	5	1214	1%	5	1214	1%	5	1214	1%
	2	9	2062	50%	8	1768	57%	9	2022	50%	8	1866	57%	11	2332	41%
	3	12	2570	42%	9	2016	56%	10	2234	50%	9	2090	56%	11	2364	46%
	4	11	2264	52%	10	2120	57%	10	2068	57%	10	2172	57%	12	2532	48%
15	**5**	11	2374	52%	11	2178	52%	10	2118	57%	11	2220	52%	10	2198	57%
	1	13	1030	0%	13	1030	0%	13	1030	0%	13	1030	0%	13	1030	0%
	2	15	1128	41%	16	1168	38%	15	1148	41%	16	1194	38%	16	1168	38%
	3	16	1186	39%	15	1134	42%	16	1166	39%	15	1134	42%	16	1166	39%
	4	15	1120	47%	15	1136	47%	16	1198	44%	16	1190	44%	18	1308	39%
16	**5**	13	1032	41%	13	1030	41%	13	1030	41%	13	1030	41%	13	1030	41%
	1	16	1216	0%	16	1216	0%	16	1216	0%	16	1216	0%	16	1216	0%
	2	18	1310	36%	18	1302	36%	18	1310	36%	17	1262	38%	20	1374	32%
	3	17	1252	34%	16	1216	37%	16	1206	37%	16	1216	37%	17	1266	34%
	4	18	1336	40%	16	1216	45%	16	1216	45%	16	1216	45%	19	1390	38%
17	**5**	17	1280	33%	16	1216	35%	16	1240	35%	16	1240	35%	16	1210	35%
	1	5	1484	1%	5	1484	1%	5	1484	1%	5	1484	1%	5	1484	1%
	2	14	2034	46%	12	1836	54%	13	1838	50%	12	1822	54%	14	1996	46%
	3	16	2056	47%	14	2044	54%	14	2042	54%	13	1986	58%	17	2302	44%
	4	17	2374	46%	14	2042	55%	14	2010	55%	15	2032	52%	15	2112	52%
18	**5**	12	1886	49%	10	1734	58%	8	1434	73%	10	1748	58%	11	1906	53%
	1	6	1020	1%	6	1020	1%	6	1020	1%	6	1020	1%	6	1020	1%
	2	18	1522	47%	17	1438	50%	17	1462	50%	16	1432	53%	19	1552	45%
	3	18	1528	50%	15	1356	61%	16	1506	57%	16	1404	57%	19	1560	48%
	4	19	1544	50%	20	1662	47%	19	1606	50%	19	1612	50%	19	1570	50%
19	**5**	16	1448	46%	13	1306	56%	13	1268	56%	13	1310	56%	13	1308	56%
	1	5	390	2%	5	390	2%	5	390	2%	5	390	2%	5	390	2%
	2	22	1034	54%	22	1020	54%	22	1020	54%	24	1104	49%	24	1088	49%
	3	24	1022	52%	23	1002	54%	22	954	57%	24	992	52%	25	1098	50%
	4	24	1038	54%	26	1120	50%	21	956	62%	24	1066	54%	26	1140	50%
20	**5**	22	970	50%	19	886	58%	18	868	61%	17	822	65%	20	948	55%
	1	7	1070	1%	7	1070	1%	7	1070	1%	7	1070	1%	7	1070	1%
	2	26	2168	45%	24	2110	49%	24	2042	49%	23	2060	51%	27	2180	44%
	3	29	2336	48%	27	2348	51%	26	2342	53%	28	2392	50%	27	2244	51%
	4	25	2096	48%	23	1998	52%	21	1808	57%	22	1988	54%	25	2132	48%
21	**5**	22	1914	47%	18	1704	57%	18	1726	57%	19	1772	54%	19	1752	54%

Table A1 continued

Instance	Class	Kontainer 1			Kontainer 2			Kontainer 3			Kontainer 4			Kontainer 5		
		K	D	U	K	D	U	K	D	U	K	D	U	K	D	U
	1	8	1184	1%	8	1184	1%	8	1184	1%	8	1184	1%	8	1184	1%
	2	26	2142	47%	22	1930	56%	23	1974	53%	21	1832	59%	25	1964	49%
	3	26	2082	50%	26	2148	50%	23	2012	57%	23	2024	57%	30	2332	43%
	4	31	2376	45%	25	2018	56%	26	2082	54%	27	2122	52%	28	2274	50%
22	5	23	2130	49%	19	1750	59%	19	1770	59%	18	1738	63%	22	1940	51%
	1	11	1294	1%	11	1294	1%	11	1294	1%	11	1294	1%	11	1294	1%
	2	26	2188	46%	26	2274	46%	24	2088	50%	25	2178	48%	27	2308	44%
	3	25	2168	53%	24	2068	55%	24	2050	55%	23	2048	58%	27	2210	49%
	4	27	2228	49%	25	2138	53%	25	2088	53%	24	2132	56%	28	2156	48%
23	5	23	1928	50%	22	1864	52%	20	1792	57%	21	1882	55%	25	2036	46%
	1	16	1658	1%	16	1658	1%	16	1658	1%	16	1658	1%	16	1658	1%
	2	26	2196	53%	26	2184	53%	27	2194	51%	27	2216	51%	27	2204	51%
	3	28	2198	47%	24	2042	54%	24	2052	54%	24	2042	54%	30	2372	44%
	4	31	2468	45%	24	2006	58%	24	2082	58%	26	2140	54%	30	2366	46%
24	5	22	1948	51%	19	1808	59%	18	1816	62%	20	1878	56%	22	1984	51%
	1	8	1478	2%	8	1478	2%	8	1478	2%	8	1478	2%	8	1478	2%
	2	35	2894	50%	34	2820	51%	34	2798	51%	35	2888	50%	38	3030	46%
	3	39	2944	46%	37	3028	48%	32	2700	56%	37	2952	48%	38	2962	47%
	4	37	3022	49%	36	3038	51%	31	2710	59%	36	2990	51%	38	3126	48%
25	5	29	2572	49%	25	2354	57%	24	2256	60%	27	2462	53%	28	2448	51%
	1	10	1362	1%	10	1362	1%	10	1362	1%	10	1362	1%	10	1362	1%
	2	34	2784	47%	29	2574	55%	32	2800	50%	30	2684	53%	33	2704	48%
	3	33	2942	50%	28	2564	59%	28	2582	59%	28	2512	59%	33	2880	50%
	4	34	2868	51%	33	2858	53%	32	2914	55%	32	2710	55%	38	3222	46%
26	5	30	2540	50%	26	2368	58%	22	2058	68%	24	2192	62%	29	2572	52%
	1	15	1830	1%	15	1830	1%	15	1830	1%	15	1830	1%	15	1830	1%
	2	35	2790	46%	32	2668	50%	31	2624	52%	34	2718	47%	37	2940	43%
	3	40	2974	45%	35	2798	51%	34	2734	53%	36	2786	50%	42	3146	43%
	4	34	2748	50%	34	2742	50%	29	2524	59%	31	2590	55%	36	2866	47%
27	5	31	2782	50%	27	2462	58%	26	2384	60%	28	2490	56%	32	2790	49%
	1	8	1700	2%	8	1700	2%	8	1700	2%	8	1700	2%	8	1700	2%
	2	34	5114	56%	34	5096	56%	33	5084	58%	36	5404	53%	38	5494	50%
	3	45	6082	46%	41	5580	51%	38	5290	55%	39	5542	53%	43	6098	48%
	4	42	5694	50%	40	5510	52%	38	5422	55%	42	5630	50%	41	5668	51%
28	5	40	6070	45%	30	4678	60%	29	4668	62%	32	4988	57%	33	5000	55%

Table A1 continued

Inst ance	Class	Kontainer 1			Kontainer 2			Kontainer 3			Kontainer 4			Kontainer 5		
		K	D	U	K	D	U	K	D	U	K	D	U	K	D	U
	1	8	1972	2%	8	1972	2%	8	1972	2%	8	1972	2%	8	1972	2%
	2	39	4366	52%	36	4210	57%	38	4294	54%	38	4272	54%	38	4376	54%
	3	47	4650	46%	39	4280	55%	40	4292	54%	41	4470	52%	43	4388	50%
	4	47	4732	51%	42	4580	57%	44	4858	54%	44	4410	54%	51	5194	47%
29	5	46	4794	45%	33	3918	63%	34	4130	61%	36	4100	58%	38	4224	55%
	1	12	1730	2%	12	1730	2%	12	1730	2%	12	1730	2%	12	1730	2%
	2	48	3782	50%	47	3812	51%	45	3674	54%	45	3632	54%	50	3888	48%
	3	53	4126	49%	52	4110	50%	49	3828	53%	49	3864	53%	53	4064	49%
	4	51	4052	51%	49	3826	53%	45	3624	57%	50	3908	52%	52	4102	50%
30	5	41	3422	51%	34	3014	62%	33	3024	64%	35	3134	60%	40	3334	52%
	1	17	2072	1%	17	2072	1%	17	2072	1%	17	2072	1%	17	2072	1%
	2	67	4930	48%	58	4296	55%	61	4652	53%	58	4394	55%	67	4984	48%
	3	69	4940	49%	63	4764	54%	62	4626	55%	65	4826	52%	71	5146	48%
	4	70	5194	52%	68	4998	53%	63	4744	57%	67	4904	54%	75	5468	48%
31	5	61	4638	48%	50	3988	59%	52	4312	56%	49	4042	60%	55	4364	53%
	1	17	2072	1%	17	2072	1%	17	2072	1%	17	2072	1%	17	2072	1%
	2	68	4984	47%	63	4690	51%	61	4718	53%	65	4782	49%	66	4788	49%
	3	65	4916	51%	61	4612	54%	59	4504	56%	64	4800	52%	64	4820	52%
	4	68	5060	50%	63	4650	53%	62	4686	54%	64	4800	53%	72	5258	47%
32	5	57	4492	50%	48	3958	59%	44	3740	65%	49	4026	58%	53	4222	54%
	1	17	2112	1%	17	2112	1%	17	2112	1%	17	2112	1%	17	2112	1%
	2	64	4944	49%	62	4606	51%	61	4530	52%	65	4896	49%	67	5010	47%
	3	71	5228	49%	63	4746	55%	62	4696	56%	64	4824	54%	69	5182	50%
	4	75	5470	47%	68	5018	52%	62	4638	57%	68	5034	52%	70	5214	51%
33	5	58	4480	50%	48	3922	61%	46	3772	63%	47	3868	62%	56	4330	52%
	1	23	1074	1%	23	1074	1%	23	1074	1%	23	1074	1%	23	1074	1%
	2	82	2542	48%	73	2302	54%	75	2388	52%	76	2456	52%	81	2458	48%
	3	87	2742	48%	77	2428	54%	74	2366	57%	81	2590	52%	86	2668	49%
	4	86	2626	50%	80	2448	54%	77	2390	56%	81	2504	53%	84	2560	51%
34	5	76	2384	48%	64	2060	57%	62	2010	58%	65	2084	56%	70	2176	52%
	1	27	1128	1%	27	1128	1%	27	1128	1%	27	1128	1%	27	1128	1%
	2	79	2908	49%	73	2700	53%	74	2748	52%	75	2776	51%	84	3042	46%
	3	95	3362	45%	78	2876	54%	75	2738	57%	80	2900	53%	91	3204	47%
	4	91	3418	49%	86	3190	52%	80	2990	56%	90	3326	50%	96	3472	47%
35	5	73	2742	51%	63	2424	59%	63	2430	59%	61	2400	61%	68	2614	54%

Table A1 continued

Instance	Class	Kontainer 1			Kontainer 2			Kontainer 3			Kontainer 4			Kontainer 5		
		K	D	U	K	D	U	K	D	U	K	D	U	K	D	U
	1	14	918	2%	14	918	2%	14	918	2%	14	918	2%	14	918	2%
	2	87	3954	47%	79	3612	51%	79	3562	51%	79	3658	51%	86	3874	47%
	3	89	3976	49%	82	3770	53%	80	3598	55%	81	3704	54%	87	3928	50%
	4	86	3744	49%	78	3428	54%	72	3228	59%	78	3494	54%	88	3828	48%
36	5	78	3504	49%	63	2872	60%	64	2972	59%	67	3046	56%	71	3226	53%

Table A2
Results obtained for the skip cutting nearest neighbor

Instance	Class	Kontainer 1			Kontainer 2			Kontainer 3			Kontainer 4			Kontainer 5		
		K	D	U	K	D	U	K	D	U	K	D	U	K	D	U
	1	3	472	1%	3	472	1%	3	472	1%	3	472	1%	3	472	1%
	2	4	488	59%	4	550	59%	4	506	59%	3	506	78%	4	516	59%
	3	5	532	49%	4	500	61%	4	526	61%	4	474	61%	4	526	61%
	4	5	588	56%	5	596	56%	4	562	70%	4	576	70%	6	592	47%
1	5	0	0	0%	0	0	0%	0	0	0%	0	0	0%	0	0	0%
	1	5	504	0%	5	504	0%	5	504	0%	5	504	0%	5	504	0%
	2	5	590	52%	4	480	65%	5	590	52%	4	480	65%	5	592	52%
	3	6	544	49%	5	638	59%	5	574	59%	5	592	59%	6	520	49%
	4	5	584	53%	4	574	66%	5	588	53%	4	574	66%	5	536	53%
2	5	4	458	49%	3	424	65%	3	428	65%	3	448	65%	4	442	49%
	1	4	540	1%	4	540	1%	4	540	1%	4	540	1%	4	540	1%
	2	6	684	57%	6	708	57%	6	654	57%	6	718	57%	6	722	57%
	3	7	776	55%	6	672	64%	6	652	64%	6	670	64%	6	678	64%
	4	5	676	62%	5	722	62%	5	720	62%	5	766	62%	6	770	52%
3	5	5	738	54%	4	672	68%	4	558	68%	4	624	68%	5	714	54%
	1	6	654	0%	6	654	0%	6	654	0%	6	654	0%	6	654	0%
	2	6	680	58%	6	716	58%	6	670	58%	6	660	58%	5	702	69%
	3	7	770	50%	6	676	58%	5	650	70%	6	630	58%	7	714	50%
	4	10	810	38%	9	766	42%	7	676	54%	9	770	42%	9	752	42%
4	5	8	666	35%	7	664	40%	8	666	35%	7	664	40%	7	664	40%
	1	4	688	1%	4	688	1%	4	688	1%	4	688	1%	4	688	1%
	2	5	742	58%	5	876	58%	4	862	72%	5	782	58%	6	774	48%
	3	5	744	60%	5	760	60%	4	714	75%	5	752	60%	5	728	60%
	4	5	888	61%	5	740	61%	5	764	61%	4	742	76%	5	912	61%
5	5	6	732	44%	5	672	53%	5	614	53%	5	636	53%	6	702	44%

Table A2 continued

Inst		Kontainer 1			Kontainer 2			Kontainer 3			Kontainer 4			Kontainer 5		
ance	**Class**	**K**	**D**	**U**	**K**	**D**	**U**	**K**	**D**	**U**	**K**	**D**	**U**	**K**	**D**	**U**
	1	6	750	0%	6	750	0%	6	750	0%	6	750	0%	6	750	0%
	2	8	752	40%	8	826	40%	8	842	40%	8	826	40%	8	808	40%
	3	6	864	63%	6	918	63%	7	898	54%	6	888	63%	7	888	54%
	4	7	790	61%	7	938	61%	7	926	61%	7	870	61%	7	938	61%
6	**5**	6	726	46%	5	856	56%	4	544	70%	4	726	70%	5	778	56%
	1	3	994	1%	3	994	1%	3	994	1%	3	994	1%	3	994	1%
	2	6	1630	59%	6	1642	59%	5	1606	71%	5	1594	71%	6	1642	59%
	3	6	1396	55%	5	1378	66%	5	1488	66%	5	1338	66%	6	1610	55%
	4	6	1442	56%	6	1296	56%	5	1326	67%	6	1404	56%	6	1502	56%
7	**5**	5	1402	58%	4	1296	73%	4	1306	73%	5	1276	58%	5	1224	58%
	1	3	994	1%	3	994	1%	3	994	1%	3	994	1%	3	994	1%
	2	5	1400	63%	5	1400	63%	4	1376	79%	5	1400	63%	5	1360	63%
	3	6	1254	59%	6	1402	59%	6	1532	59%	6	1120	59%	7	1446	50%
	4	6	1350	57%	5	1344	68%	5	1400	68%	5	1384	68%	6	1590	57%
8	**5**	4	1288	56%	4	1288	56%	4	1266	56%	3	1092	75%	4	1148	56%
	1	8	1020	0%	8	1020	0%	8	1020	0%	8	1020	0%	8	1020	0%
	2	8	1024	57%	8	1076	57%	8	1038	57%	8	1144	57%	8	1126	57%
	3	9	1126	54%	8	1140	61%	8	1098	61%	8	1122	61%	9	1070	54%
	4	7	1162	69%	7	1078	69%	7	1050	69%	7	1032	69%	8	1180	60%
9	**5**	8	1114	53%	6	972	71%	6	1038	71%	6	900	71%	6	1042	71%
	1	3	748	1%	3	748	1%	3	748	1%	3	748	1%	3	748	1%
	2	9	1688	49%	7	1486	63%	7	1486	63%	7	1390	63%	8	1596	55%
	3	7	1252	56%	7	1266	56%	6	1400	66%	6	1426	66%	7	1454	56%
	4	8	1684	64%	8	1596	64%	7	1498	73%	8	1612	64%	8	1680	64%
10	**5**	9	1590	49%	6	1328	73%	6	1146	73%	6	1394	73%	7	1666	63%
	1	3	748	1%	3	748	1%	3	748	1%	3	748	1%	3	748	1%
	2	7	1556	66%	7	1444	66%	6	1294	77%	7	1438	66%	8	1468	57%
	3	9	1562	58%	9	1862	58%	8	1562	65%	9	1772	58%	9	1604	58%
	4	10	1622	58%	9	1744	64%	9	1710	64%	9	1476	64%	10	1626	58%
11	**5**	8	1610	55%	7	1416	63%	7	1382	63%	7	1440	63%	7	1628	63%
	1	10	1122	0%	10	1122	0%	10	1122	0%	10	1122	0%	10	1122	0%
	2	10	1006	56%	8	1066	70%	9	898	62%	8	980	70%	10	1016	56%
	3	8	990	58%	7	1040	66%	7	984	66%	7	962	66%	8	982	58%
	4	10	1184	59%	9	1058	66%	10	1086	59%	10	1150	59%	10	1034	59%
12	5	9	1074	58%	8	1016	66%	8	1160	66%	8	1030	66%	9	1046	58%

IOP Conf. Series: Materials Science and Engineering **166** (2017) 012029 doi:10.1088/1757-899X/166/1/012029

Table A2 continued

Instance	Class	Kontainer 1			Kontainer 2			Kontainer 3			Kontainer 4			Kontainer 5		
		K	D	U	K	D	U	K	D	U	K	D	U	K	D	U
	1	3	4106	1%	3	4106	1%	3	4106	1%	3	4106	1%	3	4106	1%
	2	8	6752	61%	8	6774	61%	7	6306	69%	7	5772	69%	9	6494	54%
	3	8	4706	59%	7	5814	68%	7	5576	68%	7	5324	68%	8	5570	59%
	4	9	5698	60%	9	6118	60%	8	6332	67%	8	5750	67%	8	5798	67%
13	5	8	5880	59%	7	6038	67%	7	6180	67%	7	6088	67%	8	5674	59%
	1	4	1242	1%	4	1242	1%	4	1242	1%	4	1242	1%	4	1242	1%
	2	8	1920	57%	7	1802	65%	7	1918	65%	7	2058	65%	7	2014	65%
	3	8	2354	57%	7	2080	65%	7	1972	65%	7	1892	65%	8	2220	57%
	4	7	2040	61%	6	1852	71%	6	1758	71%	6	1882	71%	8	2154	53%
14	5	7	1898	56%	6	1736	65%	6	1624	65%	6	1726	65%	6	1770	65%
	1	4	1242	1%	4	1242	1%	4	1242	1%	4	1242	1%	4	1242	1%
	2	8	2156	57%	6	1908	76%	7	1822	65%	6	1704	76%	8	2014	57%
	3	9	2514	56%	7	2086	72%	7	1984	72%	7	1988	72%	8	2078	63%
	4	10	2328	57%	9	2174	64%	9	2104	64%	9	2258	64%	10	2412	57%
15	5	9	2192	64%	8	2262	72%	9	2254	64%	8	2156	72%	9	2118	64%
	1	12	1182	0%	12	1182	0%	12	1182	0%	12	1182	0%	12	1182	0%
	2	13	1068	47%	14	1156	44%	14	1224	44%	13	1106	47%	14	1146	44%
	3	14	1092	45%	13	1050	48%	14	1082	45%	14	1112	45%	15	1154	42%
	4	15	1120	47%	14	1152	50%	15	1160	47%	14	1132	50%	16	1224	44%
16	5	13	1044	41%	12	1160	44%	12	1172	44%	12	1138	44%	13	1206	41%
	1	14	1346	0%	14	1346	0%	14	1346	0%	14	1346	0%	14	1346	0%
	2	17	1350	38%	15	1356	43%	17	1390	38%	16	1344	40%	17	1280	38%
	3	15	1258	39%	15	1352	39%	15	1414	39%	15	1398	39%	16	1362	37%
	4	18	1464	40%	15	1286	48%	15	1280	48%	16	1350	45%	18	1470	40%
17	5	15	1286	37%	15	1350	37%	14	1298	40%	15	1350	37%	14	1298	40%
	1	4	1728	1%	4	1728	1%	4	1728	1%	4	1728	1%	4	1728	1%
	2	11	2184	59%	10	2388	64%	10	2202	64%	11	2190	59%	11	2184	59%
	3	12	2868	63%	11	2476	68%	11	2480	68%	11	2366	68%	12	2724	63%
	4	13	2400	60%	12	2476	64%	11	2384	70%	12	2614	64%	13	2618	60%
18	5	9	2194	65%	8	2156	73%	8	1882	73%	9	2396	65%	8	2110	73%
	1	5	902	1%	5	902	1%	5	902	1%	5	902	1%	5	902	1%
	2	13	1602	65%	13	1570	65%	13	1768	65%	14	1664	61%	14	1704	61%
	3	16	1860	57%	13	1656	70%	13	1558	70%	13	1628	70%	15	1680	61%
	4	16	2090	59%	14	1754	67%	13	1790	73%	14	2088	67%	15	2004	63%
19	5	13	1750	56%	10	1438	73%	11	1700	66%	10	1442	73%	11	1622	66%

Table A2 continued

Inst		Kontainer 1			Kontainer 2			Kontainer 3			Kontainer 4			Kontainer 5		
ance	**Class**	**K**	**D**	**U**	**K**	**D**	**U**	**K**	**D**	**U**	**K**	**D**	**U**	**K**	**D**	**U**
	1	4	528	2%	4	528	2%	4	528	2%	4	528	2%	4	528	2%
	2	17	1252	69%	17	1178	69%	17	1152	69%	17	1276	69%	17	1186	69%
	3	20	1522	63%	17	1274	74%	17	1294	74%	17	1332	74%	19	1386	66%
	4	20	1260	65%	18	1246	72%	18	1310	72%	20	1374	65%	20	1348	65%
20	**5**	18	1206	61%	15	1110	73%	15	1050	73%	15	1140	73%	19	1182	58%
	1	7	1350	1%	7	1350	1%	7	1350	1%	7	1350	1%	7	1350	1%
	2	20	2602	59%	18	2452	65%	18	2302	65%	18	2354	65%	18	2712	65%
	3	24	2806	58%	20	2798	70%	20	2640	70%	20	2696	70%	22	2982	63%
	4	18	2842	66%	18	2838	66%	16	2418	74%	17	2622	70%	18	2608	66%
21	**5**	18	2616	57%	14	2390	74%	14	2250	74%	14	2352	74%	15	2378	69%
	1	8	1384	1%	8	1384	1%	8	1384	1%	8	1384	1%	8	1384	1%
	2	20	2608	61%	18	2314	68%	17	2418	72%	19	2280	65%	19	2456	65%
	3	20	3002	65%	18	2862	72%	17	2508	77%	19	2692	69%	21	2926	62%
	4	22	2968	63%	20	3084	70%	19	2764	73%	20	3034	70%	22	3088	63%
22	**5**	18	2694	63%	15	2444	75%	16	2504	71%	15	2410	75%	17	2380	66%
	1	10	1764	1%	10	1764	1%	10	1764	1%	10	1764	1%	10	1764	1%
	2	20	2752	60%	19	2668	63%	19	2672	63%	19	2580	63%	20	2904	60%
	3	20	2780	66%	18	2664	74%	18	2678	74%	18	2518	74%	20	2700	66%
	4	21	2800	64%	18	2742	74%	19	2790	70%	18	2494	74%	21	2674	64%
23	**5**	20	2600	57%	16	2296	72%	16	2234	72%	15	2276	76%	18	2412	64%
	1	14	1886	1%	14	1886	1%	14	1886	1%	14	1886	1%	14	1886	1%
	2	21	2726	65%	20	2516	69%	21	2580	65%	19	2652	72%	21	2806	65%
	3	21	2846	62%	18	2524	73%	19	2840	69%	18	2350	73%	21	2696	62%
	4	23	2688	61%	19	2814	73%	19	2476	73%	20	2746	70%	20	2802	70%
24	**5**	18	2460	62%	16	2642	70%	15	2410	75%	16	2762	70%	17	2678	66%
	1	8	1596	2%	8	1596	2%	8	1596	2%	8	1596	2%	8	1596	2%
	2	26	3416	67%	24	3334	73%	24	3230	73%	24	3338	73%	27	3476	65%
	3	29	3654	61%	25	3448	71%	24	3396	74%	26	3408	69%	28	3720	64%
	4	27	3652	67%	27	3762	67%	25	3486	73%	25	3240	73%	27	3596	67%
25	**5**	23	3538	62%	19	3186	76%	19	3080	76%	19	3118	76%	21	3110	68%
	1	10	1402	1%	10	1402	1%	10	1402	1%	10	1402	1%	10	1402	1%
	2	23	3274	69%	21	3274	76%	22	3580	72%	21	3326	76%	24	3554	66%
	3	26	2726	63%	24	2776	69%	24	2518	69%	24	2786	69%	26	2674	63%
	4	26	3692	67%	24	3546	73%	23	3370	76%	24	3806	73%	27	3802	65%
26	5	25	3098	60%	20	2894	75%	19	2856	79%	21	2986	71%	23	3192	65%

Table A2 continued

Inst		Kontainer 1			Kontainer 2			Kontainer 3			Kontainer 4			Kontainer 5		
ance	Class	K	D	U	K	D	U	K	D	U	K	D	U	K	D	U
	1	14	2346	1%	14	2346	1%	14	2346	1%	14	2346	1%	14	2346	1%
	2	25	3444	64%	22	2918	73%	23	3254	70%	23	3330	70%	24	3366	67%
	3	29	3878	62%	25	3882	72%	25	3776	72%	25	3664	72%	30	3982	60%
	4	26	3478	66%	23	3658	74%	23	3580	74%	24	3796	71%	26	3702	66%
27	5	25	3256	63%	21	2934	74%	22	3432	71%	21	3116	74%	24	3422	65%
	1	7	1824	2%	7	1824	2%	7	1824	2%	7	1824	2%	7	1824	2%
	2	27	5574	71%	26	5296	74%	26	5320	74%	25	5000	76%	28	5980	68%
	3	33	6256	63%	28	5606	74%	29	6076	71%	28	6218	74%	32	6118	65%
	4	33	7532	63%	29	6982	72%	29	6642	72%	29	7026	72%	30	6682	70%
28	5	30	6310	60%	23	5064	79%	23	5178	79%	24	5826	75%	27	5920	67%
	1	7	2444	2%	7	2444	2%	7	2444	2%	7	2444	2%	7	2444	2%
	2	30	5244	68%	28	5272	73%	28	5686	73%	29	5262	70%	31	5654	66%
	3	34	6154	63%	29	5102	74%	28	4832	77%	29	5320	74%	33	5980	65%
	4	36	5952	66%	31	6114	77%	32	5438	75%	32	5800	75%	35	6126	68%
29	5	35	5470	60%	27	4362	77%	28	4868	74%	28	4602	74%	31	4732	67%
	1	12	2190	2%	12	2190	2%	12	2190	2%	12	2190	2%	12	2190	2%
	2	35	4728	69%	32	4170	75%	31	4044	78%	33	4290	73%	35	4722	69%
	3	39	4734	67%	34	4854	77%	37	4750	70%	35	4622	74%	38	4938	69%
	4	39	5102	66%	35	5044	74%	33	4720	78%	36	4778	72%	38	5346	68%
30	5	32	4308	66%	27	4026	78%	28	4172	75%	28	4370	75%	31	4432	68%
	1	16	2620	2%	16	2620	2%	16	2620	2%	16	2620	2%	16	2620	2%
	2	46	5624	70%	43	5282	75%	42	5300	76%	42	5412	76%	46	5542	70%
	3	51	6488	67%	46	5998	74%	45	6114	76%	46	5664	74%	49	5726	70%
	4	53	6484	68%	48	6242	75%	46	6094	78%	49	5892	74%	55	6240	66%
31	5	51	4770	57%	39	4104	75%	39	4232	75%	41	4264	71%	46	4684	64%
	1	16	2620	2%	16	2620	2%	16	2620	2%	16	2620	2%	16	2620	2%
	2	44	5348	73%	42	5488	76%	41	5326	78%	43	5908	75%	45	5668	71%
	3	49	6326	67%	43	5896	77%	43	5700	77%	43	5782	77%	48	6068	69%
	4	49	6218	69%	44	6044	77%	43	5814	78%	45	5810	75%	50	6102	67%
32	5	49	4894	58%	39	4308	73%	38	4004	75%	40	4142	71%	46	4532	62%
	1	17	2112	1%	17	2112	1%	17	2112	1%	17	2112	1%	17	2112	1%
	2	49	4574	64%	46	4488	69%	45	4560	70%	45	4412	70%	50	4782	63%
	3	53	6710	65%	46	5712	75%	46	5874	75%	46	5864	75%	51	5928	68%
	4	52	6662	68%	47	6588	76%	47	6038	76%	47	6656	76%	54	6524	66%
33	5	48	4612	61%	40	4260	73%	39	4220	75%	40	4178	73%	45	4350	65%

Table A2 continued

Instance	Class	Kontainer 1			Kontainer 2			Kontainer 3			Kontainer 4			Kontainer 5		
		K	D	U	K	D	U	K	D	U	K	D	U	K	D	U
	1	22	1192	1%	22	1192	1%	22	1192	1%	22	1192	1%	22	1192	1%
	2	62	2356	63%	58	2194	68%	57	2220	69%	58	2282	68%	63	2426	62%
	3	69	2538	61%	60	2266	70%	61	2360	69%	63	2400	67%	67	2546	63%
	4	68	2474	63%	62	2482	70%	61	2404	71%	63	2494	68%	67	2646	64%
34	5	62	2364	58%	51	2106	71%	49	2076	74%	52	2134	70%	57	2286	63%
	1	26	1354	1%	26	1354	1%	26	1354	1%	26	1354	1%	26	1354	1%
	2	57	3740	67%	50	3524	77%	51	3494	75%	51	3676	75%	55	3632	70%
	3	67	4222	63%	58	4138	73%	56	3830	76%	58	4178	73%	64	4062	66%
	4	64	4464	70%	59	4226	76%	56	4118	80%	60	4220	75%	67	4516	67%
35	5	58	3866	64%	46	3812	80%	47	3364	79%	46	3834	80%	54	3840	68%
	1	14	1062	2%	14	1062	2%	14	1062	2%	14	1062	2%	14	1062	2%
	2	59	3988	69%	53	3692	77%	53	3514	77%	54	3562	75%	56	3758	73%
	3	66	4188	66%	60	3858	73%	58	3840	75%	58	3906	75%	65	4126	67%
	4	64	4112	66%	55	3822	77%	55	3658	77%	56	3706	75%	63	4064	67%
36	5	64	3552	59%	51	2982	74%	50	2964	76%	53	3056	71%	58	3300	65%

Integrating data mining technique and AHP in market analysis to propose new product development in real estate

Yunita[1], **M. Galinium**[1], **Lukas**[2]

[1]Faculty of Engineering and IT Swiss German University, EduTown BSD City, Tangerang, INDONESIA
[2]Faculty of Engineering, Universitas Katolik Indonesia Atma Jaya, Jakarta, INDONESIA

E-mails: unita.st@gmail.com, maulahikmah.galinium@sgu.ac.id, lukas@atmajaya.ac.id

Abstract. New product development in real estate industry is a challenging process since it is related to long term concept and high cost. A newly proposed product development should meet customer need and their preferences which appropriate with customer buying power and company value. This research use data mining for profiling customer transaction and Analytic Hierarchy Process (AHP) method for product selection in new product development. This research utilizes Weka as data mining open source software to profiling data customers. The analysis correlated product preferences and profiling demography such as city, age, gender and occupation. Demography profiles gives description buying power and product preferences. The products proposed are based on customer profiles and rank of the product by AHP method. The product with the highest score will be proposed as new product development. Case studies of this research are real estate projects in Serang, Makassar, and Balikpapan. Makassar and Balikpapan are the project that already gained success and Serang is new project which new products development will be proposed to launch. Based on profiling and product preference of customer in Balikpapan, Makassar, and prospectus of Serang markets, new products development that will be proposed are house type of 120/200 m^2 with price around Rp1.300.000.000 and house type of 71/120 m^2 with price around Rp800.000.000. The markets of Serang and Balikpapan have similarities in profiles as urban city so the new products development will adopt the succeed story of Balikpapan project.

1. Introduction

Nowadays, almost all company operations using information technology to manage all transaction data. Currently information technologies not only as a business support in business operation but also become one of strategic planning to gain business success. Data records which become huge should give added value in management decision in making knowledge-driven decisions to improve current business or develop new business. To capture the market in new business development the company must know what they know in this digital era.

A long with database development, a treasure is found in data tomb. Data become precious when knowledge is found. The process called Knowledge Data Discovery; it is a nontrivial process of identifying valid, novel, potentially useful and ultimately understandable pattern in data [1].

In new business development, marketing is one of organization which identifies unfulfilled needs and preferences, define and measures their interest, determines which target market the organization can best serve, decides on appropriate products, services and program to serve these markets, and calls upon everyone in the organization to think and serve the customer [2]. In order to develop new product the marketing function needs to be fulfilled wisely. Data transaction can be one of resources in gaining knowledge from pattern of customer data. Therefore, based on knowledge of data discovery, the new product can be proposed to fit with the market. In the processes which involve multi criteria decision to decide appropriate products, another technique is needed as decision support system. AHP technique decomposes a complicated problem into a multilevel hierarchical structure of objective, criteria and alternatives [3].

In property industry, a concept of city development is crucial. Creating the new product development that matches with customer preference becomes a big challenge since it related with big investments in a long term. In the competitive market, data mining will give an insight for management to help find the profile of customers and find their preference, and selecting the new product development appropriate with their preference.

2. Data Mining

Data mining methodology used in this research is CRISP-DM that stands for CRoss-Industry Standard Process for data mining process was developed by Daimler Chrysler, SPSS and NCR. CRISP-DM methodology consists on a cycle that comprises six stages as shown in Figure 1 [4]:

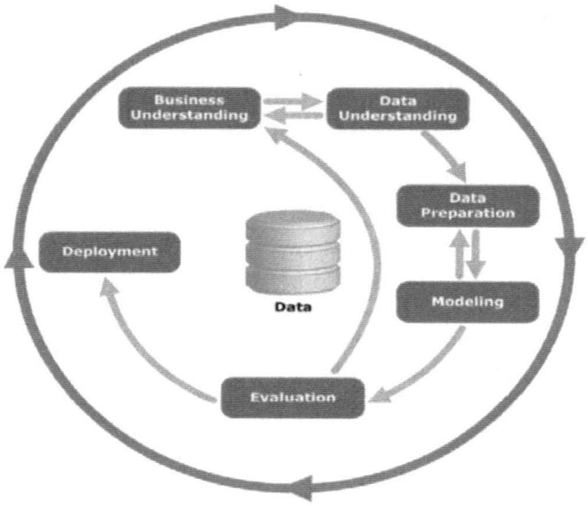

Figure 1. CRISP-DM Methodology (Chapman *et al.*, 2000)

1. Business understanding – First phase to understanding business objectives and requirement of business perspective to define data mining task in achieving the objectives.
2. Data understanding – phase to collect data and understanding the data to identifying data quality and discover data insights.
3. Data preparation – phase to prepare raw data set to final data set which clean and qualified to process.
4. Modelling – phase to apply modelling techniques that appropriate to data mining task.
5. Evaluation – phase to evaluate the model whether the model is properly achieves the business objective.
6. Deployment – phase to deploy the result to another process.

3. AHP

AHP is a multi –criteria decision making method developed by Saaty and is one of the best known which allows both qualitative and quantitative approaches to solve complex decision problems [3].

The method is modelling the complex problem into a hierarchical structure which consists of the goal, criteria and sub criteria. The subjective evaluations are converted into score on 1-9 scaling method and calculate the weight then we can rank the values.

4. Case studies

Case studies of this research are real estate projects in Serang, Makassar, and Balikpapan. Makassar and Balikpapan are the project that already gain succeeds and Serang is new project which new products development will be proposed to launch. Data used in this research is shown in Table 1 as follow:

Table 1. Data sampling

NO	YEAR TRANSACTION	DATASET 1 SERANG	DATA SET 2 MAKASSAR	DATA SET 3 BALIKPAPAN
1	2008	0	0	25
2	2009	80	0	33
3	2010	123	105	88
4	2011	85	203	105
5	2012	68	193	60
6	2013	76	159	28
7	2014	99	92	8
8	2015	40	153	3
	Total Data	**571**	**905**	**350**

To propose new product development in real estate project, the methodologies used in this research are described in Figure 2, as follows:

Phase 1:	
CRISP - Data Mining	: Finding profile of customer in Serang Finding product preference in Balikpapan & Makassar
Phase 2:	
Literature Review	: Market analysis in market trend and need Criteria new product proposed
Phase 3:	
AHP	: Ranking new product proposed

Figure 2. Phases of proposed methodology

4.1. Phase 1 – CRISP - Data Mining

Phase of CRISP – Data mining for this research as follow:

1. Business understanding, the data mining technique used to profiling the customer in Serang and finding product preferences in Makassar and Balikpapan.
2. Data understanding, to reach the goal the data used has been adjusted. The data field that not needed are eliminated
3. Data preparation, to get demography information such as province, age, gender we break the attribute of ID number (KTP) which has 16 digits as follow: ABCDEFGHIJKLMNOP.
   AB : Defines province where the citizen live
   CD : Defines city/kabupaten where the citizen live
   EF : Defines kecamatan where the citizen live
   GH : Defines birth date and show the gender too, if the citizen is female then the birth date will date will be added by 40.
   KL : Define year of citizen birth
   MNOP : Define the sequential number based on same birth date in area where the KTP issued.

4. Modelling, the data mining modelling is using Weka with data mining task are classification and clustering. The data modelling and analysis results are as shown in Table2.

Table 2. Data modelling analysis

Data Analysis	Serang	Makassar	Balikpapan
Customer Profile	Status: 74.44% married 23.81% single 1.75% widow ID town variance from many regions in Indonesia	Makassar customer are originally from Makassar. Their ID presents their economic dweller, and they prefer to buy property in their home town.	Urban city come from many regions in Indonesia
Product preference	First buyer for single home with price range Rp160 – 180 mio with occupation Karyawan or better type price range Rp190 – 300 mio Up grader for married, bigger house price range Rp 300 mio – Rp 700 mio	Home with price range Rp 200 – 400 mio	First buyer: price around Rp500 mio LB less than 102 m (range 49 – 113m), Upgrader: price around Rp800 mio – Rp6 bio LT is less than 140 and LB more than 102 m (range 135 – 171 m) LT more than 140 m.

5. Evaluation, the model are evaluated with business development team since data represented need to be validated, and data mining is an iterative process.
6. Deployment, the model are deployed to next process AHP.

4.2. Literature Review

Result of Indonesia Property Sentiment Survey H2-2015 by Rumah123.com is 65% people ages 21-30 are motivated to own their house [5]. The unique is they buy to invest and then upgrade the house. 29% of first buyer is young people and the up grader is 6% ages 31-40. The main indicator in this survey is income. People with income less than Rp100.000.000 in a year tend to find house with price less than Rp300.000.000. But people with income more than Rp1.000.000.000/year are often to looking for the property with that range price, because it's easy to reinvest. Based on Cushman and Wakefield Indonesia in 2015, customers preference in middle of 2015 are house type 45/60 and 120/115 with price range Rp600.000.000 – Rp1.200.000.000 [6].
Based on customers profiles of Serang analysis, Serang customers have buying power of property with price range Rp259.325.000 and up to Rp737.361.000, the citizens are new urban from different province in Indonesia. Their age more than 29 years old, worked as PNS, worker in chemical industry at Cilegon, and international company in Dubai, Europe, and Asia. In Serang, there are few developers that provides suitable home for up grader.

4.3. AHP

With the consideration of customer profile and literature review new product development are defined and built as product selection matrix shown in Table 3.

Table 3. Product selection matrix

Type 1 • Price Range: Rp300.000.000 • LB/LT: 27/60 m^2 • Design: Modern minimalis • Facilities: Cluster Gate	Type 2 • Price Range: Rp400.000.000 • LB/LT: 36/72 m^2 • Design: Modern minimalis • Facilities: Cluster Gate
Type 3 • Price Range: Rp800.000.000 • LB/LT: 71/120 m^2 • Design: Modern Classic • Facilities: Cluster Gate, neighbourhood CCTV	Type 4 • Price Range: Rp 1.300.000.000 • LB/LT: 120/200 m^2 • Design: Modern Classic • Facilities: Cluster Gate, neighbourhood CCTV

In each group, with the business development member team, scoring the 1- 9 scaling method, then count the weight.

The highest weight score is Type 4 with score 0.586 and for option the second type is Type 3 with score 0.224 as shown in Table 4. In Serang, properties available are small to medium type by local developers, there are opportunities to market bigger house and complete facilities.

Table 4. Pair comparison of houses type

House types	T1	T2	T3	T4	WEIGHT
Type 1 (T1)	1	2	0.5	0.2	0.118
Type 2 (T2)	0.5	1	0.2	0.2	0.073
Type 3 (T3)	2	5	1	0.2	**0.224**
Type 4 (T4)	5	5	5	1	**0.586**
CR =0.088					

Price element is the element which has highest weight with score 0.548 and (LB/LT) has the second element with score 0.256 as shown in Table 5. For first buyer segment, price will be the important element for consideration and then the building area and land area (LB/LT). They will accept the detail product offered.

Table 5. Pair wise comparison Type 1

Type 1	Price T1	LB/LT T1	Design T1	Facilities T1	WEIGHT
Price T1	1	3	5	7	**0.548**
LB/LT T1	1/3	1	3	5	**0.256**
Design T1	1/5	1/3	1	5	0.145
Facilities T1	1/7	1/5	1/5	1	0.052
CR=0.093					

Comparing criteria element in Type 2 shown in Table 6, which has more expensive than Type 2, the highest weight important element is Facilities with score 0.520 and the second important is Design with score 0.297. For first buyer who have more money or up grader they will start consider the facilities provided.

Table 6. Pair wise comparison Type 2

Type 2	Price T2	LB/LT T2	Design T2	Facilities T2	WEIGHT
Price T2	1	1/3	1/5	1/7	0.058
LB/LT T2	3	1	1/3	1/5	0.124
Design T2	5	3	1	½	**0.297**
Facilities T2	7	5	2	1	**0.520**
CR=0.026					

Comparing criteria element in Type 3 shown in Table 7, which has more expensive than Type 1 and Type 2, the highest weight important element is Facilities with score 0.589 and the second important is Design with score 0.216. For market segment with high price, the facilities become high consideration.

Table 7. Pair wise comparison Type 3

Type 3	Price T3	LB/LT T3	Design T3	Facilities T3	WEIGHT
Price T3	1	½	1/3	1/5	0.080
LB/LT T3	2	1	1/3	1/5	0.116
Design T3	3	3	1	1/5	**0.216**
Facilities T3	5	5	5	1	**0.589**
CR=0.081					

Comparing criteria element in Type 4 shown in Table 8, which have more expensive than Type 1 and Type 2 and Type 3. The highest weight important element is LB/LT with score 0.473 and the second important is Design with score 0.298. The consideration for highest price product will more detail, where markets want more in space, facilities and design.

Table 8. Pair wise comparison Type 4

Type 4	Price T4	LB/LT T4	Design T4	Facilities T4	WEIGHT
Price T4	1	1/5	1/3	½	0.087
LB/LT T4	5	1	2	3	**0.473**
DesignT4	3	½	1	3	**0.298**
Facilities T4	2	1/3	1/3	1	0.142
CR=0.024					

Table 9. Weight of selected product

	Weight of each group	Elements	Weight of each element	Total weight
TYPE 4	**0.586**	• Price Range: Rp 1.300.000.000	0.087	0.050
		• LB/LT: 120/200 m^2	**0.473**	**0.277**
		• Design: Modern Classic	0.298	0.175
		• Facilities: Cluster Gate, neighbourhood CCTV	0.142	0.083
TYPE 3	**0.224**	• Price Range: Rp 800.000.000	0.080	0.017
		• LB/LT: 71/120 m^2	0.116	0.026
		• Design: Modern Classic	0.216	0.048
		• Facilities: Cluster Gate, neighbourhood CCTV	**0.589**	**0.132**
TYPE 1	0.118	• Price Range: Rp 300.000.000	**0.548**	**0.065**
		• LB/LT: 37/60 m^2	0.256	0.030
		• Design: Modern minimalist	0.145	0.017
		• Facilities: Cluster Gate	0.052	0.006
TYPE 2	0.073	• Price Range: Rp 400.000.000	0.058	0.004
		• LB/LT: 45/60 m^2	0.124	0.009
		• Design: Modern minimalist	0.297	0.022
		• Facilities: Cluster Gate	**0.520**	**0.038**

Finally the overall priority scores are calculated as shown in Table 9. Type 4 is the proposed new product developments which have the highest weight and the second type option is Type 3. The opportunity of exclusive product with brand, facilities and design price range Rp800 mio –

Rp1.3 bio will be consider for new product development proposed in Serang.

5. Conclusion

The research shows that data mining process finds the profile of Serang customers and products preference in Balikpapan and Makassar project. Serang and Balikpapan has similarity as urban city, the market growth since there is opportunity in industry that attracted urban people as economic dweller to live in new city. Price range and product type of Makassar project has few variances customizes buying power of the market. Serang as new urban city need an exclusive product, with buying power up to Rp750.000.000, big house in type 3 and 4 can be proposed as new product development.

From experiment that has been done, data transaction contains valuable information that can be use in reselling, target marketing, and customer relationship management for future works.

However new product development is a complex process, since there many factor that not captured in data. Economic situation, government regulation, and customer behaviours in changed markets. Other technique is needed to complement the process.

References

[1] Fayyad, U.M., Smyth, P. and Uthurusamy, R. eds., 1996. *Advances in knowledge discovery and data mining* (Vol. 21). Menlo Park: AAAI press.

[2] Kotler, P., 2011. *Marketing insights from A to Z: 80 concepts every manager needs to know*. John Wiley & Sons.

[3] Saaty, T.L., 1988. What is the analytic hierarchy process? In *Mathematical models for decision support* (pp. 109-121). Springer, Berlin.

[4] Chapman, P., Clinton, J., Kerber, R., Khabaza, T., Reinartz, T., Shearer, C. and Wirth, R., 2000. CRISP-DM 1.0 Step-by-step data mining guide.

[5] Kompas. (2015, November 3). Rumah Rp300 Jutaan Diincar Anak Muda Usia 21-30 Tahun. July 2016. http://properti.kompas.com/read/2015/11/03/150000021/ Rumah.Rp.300.Jutaan. Diincar.Anak.Muda.Usia.21-30.Tahun

[6] Peluang Properti. (2015, October 15). Rumah yang jadi incaran para konsumen. July 2016. http://www.peluangproperti.com/berita/kpr/2015-10/5898/rumah-yang-jadi-incaran-para-konsumen#sthash.qRUAIaLB.dpuf

[7] Jain, S., Aalam, M.A. and Doja, M.N., 2010. K-means clustering using weka interface. In *Proceedings of the 4th National Conference* (pp. 25-26).

[8] Rahadi, R.A., Wiryono, S.K., Koesrindartoto, D.P. and Syamwil, I.B., 2012. Relationship between consumer preferences and value propositions: a study of residential product. *Procedia-Social and Behavioral Sciences*, *50*, pp.865-874.

[9] Rahadi, R.A., Wiryono, S.K., Koesrindartoto, D.P. and Syamwil, I.B., 2013. Attributes influencing housing product value and price in Jakarta metropolitan region. *Procedia-Social and Behavioral Sciences*, *101*, pp.368-378.

[10] Singh Swati, Dubey Gaurav, "*Finding interest people in purchasing real estate by using data mining techniques*". International Journal of Recent Technology and Engineering(IJRTE) ISSN:2277-3878, Volume X, Issue X.

[11] Ho, H.P., Chang, C.T. and Ku, C.Y., 2015. House selection via the internet by considering homebuyers' risk attitudes with S-shaped utility functions. *European Journal of Operational Research*, *241*(1), pp.188-201.

[12] Sener, H.Y., 2014. Determining New Markets Using Analytic Hierarchy Process: Case Study in Güral Porcelain. *International Journal of Marketing Studies*, *6*(5), p.149.

[13] Padhy, N., Mishra, D. and Panigrahi, R., 2012. The survey of data mining applications and feature scope. *arXiv preprint arXiv:1211.5723*.

[14] Chimbalkar, A., Patil, S., Jain, M., Kate, G. and Dwivedi, P., 2014. Finding n-Most Demanding Products for Real-Estate Industry. *International Journal of Computer Applications*, *93*(12).

IOP Conf. Series: Materials Science and Engineering **166** (2017) 012031 doi:10.1088/1757-899X/166/1/012031

Analysis of data mining classification by comparison of C4.5 and ID algorithms

R. Sudrajat[1], I. Irianingsih[1], D. Krisnawan[2]

[1]Department of Computer Science, Padjadjaran University, Sumedang 45363, Indonesia
[2]Department of Mathematics, Padjadjaran University, Sumedang 45363, Indonesia

E-mails: r.sudrajat@unpad.ac.id, iinirianingsih@gmail.com, dillan.k@gmail.com

ABSTRACT. The rapid development of information technology, triggered by the intensive use of information technology. For example, data mining widely used in investment. Many techniques that can be used assisting in investment, the method that used for classification is decision tree. Decision tree has a variety of algorithms, such as C4.5 and ID3. Both algorithms can generate different models for similar data sets and different accuracy. C4.5 and ID3 algorithms with discrete data provide accuracy are 87.16% and 99.83% and C4.5 algorithm with numerical data is 89.69%. C4.5 and ID3 algorithms with discrete data provides 520 and 598 customers and C4.5 algorithm with numerical data is 546 customers. From the analysis of the both algorithm it can classified quite well because error rate less than 15%.

1. Introduction

Investment is currently growing very rapidly, public is interesting to do a form of investment such as deposits and mutual funds. On the other hand, the development of information technology has developed rapidly, many science informatics used in the banking sector, one of them is data mining. Data mining is used to explore large amounts of data into patterns/rules so as to simplify large data into parts more modest but has a precious value. C4.5 and ID3 algorithms are algorithms used in data mining to create decision tree, where decision tree is a classification method with an accurate prediction. Decision tree method to change the very large data into a decision tree that represents the rule.

1.1 Decision Tree

Classification by decision tree, a classification that is very reliable as a means of classification and prediction. In general, a decision tree is a modeling picture of a subject that is composed of a series of decisions that provide a solution. Classification rules can be easily presented in easy language so that users can understand it, or in its database as Microsoft Excel, Microsoft Access and SQL. In some applications, and prediction accuracy of classification is important, for example, predict potential customers in the business market. There are various algorithms to build a decision tree, the most popular is the ID3 and C4.5.

1.2 C4.5 Algorithm

C4.5 algorithm is an algorithm to form a decision tree by counting the value of the gain, where the biggest gains are to be used as an initial node or the root node. C4.5 algorithms step in building a decision tree as follows:

- Select the attribute with the largest gain value as the root.
- Create a branch for each value.
- For the case of the branches.
- Repeat the process for each branch until all cases the branches have the same class.

Entropy and gain search formula as follows:

$$Gain(S) = Entropy(S) - \sum_{i=1}^{n} \frac{|S_i|}{|S|} Entropy(S_i)$$

where

S = the set of cases
n = number of partitions
$|S_i|$ = number of cases in the partition i
$|S|$ = number of cases in S.

$$Entropy(S) = \sum_{i=1}^{n} -p_i \log_2 p_i$$

where

S = The set of cases
n = number of partitions
p_i = proportion of S_i to S.

1.3 ID3 Algorithm
ID3 (Iterative Dichotomiser3) algorithm is used to form a decision tree by calculating the entropy values, created by Ross Quinlan. Step ID3 algorithm to construct a decision tree as follows:

- Select all the existing attributes and calculate the entropy of each attribute.
- Select an attribute with the smallest entropy, then these attributes be the root node.
- Create an attribute node with the smallest entropy.
- Repeat the process for each branch until all cases the branches have the same class.

2. Analysis and System Design

2.1 Analysis Classification System
In the analysis stage of this system, which will be the topic of discussion is about what the problems faced as an analysis of the data and the accuracy of the analysis of algorithms C4.5 and ID3.

2.1.1 Data Analysis The data used in the making of this program is the customer data a bank with attribute goal is PEP (Portfolio Evaluation Plan) or the planned investment, the investment data in the form of excel data which attributes the following selection *id, age, sex, region, income, married, children, car, save act, current act and mortgage*, and the attributes of the destination is pep. Total investment data as many as 600 data, the data is stored in csv (Comma Separated Values) format.

2.1.2 Analysis Software Some of the software used in this study as follows: (1) The operating system used Microsoft Windows Vista Home Premium, (2) Net Beans Java programming as the software used to build the program, (3) Microsoft Excel as software that serves to change the data in .xls to .csv.

2.2 Designing Data Classification
Classification of data that will be discussed is divided into three parts, namely the classification of data using ID3 algorithm with discrete data, the algorithm C4.5 and C4.5 algorithms discrete data with numerical data. Data used in the calculation of investment manual 10 records.

Table 1. Experiment Data 1

ID	AGE	SEX	REGION	INCOME	MARRIED	CAR
1	MIDDLE	F	INNER_CITY	LOW	NO	NO
2	MIDDLE	F	TOWN	LOW	YES	NO
3	MIDDLE	F	RURAL	LOW	NO	YES
4	YOUNG	M	INNER_CITY	LOW	YES	NO
5	OLD	F	RURAL	HIGH	YES	NO
6	OLD	F	TOWN	MID	YES	NO
7	YOUNG	M	RURAL	LOW	NO	NO
8	OLD	M	TOWN	MID	YES	NO
9	MIDDLE	F	SUBURBAN	MID	YES	YES
10	OLD	M	INNER_CITY	HIGH	YES	NO

Table 2. Experiment Data 2

CHILDREN	SAVE ACT	CURRENT ACT	MORTGAGE	PEP
ONE	NO	NO	NO	YES
ONE	NO	YES	NO	YES
NOL	NO	YES	NO	YES
NOL	YES	YES	NO	YES
NOL	YES	NO	NO	NO
TWO	YES	YES	NO	YES
NOL	NO	YES	NO	YES
NOL	NO	YES	YES	YES
TWO	NO	NO	NO	NO
TWO	YES	YES	YES	YES

Table 3. Gain Value Calculation and Entropy

Attribute	Sub-attribute	Number of Cases (S)	NO (S1)	YES (S2)	Entropy	Gain
Total		10	2	8	0.7219280946	
AGE						0.0729055950
	MIDDLE	4	1	3	0.8112781245	
	YOUNG	2	0	2	0	
	OLD	4	1	3	0.8112781245	
SEX						0.1709505941
	FEMALE	6	2	4	0.9182958341	
	MALE	4	0	4	0	
REGION						0.1709505942
	INNER_CITY	3	0	3	0	
	TOWN	3	1	2	0.9182958341	
	RURAL	3	1	2	0.9182958341	
	SUBURBAN	1	0	1	0	
INCOME						0.2464393444
	Low	5	0	5	0	
	Mid	3	1	2	0.9182958341	
	High	2	1	1	1	
MARRIED						0.1177436965
	YES	7	2	5	0.8631205687	
	NO	3	0	3	0	
CHILDREN						0.0854752971
	NOL	5	1	4	0.7219280946	
	ONE	2	0	2	0	
	TWO	3	1	2	0.9182958341	

Table 4. Calculation of Node Root for C4.5 Algorithm

Attribute	Sub-attribute	Number of Cases (S)	NO (S1)	YES (S2)	Entropy	Gain
INCOME-MID		3	1	2	0.9182958341	
AGE						
	MIDDLE	1	1	0	0	-
	YOUNG	0	0	0	0	
	OLD	2	0	2	0	
SEX						0.2516291674
	FEMALE	2	1	1	1	
	MALE	1	0	1	0	
REGION						0.2516291674
	INNER_CITY	0	0	0	0	
	TOWN	2	1	1	1	
	RURAL	0	0	0	0	
	SUBURBAN	1	0	1	0	
MARRIED						0
	YES	3	1	2	0.9182958341	
	NO	0	0	0	0	
CHILDREN						0.2516291674
	NOL	1	0	1	0	
	ONE	0	0	0	0	
	TWO	2	1	1	1	
	THREE	0	0	0	0	

Table 5. Calculation of Node Root for ID3 Algorithm

Attribute	Sub-Attribute	Number of Cases (S)	NO (S1)	YES (S2)	Entropy
CHILDREN-NOL		5	1	4	0.7219280946
Age					
	Middle	1	0	1	0
	Young	2	0	2	0
	Old	2	1	1	1
Sex					
	Female	2	1	1	1
	Male	3	0	3	0
Region					
	Inner_city	1	0	1	0
	Town	1	0	1	0
	Rural	3	1	2	0
	Suburban	0	0	0	0
Married					
	Yes	3	1	2	0.9182958341
	No	2	0	2	0
Income					
	Low	3	0	3	0
	Mid	1	0	1	0
	High	1	1	0	0

From the calculation of Table 4 it is obtained the decision tree as follows:

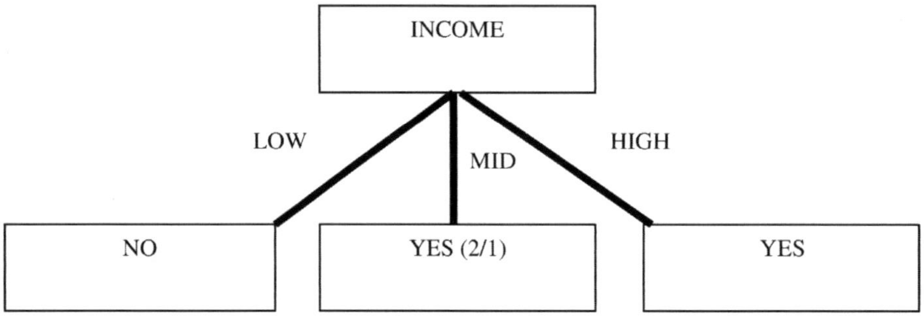

Figure 1. C4.5 Decision Tree Algorithm

Table 6. Calculation of Node 1 for Algorithm ID3

Attribute	Sub_ Attribute	Number of Cases (S)	NO (S1)	YES (S2)	Entropy
CHILDREN-NOL-		3	1	2	0.9182958341
MARRIED-YES					
Age					
	Middle	0	0	1	0
	Young	1	0	1	0
	Old	2	1	1	1
Sex					
	Female	1	1	0	0
	Male	2	0	2	0
Region					
	Inner_city	1	0	1	0
	Town	1	0	1	0
	Rural	1	1	0	0
	Suburban	0	0	0	0
Income					
	Low	1	0	1	0
	Mid	1	0	1	0
	High	1	1	0	0

From the calculation of Table 6 it is obtained the decision tree as follows:

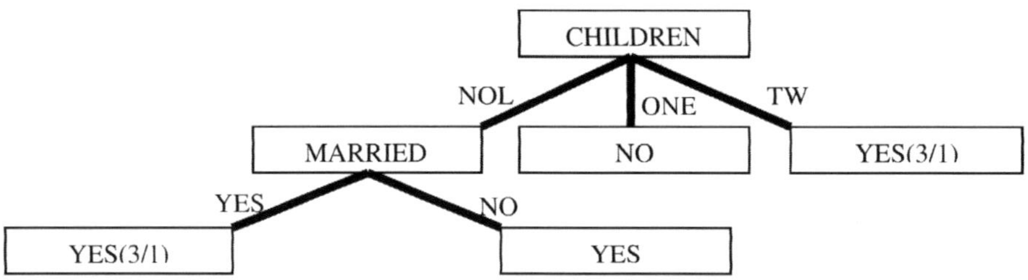

Figure 2. Decision Tree ID3 Algorithm

3. Results and Discussion

In this experiment, use the data as much as 100 records, the data is tested again using ID3 algorithm with discrete data, the algorithm C4.5 and C4.5 algorithms discrete data with continuous data. The results of these experiments obtained a decision tree as follows:

```
21:23:40 - trees.Id3
children = ONE
|   age = MIDDLE: YES
|   age = YOUNG
|   |   married = NO: YES
|   |   married = YES
|   |   |   sex = FEMALE: NO
|   |   |   sex = MALE
|   |   |   |   save_act = NO: NO
|   |   |   |   save_act = YES: YES
|   age = OLD: YES
children = THREE
|   age = MIDDLE
|   |   save_act = NO: NO
|   |   save_act = YES: YES
|   age = YOUNG: NO
|   age = OLD: YES
children = NOL
|   save_act = NO
|   |   married = NO: YES
|   |   married = YES
|   |   |   age = MIDDLE: NO
|   |   |   age = YOUNG: NO
|   |   |   age = OLD: YES
|   save_act = YES
|   |   region = INNER_CITY
|   |   |   car = NO
|   |   |   |   age = MIDDLE: NO
|   |   |   |   age = YOUNG
|   |   |   |   |   sex = FEMALE: NO
|   |   |   |   |   sex = MALE
|   |   |   |   |   |   current_act = NO: NO
|   |   |   |   |   |   current_act = YES: YES
|   |   |   |   age = OLD
|   |   |   |   |   sex = FEMALE: YES
|   |   |   |   |   sex = MALE: NO
|   |   |   car = YES: NO
|   |   region = TOWN: NO
|   |   region = RURAL: NO
|   |   region = SUBURBAN
|   |   |   married = NO: YES
|   |   |   married = YES: NO
children = TWO
|   income = LOW
|   |   save_act = NO
|   |   |   age = MIDDLE: NO
|   |   |   age = YOUNG: YES
|   |   |   age = OLD: null
|   |   save_act = YES: NO
|   income = MID
|   |   region = INNER_CITY
|   |   |   age = MIDDLE: NO
|   |   |   age = YOUNG: NO
|   |   |   age = OLD: YES
|   |   region = TOWN
|   |   |   sex = FEMALE: YES
|   |   |   sex = MALE
|   |   |   |   age = MIDDLE: null
|   |   |   |   age = YOUNG: YES
|   |   |   |   age = OLD: NO
|   |   region = RURAL: YES
|   |   region = SUBURBAN: NO
|   income = HIGH
|   |   married = NO: YES
|   |   married = YES
|   |   |   car = NO: YES
|   |   |   car = YES: NO
```

Figure 3. Decision Tree ID3 Algorithm with Discrete Data

```
22:06:05 - trees.J48
------------------

children = ONE: YES (19.0/2.0)
children = THREE
|   age = MIDDLE
|   |   save_act = NO: NO (2.0)
|   |   save_act = YES: YES (2.0)
|   age = YOUNG: NO (5.0)
|   age = OLD: YES (1.0)
children = NOL
|   save_act = NO
|   |   married = NO: YES (5.0)
|   |   married = YES
|   |   |   age = MIDDLE: NO (2.0)
|   |   |   age = YOUNG: NO (4.0)
|   |   |   age = OLD: YES (2.0)
|   save_act = YES: NO (28.0/3.0)
children = TWO
|   income = LOW: NO (14.0/1.0)
|   income = MID
|   |   age = MIDDLE: NO (2.0)
|   |   age = YOUNG: YES (2.0/1.0)
|   |   age = OLD: YES (7.0/2.0)
|   income = HIGH: YES (5.0/1.0)
```

Figure 4. Decision Tree C4.5 Algorithm with Discrete Data

```
22:12:01 - trees.J48

children <= 0
|   save_act = NO
|   |   married = NO: YES (5.0)
|   |   married = YES
|   |   |   age <= 51: NO (6.0)
|   |   |   age > 51: YES (2.0)
|   save_act = YES: NO (28.0/3.0)
children > 0
|   children <= 1
|   |   age <= 31: NO (2.0)
|   |   age > 31: YES (16.0)
|   children > 1
|   |   income <= 30085.1: NO (28.0/3.0)
|   |   income > 30085.1: YES (13.0/1.0)
```

Figure 5. Decision Tree C4.5 Algorithm with Continuous Data

Table 7. Data classification results

	C4.5 Algorithm						Total	ID3 Algorithm						Total
YES	37	42	43	41	42	39	244	41	49	44	45	49	44	272
NO	53	50	52	40	40	41	276	59	52	56	56	52	51	326
						Total	520						Total	598
	Unclassified						80	Unclassified						2

Table 8. Classified Customers with Discrete Data

	C4.5 Algorithm						Total
YES	35	43	44	43	43	37	245
NO	56	46	52	50	52	45	301
						Total	546
	Unclassified						54

Table 9. Classified Customers with Continuous Data

Trial	ID3 with discrete data (%)	Failure (%)	C4.5/J48 with discrete data (%)	Failure (%)	C4.5/J48 with numeric data (%)	Failure (%)
1	100	0	93	7	93	7
2	100	0	91	9	88	12
3	99	1	94	6	95	5
4	100	0	80	20	92	8
5	100	0	81	19	84	16
6	100	0	84	16	86	14
MEAN	99,83	0.16	87,16	12,83	89,69	10,33

4. Conclusion

1. ID3 Algorithm excellent classifying investment data with a discrete form for 99.83% of the data can be classified by degree of error 0:16%, while the C4.5 algorithm to classify well enough to form discrete investment data for 87.16% of the data can be classified by

an error rate of 12.83% and 89.69% in numerical form with an error rate of 10.33%. Algorithm ID3 classifying investment data better than C4.5 algorithm.
2. ID3 algorithm to classify as many as 598 discrete data to unclassified 2, the algorithm C4.5 with as many as 520 discrete data classifying the unclassified 80 and C4.5 algorithms with continuous data to classify as many as 546 Unclassified 54 customers.
3. The decision tree obtained in this study, has a decision tree ID3 algorithm is more complex than the C4.5 algorithm, so it can be viewed accuracy of ID3 algorithm can better classify. The more complex a decision tree algorithm, the better classifying data.

References

[1] Berry, Michael J.A. and Gordon S. Linoff. 2004. Data Mining Techniques for Marketing. Second Edition. Wiley Publishing, Inc.
[2] Chen, M., Han, J. and Yu, P. 1996. Data Mining: An Overview from a Database Perspective. IEEE Trans. Knowledge and Data Engineering. 8(6): 866-881.
[3] Larose, Daniel T. 2005. Discovering Knowledge in Data: An Introduction to Data Mining. John Willey & Sons, Inc.
[4] Mobasher, Bamshad. 2005. Clasification via Decision Trees in WEKA. http://maya.cs.depaul.edu/classes/ect584/weka/classify.html.
[5] Moertini, Veronica S. towards the Use of C4.5 Algorithm for Classifying Banking Dataset. http://home.unpar.ac.id/~integral/Volume%208/Integral%208%20No.%202/C45%20 Algorihm.
[6] Sudjana. 2010. Klasifikasi Data Nasabah Sebuah Asuransi Menggunakan Algoritma C4.5. http://journal.uii.ac.id/index.php/Snati/article/view/1923.
[7] Sunni, Ismail. 2011. Perbedaan Algoritma ID3, C4.5 dan J48. http://webcache.googleusercontent.com/search?q=cache:http://codemath.wordpress.com/2011/06/20/perbedaan-algoritma-id3-c4-5-dan-j48.
[8] Quinlan, Ross. 2000. ID3 Algorithm. http://www.soc.napier.ac.uk/~peter/vldb/dm/node11.html.

Adaptive spline autoregression threshold method in forecasting Mitsubishi car sales volume at PT Srikandi Diamond Motors

D. Susanti, E. Hartini and A. Permana

Department of Mathematic, Padjadjaran University Sumedang 45363, Indonesia

E-mails: dwi.susanti@unpad.ac.id, euis_hartini@yahoo.co.id, permanaarif009@gmail.com

Abstract. Sale and purchase of the growing competition between companies in Indonesian, make every company should have a proper planning in order to win the competition with other companies. One of the things that can be done to design the plan is to make car sales forecast for the next few periods, it's required that the amount of inventory of cars that will be sold in proportion to the number of cars needed. While to get the correct forecasting, on of the methods that can be used is the method of Adaptive Spline Threshold Autoregression (ASTAR). Therefore, this time the discussion will focus on the use of Adaptive Spline Threshold Autoregression (ASTAR) method in forecasting the volume of car sales in PT.Srikandi Diamond Motors using time series data.In the discussion of this research, forecasting using the method of forecasting value Adaptive Spline Threshold Autoregression (ASTAR) produce approximately correct.

1. Introduction

This research method used is the method of Adaptive Threshold Autoregression Spline. Adaptive Threshold Autoregression Spline method (ASTAR) can be used in various types of data so that its application be varied. Thesis research on the development of Stevens, J.G. (1991) and the simulation data using the volume of sales of Mitsubishi cars at PT Srikandi Diamond Motors, with variable used is the number of car sales. The data obtained were tested using normality test for measuring the data obtained so that it can be used in parametric statistics (inferential statistics).

2. Model Identification

Identification of the model is the data that will be tested is stationary or not. There are two ways that can be used in stationary test, graphs plot the data (time plot) and the unit root test commonly referred to Augmented Dickey Fuller test (ADF). In the graph plots the data, the way we need to do is look at the visible time series data is stationary or not. While using a stationary test unit root test. If data is stationary, then the time series model that can be used, namely the AR(p), MA(q), ARMA(p,q). Meanwhile, if the data is not stationary, it must go through the process different advance until the data becomes stationary. Next, use the Order Identification Time Series model by analysing the data plot of the autocorrelation function (ACF) and partial autocorrelation function (PACF). As for the use, for the AR using the partial autocorrelation function (PACF) while for the MA process using the autocorrelation function (ACF). Then do the Time Series Model Parameter Estimation order to obtain *p*, *d* and *q* of a time series model, the next estimate of the model parameters in the get. One method that can be used in the estimation of the parameters is the least squares method with the help of software E Views. If you get the estimated value of the desired parameter, followed by *t*-test to determine the parameters of significant or not. In addition to the *t*-test, performed statistical tests also *F*-test to determine the significance of all independent variables as a unit to. To determine the best time series model can be determined by comparison of the value of Akaike Information Criterion (AIC) and Schwarz Information Criterion (SIC). If the value of AIC and SIC a model of the smaller and the better model.

3. Parameter Estimation Model ASTAR

In estimating model parameters Adaptive Threshold Autoregression Spline (ASTAR), by calculating the autocorrelation parameter values and a constant value of the time series model. Parameter autocorrelation is a characteristic that states the correlation coefficient of a population. The general shape parameter p order autocorrelation to have the following equation:

$$\theta_i = \frac{n(\sum(Y_{t-1}Y_t)) - \sum(Y_{t-1})\sum(Y_t)}{n(\sum(Y_{t-1}{}^2) - (\sum Y_{t-1})^2)}. \tag{1}$$

4. Method Spline Adaptive Threshold Autoregression (ASTAR)

Adaptive Threshold Autoregression Spline method (ASTAR) is a nonlinear time series modeling threshold with the predictor is a lag value (Y_{t-d}) as in the time series [3]. The method can form a model with a limit cycles when the data time series model shows the periodic behavior. Adaptive Threshold Autoregression Spline method (ASTAR) is the development of methods Autoregressive (AR). In the journal, entitled A New Approach to Adaptive Spline Threshold Autoregression Toprak and Taylan suggested a continuity between the methods of autoregressive (AR) and Spline Adaptive Threshold Autoregression (ASTAR). The general form of the equation with the order p AR method is as follows:

$$AR(p) = Y_t = c + \theta_1 Y_{t-1} + \theta_2 Y_{t-2} + \cdots + \theta_p Y_{t-p} + \varepsilon_t. \tag{2}$$

The general form of the equation with the ordo p ASTAR method is as follows:

$$\begin{aligned} ASTAR(p) &= Y_t = c + \theta_1(\lambda_1 Y_{t-1}) + \theta_2(\lambda_2 Y_{t-2}) + \cdots + \theta_p(\lambda_p Y_{t-p}) + \varepsilon_t \\ Y_t &= c + \sum_{i=1}^{p} \theta_i(\lambda_i Y_{t-i}) + \varepsilon_t. \end{aligned} \tag{3}$$

5. Result and Discussion

The data used is the volume of car sales and inventory PT Srikandi Diamond Motors in June 2013 until May 2015 (Table 1).

First to identify the time series model with stationary identification data to determine the sales data of Mitsubishi cars in Diamond Motors PT Srikandi whether stationary or not, with a Unit Root Tests using the Augmented Dickey-Fuller (ADF) with the help of software E-Views 7 of SPSS results are in Table 2.

From Table 2, it is known that the statistical value of t = -3.081740 < 0.1347 -3.632896 and probability value > 0.05. Data is not stationary, subsequent differentiation process is carried out first (first difference), the result is Table 3.

Table 1. Sales Volume Data and Car Inventory

Year	Period	Month	Sale	Stock
2013	1	June	162	300
	2	July	186	300
	3	August	144	300
	4	September	151	300
	5	October	141	300
	6	November	154	300
	7	December	160	250
2014	8	Januar	168	250
	9	February	142	250
	10	March	160	250
	11	April	169	250
	12	May	151	250
	13	June	140	250
	14	July	84	250
	15	August	99	250
	16	September	124	250
	17	October	75	175
	18	November	109	175
	19	December	110	175
2015	20	January	116	175
	21	Februari	95	175
	22	March	112	175
	23	April	131	175
	24	May	103	175

Table 2. Results of Augmented Dickey-Fuller (ADF test)

Null Hypothesis: PENJUALAN has a unit root
Exogenous: Constant, Linear Trend
Lag Length: 0 (Automatic - based on SIC, maxlag=4)

		t-Statistic	Prob.*
Augmented Dickey-Fuller test statistic		-3.081740	0.1347
Test critical values:	1% level	-4.440739	
	5% level	-3.632896	
	10% level	-3.254671	

**MacKinnon (1996) one-sided p-values.*
Augmented Dickey-Fuller Test Equation
Dependent Variable: D(PENJUALAN)
Method: Least Squares
Date: 06/04/16 Time: 05:09
Sample (adjusted): 2013M07 2015M04
Included observations: 22 after adjustments

Variable	Coefficient	Std. Error	t-Statistic	Prob.
SALE(-1)	-0.732093	0.237558	-3.081740	0.0061

C	*123.1955*	*42.67979*	*2.886508*	*0.0095*
@TREND(2013M06)	*-2.293138*	*1.108578*	*-2.068540*	*0.0525*

R-squared	*0.338053*	*Mean dependent var*	*-1.409091*
Adjusted R-squared	*0.268374*	*S.D. dependent var*	*24.79566*
S.E. of regression	*21.20900*	*Akaike info criterion*	*9.072852*
Sum squared resid	*8546.611*	*Schwarz criterion*	*9.221631*
Log likelihood	*-96.80137*	*Hannan-Quinn criter.*	*9.107900*
F-statistic	*4.851597*	*Durbin-Watson stat*	*1.840305*
Prob(F-statistic)	*0.019853*		

Table 3. Results First Difference

Null Hypothesis: D(PENJUALAN) has a unit root
Exogenous: Constant, Linear Trend
Lag Length: 0 (Automatic - based on SIC, maxlag=4)

		t-Statistic	*Prob.**
Augmented Dickey-Fuller test statistic		*-6.628693*	*0.0001*
Test critical values:	*1% level*	*-4.467895*	
	5% level	*-3.644963*	
	10% level	*-3.261452*	

**MacKinnon (1996) one-sided p-values.*
Augmented Dickey-Fuller Test Equation
Dependent Variable: D(PENJUALAN,2)
Method: Least Squares
Date: 06/04/16 Time: 05:15
Sample (adjusted): 2013M08 2015M04
Included observations: 21 after adjustments

Variable	*Coefficient*	*Std. Error*	*t-Statistic*	*Prob.*
D(SALE(-1))	*-1.392185*	*0.210024*	*-6.628693*	*0.0000*
C	*-12.75254*	*11.38217*	*-1.120396*	*0.2773*
@TREND(2013 M06)	*0.766643*	*0.845363*	*0.906881*	*0.3765*

R-squared	*0.712181*	*Mean dependent var*	*-0.238095*
Adjusted R-squared	*0.680201*	*S.D. dependent var*	*41.47277*
S.E. of regression	*23.45316*	*Akaike info criterion*	*9.279452*
Sum squared resid	*9900.916*	*Schwarz criterion*	*9.428669*
Log likelihood	*-94.43424*	*Hannan-Quinn criter.*	*9.311836*
F-statistic	*22.26966*	*Durbin-Watson stat*	*1.960010*
Prob(F-statistic)	*0.000014*		

From Table 3, it is known that the statistical value $t = -6.628693 > -3.644963$ 0.0001 and a probability value of <0.05. The results show that the data is stationary on the differentiation of the first stage (1st difference). Once the data is stationary, it will further analyse the data plot of the autocorrelation function (ACF) and partial autocorrelation function (PACF) as shown in Table 4 below.

Table 4. Plot ACF and PACF

Correlogram of PENJUALAN

Date: 06/15/16 Time: 19:30
Sample: 2013M06 2015M04
Included observations: 23

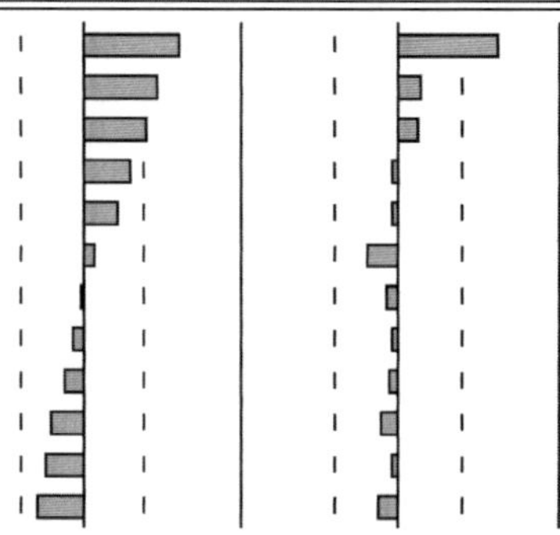

Autocorrelation	Partial Correlation		AC	PAC	Q-Stat	Prob
		1	0.644	0.644	10.832	0.001
		2	0.494	0.136	17.516	0.000
		3	0.431	0.120	22.860	0.000
		4	0.328	-0.028	26.119	0.000
		5	0.243	-0.025	28.012	0.000
		6	0.080	-0.198	28.229	0.000
		7	-0.015	-0.078	28.237	0.000
		8	-0.067	-0.037	28.409	0.000
		9	-0.135	-0.051	29.152	0.001
		10	-0.216	-0.102	31.223	0.001
		11	-0.259	-0.044	34.445	0.000
		12	-0.316	-0.117	39.652	0.000

In Table 4, plots of ACF disconnected (cuts off) ranging from lag to 1, so there is an indication to model data using the model MA (1), and plot PACF disconnected (cuts off) ranging from lag to 1, can also model the model data AR (1). Possible models are good enough to model Mitsubishi car sales data is an AR (1) or MA (1). In the next phase of parameter estimates are calculated for both models. The equation used in the estimation of the parameters are as follows:

$$Model\ AR(1): r_t = c + \phi r_{t-1} + \varepsilon_t$$

$$Model\ MA(1): r_t = \mu + \theta \varepsilon_{t-1} + \varepsilon_t$$

The result of estimation of the model parameters AR (1) and MA (1) are:

Table 5. Parameter Estimation models AR (1)

Dependent Variable: SALE
Method: Least Squares
Date: 06/15/16 Time: 20:39
Sample (adjusted): 2013M07 2015M04
Included observations: 22 after adjustments
Convergence achieved after 3 iterations

Variable	*Coefficient*	*Std. Error*	*t-Statistic*	*Prob.*
C	*130.2205*	*13.83645*	*9.411414*	*0.0000*
AR(1)	*0.644284*	*0.164777*	*3.910046*	*0.0009*

R-squared	*0.433242*	*Mean dependent var*	*132.7727*
Adjusted R-squared	*0.404905*	*S.D. dependent var*	*29.66147*
S.E. of regression	*22.88158*	*Akaike info criterion*	*9.185050*
Sum squared resid	*10471.33*	*Schwarz criterion*	*9.284235*
Log likelihood	*-99.03555*	*Hannan-Quinn criter.*	*9.208415*

F-statistic	*15.28846*	*Durbin-Watson stat*	*2.261518*
Prob(F-statistic)	*0.000868*		
Inverted AR Roots	*.64*		

In the table 5, it can be seen that the probability (*t*-statistic) of parameter c and φ respectively are 0.0000 and 0.0009 is smaller than α = 0.05 so H_0 t-test for the AR model (1) is rejected. Then for the *F*-test, it can be seen that the value of probability (*F*-statistic) = 0.000868 < α, so H0 *F*-test was rejected for the AR model (1). Thus, it can be said that all the independent variables in the model AR (1) significantly to the model. The next table shows the results of the estimation model MA (1) as follows:

Table. 6 Parameter Estimation models MA (1)

Dependent Variable: SALE
Method: Least Squares
Date: 06/15/16 Time: 20:59
Sample: 2013M06 2015M04
Included observations: 23
Convergence achieved after 7 iterations
MA Backcast: 2013M05

Variable	*Coefficient*	*Std. Error*	*t-Statistic*	*Prob.*
C	*134.0130*	*7.967499*	*16.81996*	*0.0000*
MA(1)	*0.522741*	*0.185667*	*2.815483*	*0.0104*

R-squared	*0.302369*	*Mean dependent var*	*134.0435*
Adjusted R-squared	*0.269149*	*S.D. dependent var*	*29.61338*
S.E. of regression	*25.31644*	*Akaike info criterion*	*9.383726*
Sum squared resid	*13459.36*	*Schwarz criterion*	*9.482465*
Log likelihood	*-105.9128*	*Hannan-Quinn criter.*	*9.408558*
F-statistic	*9.101883*	*Durbin-Watson stat*	*1.638561*
Prob(F-statistic)	*0.006563*		
Inverted MA Roots	*-.52*		

In Table 6, it can be seen that the probability (*t*-statistic) of parameter c and φ respectively are 0.0000 and 0.0104 is smaller than α = 0.05 so H0 *t*-test for models MA (1) was rejected. Then for the *F*-test, it can be seen that the value of probability (F-statistic) = 0.006563 < α, so H0 *F*-test was rejected for models MA (1). Thus, it can be said that all the independent variables in the model MA (1) significantly to the model. To determine the best models of both models can use comparative value of *Akaike Information Criterion* (AIC) and Schwarz Information Criterion (SIC). Based on the Tables 5 and 6, obtained Table 7 as follows:

Table 7. Comparison of AIC and SIC

Time Series Model	AIC	SIC
AR(1)	9.185050	9.284235
MA(1)	9.383726	9.482465

In Table 7, note that the value in the AIC and SIC AR model (1) is smaller than the model MA (1), so the model AR (1) it is better to use in the data model Mitsubishi car sales in PT Srikandi Diamond

Motors. After identification of the time series model, do the model parameter estimation Adaptive Threshold Autoregression Spline (ASTAR). Parameter estimation is performed to find the value of θ_i and c as autocorrelation parameters and constants, the results obtained in table 8 below:

Table 8. Frequency with Time Difference 1 Month

t	Y_{t-1}	Y_t	Y_tY_{t-1}	Y_{t-1}^2	Y_t^2
2	162	186	30132	26244	34596
3	186	144	26784	34596	20736
4	144	151	21744	20736	22801
5	151	141	21291	22801	19881
6	141	154	21714	19881	23716
7	154	160	24640	23716	25600
8	160	168	26880	25600	28224
9	168	142	23856	28224	20164
10	142	160	22720	20164	25600
11	160	169	27040	25600	28561
12	169	151	25519	28561	22801
13	151	140	21140	22801	19600
14	140	84	11760	19600	7056
15	84	99	8316	7056	9801
16	99	124	12276	9801	15376
17	124	75	9300	15376	5625
18	75	109	8175	5625	11881
19	109	110	11990	11881	12100
20	110	116	12760	12100	13456
21	116	95	11020	13456	9025
22	95	112	10640	9025	12544
23	112	131	14672	12544	17161
24	131	103	13493	17161	10609
TOTAL	3083	3024	417862	432549	416914

Based on Table 8, it is known that $\sum Y_t = 3024$, $\sum Y_{t-1} = 3083$, $\sum Y_{t-1}Y_t = 417862$, and $\sum {Y_{t-1}}^2 = 432549$.

$$\theta_i = \frac{n(\sum(Y_{t-1}Y_t)) - \sum(Y_{t-1})\sum(Y_t)}{n(\sum({Y_{t-1}}^2)-(\sum Y_{t-1})^2)} = \frac{(23(417862))-(3083(3024))}{(23(432549))-(3083^2)} = \frac{9610826-9322992}{9948627-9504889} = \frac{287834}{443738}$$

$$= 0.6486575 \approx 0.65$$

$$c = \bar{Y}_t + \theta_i \bar{Y}_{t-1} = \left(\frac{3024}{23}\right) + 0.65\left(\frac{3083}{23}\right) = 44.52994785 \approx 44.53$$

Thus, the value of sales data parameter is almost equal to a constant value of 0.65 and sales data is almost equal to 44.53, with the construction of the probability of normal table values obtained broad areas as follows:

Table 9. Wide Area (λ)

No	Lower Bound	Z	Square Area (λ)
1	74.5	-0.60	0.2743
2	93.5	-0.40	0.3446
3	112.5	-0.21	0.4168
4	131.5	-0.01	0.4960
5	150.5	0.19	0.5753
6	169.5	0.39	0.6517

According to the table 9, the value of the area with a lower limit of 74.5 , 93.5 , 112.5 , 131.5 , 150.5 , 169.5 respectively are 0.2743 , 0.3446 , 0.4168 , 0.4960 , 0.5753 , and 0.6517. Preliminary analysis of data were normally distributed and stationary, and has obtained parameter values and constants, we then determined Spline models Adaptive Threshold Autoregression (ASTAR), using the formula:

$$\begin{aligned} Y_t &= c + \theta_1(\lambda_1 Y_{t-1}) + \theta_2(\lambda_2 Y_{t-2}) + \cdots + \theta_p(\lambda_p Y_{t-p}) + \varepsilon_t \\ &= 44.53 + 0.65(\lambda_1 Y_{t-1}) + 0.65(\lambda_2 Y_{t-2}) + 0.65(\lambda_3 Y_{t-3}) + \varepsilon_t. \end{aligned} \quad (4)$$

Equation (4) will be used in forecasting the volume of sales of Mitsubishi cars PT Srikandi Diamond Motors with independent variable volume sales of 2 or 3 periods before time *t*. The results of the forecast volume of car sales in the period to 25, 26, and 27 are as follows:

The forecast for the period to 25 (June 2015) using two independent variables the previous period, the sales volume for the period 23 (April 2015), and the period to 24 (May 2015).

$$Y_{25} = 44.53 + 0.65(\lambda_1 Y_{24}) + 0.65(\lambda_2 Y_{23}) = 44.53 + 0.65(\lambda_1(103)) + 0.65(\lambda_2(131))$$
$$= 44.53 + 0.65(0.3446(103)) + 0.65(0.4168(131)) = 103.09 \approx 104$$

The forecast for the period to 26 (July 2015) using two independent variables the previous period, the volume of sales in the period to 24 (May 2015), and the period of the 25th (June 2015).

$$Y_{26} = 44.53 + 0.65(\lambda_1 Y_{25}) + 0.65(\lambda_2 Y_{24})$$
$$= 44.53 + 0.65(\lambda_1(103)) + 0.65(\lambda_2(103))$$
$$= 44.53 + 0.65(0.3446(103)) + 0.65(0.3446(103)) = 90.67194 \approx 91$$

The forecast for the period to 27 (August 2015) using two independent variables the previous period, the volume of sales in the period to 25 (June 2015), and the period to 26 (July 2015).

$$Y_{27} = 44.53 + 0.65(\lambda_1 Y_{26}) + 0.65(\lambda_2 Y_{25})$$
$$= 44.53 + 0.65(\lambda_1(91)) + 0.65(\lambda_2(103))$$
$$= 44.53 + 0.65(0.2743(91)) + 0.65(0.3446(103)) = 93.692375 \approx 94$$

Thus, the results obtained forecasting sales volume car with two independent variables previous period, i.e. the period to 25 104 cars, a period of 26 to as many as 91 cars, and a period of 27 to as many as 94 cars:

The forecast for the period to 25 (June 2015) by using three independent variables previous period, the sales volume for the period 22 (March), the period to 23 (April 2015), and the period to 24 (May 2015):

$$Y_{25} = 44.53 + 0.65(\lambda_1 Y_{24}) + 0.65(\lambda_2 Y_{23}) + 0.65(\lambda_2 Y_{22})$$
$$= 44.53 + 0.65(\lambda_1(103)) + 0.65(\lambda_2(131)) + 0.65(\lambda_3(112))$$
$$= 44.53 + 0.65(0.3446(103)) + 0.65(0.4168(131)) + 0.65(0.3446(112))$$
$$= 128.17837 \approx 129$$

Forecast period to 26 (June 2015) by using three independent variables previous period, the sales volume for the period 23 (April 2015), the period to 24 (May 2015), and the period of the 25th (June 2015):

$$Y_{26} = 44.53 + 0.65(\lambda_1 Y_{25}) + 0.65(\lambda_2 Y_{24}) + 0.65(\lambda_3 Y_{23})$$
$$= 44.53 + 0.65(\lambda_1(128)) + 0.65(\lambda_2(103)) + 0.65(\lambda_3(131))$$
$$= 44.53 + 0.65(0.4168(128)) + 0.65(0.3446(103)) + 0.65(0.4168(131))$$
$$= 137.76925 \approx 138$$

Forecast period to 27 (June 2015) by using three independent variables previous period, the volume of sales to 24 (May 2015), the period of the 25th (June 2015), and the period to 26 (July 2015).

$$Y_{27} = 44.53 + 0.65(\lambda_1 Y_{26}) + 0.65(\lambda_2 Y_{25}) + 0.65(\lambda_3 Y_{24})$$
$$= 44.53 + 0.65(\lambda_1(138)) + 0.65(\lambda_2(128)) + 0.65(\lambda_3(103))$$
$$= 44.53 + 0.65(0.4960(138)) + 0.65(0.41681(128)) + 0.65(0.3446(103))$$
$$= 132.38177 \approx 133$$

Thus, the results obtained forecasting sales volume by 3 independent variables previous period, the period to 25 as many as 129 cars, a period of 26 to 138 cars, and a period of 27 to as many as 133 cars.

5.1. Mean Y_t Variables 2 and 3 Period Prior t

Y_t value calculation results. Y_t average value can be determined that is a suspected closest to the actual value, the value of the average for the month of June, July and August 2015 as follows:

1. Mean value for period 25 (June 2015).

$$\bar{Y}_{25} = \frac{Y_{25(2)} + Y_{25(3)}}{2} = \frac{104+129}{2} = \frac{233}{2} = 116.5 \approx 117$$

2. Mean value for period ke 26 (July 2015).

$$\bar{Y}_{26} = \frac{Y_{26(2)} + Y_{26(3)}}{2} = \frac{91+138}{2} = \frac{229}{2} = 114.5 \approx 115$$

3. Mean value for period ke 27 (August 2015).

$$\bar{Y}_{27} = \frac{Y_{27(2)} + Y_{27(3)}}{2} = \frac{94+133}{2} = \frac{227}{2} = 113.5 \approx 114$$

Thus, the value of the average volume of car sales, i.e. the period to 25 as many as 117 cars, a period of 26 to 115 cars, and a period of 27 to as many as 114 units.

5.2. Comparison of Forecasting and Actual Value

Comparison of forecasting ($Y_{t_{(2)}}$, $Y_{t_{(2)}}$) and the actual value (Y_t) as follows:

Tabel 10. Comparison of Forecasting and Actual Value

No	Period	$Y_{t_{(2)}}$	$Y_{t_{(3)}}$	$\bar{Y}_t$	Y_t
1	25 (June)	104	129	117	107
2	26 (July)	91	138	115	126
3	27 (August)	94	133	114	114

From Table 10, it appears that the average value has a value close to the true value. Where the value of the forecast for the period to 25 (June 2015), 26 (July 2015), and 27 (August 2015) have flats (Y_t) 117, 115, 114, while the actual value in the period to 25 (June 2015), 26 (July 2015), and 27 (August 2015) respectively 107, 126, 114.

For more details, can be seen in Figure 4 below:

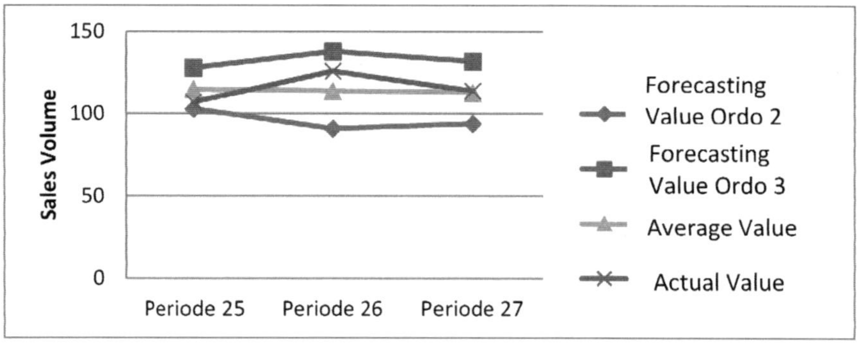

Figure 4. Comparison Graph Forecasting Value and Actual Value

Based on Figure 4, the actual value is between two values forecast by the order and the value forecast by the order 3. It shows Spline method Adaptive Threshold Autoregression (ASTAR) suitable to be applied in forecasting the volume of sales of Mitsubishi cars PT Srikandi Diamond Motors with single-element data. The value of error ratio predictive value and the actual value in the period of June, July and August 2015 as follows:

$$E_t = \frac{\left|Y_{t_{actual}} - Y_{t_{forecast}}\right|}{Y_{t_{actual}}} \times 100\%$$

Value of error for the period to 25 (June 2015):

$$E_{25} = \frac{|Y_{25} - Y_{(25)}|}{Y_{25}} \times 100\% = \frac{|107-117|}{107} \times 100\% = \frac{10}{107} \times 100\% = 0.0935 \times 100\% = 9.35\ \%$$

Value of error for the period to 26 (July 2015):

$$E_{26} = \frac{|Y_{26} - Y_{(26)}|}{Y_{26}} \times 100\% = \frac{|126-115|}{126} \times 100\% = \frac{11}{126} \times 100\% = 0.0873 \times 100\% = 8.73\ \%$$

Value of error for the period to 27 (August 2015)

$$E_{27} = \frac{|Y_{27} - Y_{(27)}|}{Y_{27}} \times 100\% = \frac{|114-114|}{114} \times 100\% = \frac{0}{114} \times 100\% = 0 \times 100\% = 0\ \%$$

Thus, the error value comparison of forecast and actual value in the period to 25, 26 and 27 respectively is 9:35%, 8.73%, and 0%.

For more details can be seen in Figure 5 as follows:

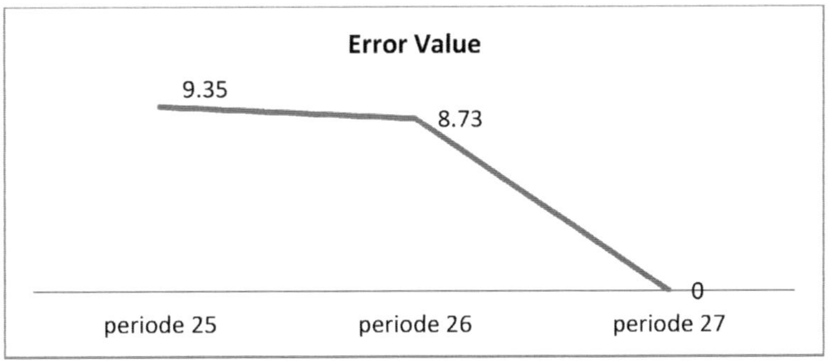

Figure 5. Compare to Error Value

In figure 5 shows the value uniformity errors that occur from 3 to forecast period. The error value occurred in the period 25 and the value of the smallest error occurred in the period of 27. The results forecast approaching its true value, and includes a maximum estimated value of sales and minimal sales so as to facilitate in determining the amount of supply of cars needed in some future period and can be used for all types of data both data trend, seasonal, or cyclical,

6. Conclusion

Based on the results of data processing and discussion can be concluded that:
Based on our forecasting method Spline Adaptive Threshold Autoregression (ASTAR) for Mitsubishi car sales volume PT Srikandi Diamond Motors in the period to 25, 26, and 27, respectively 115, 114, and 113 cars with a maximum of alleged number 128, 138, and 132 cars and the alleged minimal amount of 103, 91, and 94, and the forecast can only be applied to data that is stationary. Great forecasting error ratio value and the actual value for the volume of car sales Mitsubishi Diamond Motors PT Srikandi period to 25, 26 and 27 respectively at 7:48% 9:52% and 0.88%.

References

[1] Arsyil, H.S. 2013. *Proses Differencing pada Analisis Runtun Waktu.* [Online]. In https://statistikawanku.wordpress.com/2013/03/29/proses-differencing-pada-analisis-runtun-waktu/ [25 September 2015].
[2] Hogg, M. dan Craig. 2007. *Introduction to Mathematical Statistics.* India: Dorling Dharma Patria.

[3] Lewis, P. W. dan Stevens, J. C. 1991. Nonlinear Modelling of Times Series Using Multivariate Adaptive Regression Splines.
[4] PQ Bandung. 2013-2015. *Laporan Penjualan*. Bandung: PT Srikandi Diamond Motors.
[5] Putra, W. 2013. *Analisis Statistika*. [Online]. In http://analisis-statistika.blogspot.co.id/2013/03/mengenal-distribusi-normal-dan-cara.html [8 December 2015].
[6] Stevens, J.G. (1991). An Investigation of Multivariate Adaptive Regression Splines for Modeling and Analysis of Univariate and Semi-Multivariate Time series Systems. Ph.D. Thesis, Naval Postgraduate School, California, USA.
[7] Subagyo, P. 1986. *Forecasting Konsep Dan Aplikasi*. Yogyakarta: Penerbit BPFE Yogyakarta.
[8] Supranto, J. 1981. *Metode Peramalan Kuantitatif Untuk Perencanaan*. Jakarta: Penerbit Gramedia.
[9] Secil Toprak dan Pakize Taylan, Journal of the American Statistical Association, 2012 : *A New Approach to Adaptive Spline Threshold Autoregression*

IOP Conf. Series: Materials Science and Engineering **166** (2017) 012033 doi:10.1088/1757-899X/166/1/012033

Model for determining logistic distribution center: case study of Mount Merapi eruption disaster

T.J. Ai and S.S. Wigati

Department of Industrial Engineering, Universitas Atma Jaya Yogyakarta, Jl. Babarsari No. 44 Yogyakarta 55281, Indonesia

Emails: jinai@mail.uajy.ac.id, yayan@mail.uajy.ac.id

Abstract. As one of the most active volcano in the earth, Mount Merapi is periodically erupted and it is considered as a natural disaster for the surrounding area. Kabupaten Sleman as one of the nearest location to this mount has to be always prepared to this disaster. The local government already set three different groups of region, in which potentially affected by Mount Merapi eruption, called KRB I, KRB II, and KRB III. Region KRB III is the closest area to the mount crater and most often affected by the eruption disaster. Whenever KRB III is affected, people live in that area usually being transfer to the next region set that is KRB II. The case presented in this paper is located at the KRB II region, which is the second closest region to the mount crater. A humanitarian distribution system has to be set in this region, since usually this region is became the location of shelters for KRB III population whenever a 'big' eruption is happened. A mathematical model is proposed in this paper, for determining the location of distribution center, vehicle route, and the amount of goods delivered to each customer. Some numerical illustration are presented in order to know the behavior of the proposed model.

1. Introduction

Disaster is a natural phenomenon that cannot be avoided. During 2003 - 2012, the worldwide average annual disaster frequency was 388, and the average annual of people killed by disasters was 106,654 [1]. Disaster not only killed of peoples, but also in the number of injured, displaced, material losses, environmental damage and psychological impact. Losses in term of financial and psychological were a matter that cannot be recovered in a short time. Therefore, preparedness is important to reduce losses, both material and immaterial.

Preparedness is an important factor to overcome disaster. Experience from any disaster events that have occurred, could be used as the basis for prevention and management of disasters. An increase in the number and impact of disasters should be a concern in disaster response operations to more efficiently and effectively [2]. The magnitude of mortality and financial losses that occur indicate the absence of good preparedness in the face of disaster. Humanitarian logistics is one kind of preparedness to be done. This activity, or usually called disaster relief operations (DRO), includes assessing demand, procurement, prioritization, receiving goods, selection, store, search and delivery. This activity is somewhat similar to commercial supply chain management [3].

As well as commercial logistics system, the humanitarian logistics system includes several stages of the disaster, preparation, planning, procurement, transportation, storage, and distribution [4]. The main difference is the main focus. In the commercial logistics system, the final consumer, which is the input source of funds of the entire supply chain, becomes the main focus. However, in the humanitarian logistics system, the end consumer does not have participation in transaction activity, neither control of the supply. When applied to the activity of humanity, the supply chain must be flexible and able to respond quickly to events that cannot be predicted effectively and efficiently. Disaster logistics supply distribution system in contrast to the more commercial logistics focuses on the problem of costs and benefits [5]. Supply logistics of disaster in order to have a fast response and appropriate allocation in response to a disaster.

Research related to the logistics distribution model during disaster by determining the distribution center is based on a risk map has not been done. This model is necessary to help all the parties conducting disaster preparedness. The data required in the form of risk maps are now widely made and continue to be made to areas with disaster potential.

2. Case Study

Mount Merapi is one of the most active volcanoes in the earth. The volcano periodic activities that can be a natural disaster for the surrounding area is its eruption. Since the eruption is frequently happened, Kabupaten Sleman, which is located in Yogyakarta Province, Republic of Indonesia, has to be prepared to this disaster, since some regions of Kabupaten Sleman are placed next to this mountain. The local government already sets three different groups of region, in which potentially affected by Mount Merapi eruption, called KRB I, KRB II, and KRB III. It is noted that KRB III is the closest area to the mount crater and most often affected by the eruption disaster. Whenever KRB III is affected, people live in that area usually being transfer to the next region set that is KRB II.

Our case are focus on the KRB II, in which during the disaster people live in this region combined with the people from KRB III are gathered into some centralized shelters for effective disaster mitigation. For example, if the scale of eruption is increasing and the KRB II region became affected, the transfer of people to safer region, i.e. KRB I, can be coordinated and performed effectively. Another example is the distribution of humanitarian aid to the affected people can be conducted efficiently. Based on several occasion of this eruption, it is required to have a big depot or logistic distribution center for serving the shelters in KRB II. This depot responsibility is to handle the humanitarian aid received from many different stakeholders and distribute the aid to all shelters. For effectively coordinate the shelters, the government set village administrator as the coordinator all shelter within their village. Hence, the common flows of humanitarian aid in this stage are depot – village administrator office – shelter. The current village administrator offices in the KRB II are described in Table 1, together with population estimation of each village.

Table 1. Village, GPS coordinate of the administrator office, and demand estimation in KRB II

No	Village	GPS Coordinate	Population Estimation
2	Girikerto	(-7.6242265, 110.3911628)	946
3	Wonokerto	(-7.6207774, 110.3829727)	590
4	Purwobinangun	(-7.6492907, 110.3923309)	814
5	Candibinangun	(-7.6633180, 110.3993969)	580
6	Hargobinangun	(-7.6397432, 110.4258035)	3912
7	Wukirsari	(-7.6577791, 110.4438001)	1826
8	Argomulyo	(-7.6594539, 110.4563488)	4254
9	Kepuharjo	(-7.6251696, 110.4524367)	7104
10	Umbulharjo	(-7.6257874, 110.4332686)	2040
11	Sindumartani	(-7.6915330, 110.4761452)	599

Other problems that usually rose during the distribution of humanitarian aid are related to the route of distribution vehicles and the distribution service level. For this case, the local government is allocated only single distribution vehicle for serving the KRB II area. Therefore, the scheduling of this vehicle is very important so that all the villages in the area can be covered within the operational time. Related to the service level, it is common in this situation that the supply of humanitarian aid available to be distributed is less than the total demand of the aid for all villages. As consequences, the service level may less than 100%. There is a necessity to make the service level among village are balanced so that every village are experienced more or less the same quality of fondness from various stakeholders during the time of disaster. In an extreme situation, distribution that make 100% service level to village A and 0% service level to village B can lead to chaotic condition since the people in village B are felt abandoned, jealous to the people in village A, and think that the government is unfair to them. By creating balance of the service level, therefore, the last illustration can be avoided.

3. Mathematical Model

Based on the case described above, we can formulate a mathematical model for determine the location of logistic distribution center, the vehicle route, and the amount of goods delivered to each customer location.

In order to determine the logistic distribution center, we are using center of gravity approach. The formula to obtain the center are

$$\bar{x} = \frac{\sum x_i \varphi_i}{\sum D_i} \quad (1)$$

$$\bar{y} = \frac{\sum y_i \varphi_i}{\sum D_i} \quad (2)$$

where

$\bar{x}$: x coordinate of the logistic distribution center
$\bar{y}$: y coordinate of the logistic distribution center
x_i : x coordinate of village administrator office i
y_i : y coordinate of village administrator office i
φ_i : population estimation of village i

A mixed integer linear programming is formulated in order to simultaneously the vehicle route and the amount of goods delivered to each customer location. It is defined that the customers are located in each village administrator office. Several assumptions for this formulation are:

- Number of customer location is known
- Each customer location can be visited once in each vehicle route
- Vehicle has access from the distribution center to every customer location and from every customer location to other customer locations
- Every vehicle route starts from and ends at the distribution center
- The distance is symmetric, i.e. the distance from customer i to customer j is equal to the distance from customer j to customer i. Hence, the travel time is also symmetric
- Whenever vehicle is completing its service to a certain route, it can be assigned to serve another route as long as its total time is not exceeding the service time availability
- The amount of goods delivered to each customer is not exceeding its corresponding customer demand
- Travel time is deterministic static
- Loading and unloading time in every location is similar and constant

The model is formulated with two set of decision variables:

- Q_{jr}, amount of goods delivered to customer j by vehicle r
- X_{ijr}, a binary variable indicating the vehicle route, in which $X_{ijr} = 1$ whenever vehicle r travel to location j right after location i and $X_{ijr} = 0$ for other conditions

The mixed integer linear programming is formulated using two sets definition which are the set of location J and set of vehicle route R. The member of set J are the distribution center (set as j=1) and all the customer location (set as j=2,3, ...)

The model is presented below:

$$\text{Min } \sum_{j\in J}\left(D_j - \sum_{r\in R} Q_{jr}\right) \quad (3)$$

Subject to

$$\sum_{r\in R} Q_{jr} \leq D_j, \quad \forall j \in J \quad (4)$$

$$\sum_{r\in R}\sum_{j\in J} Q_{jr} \leq S \quad (5)$$

$$\sum_{j\in J} B.Q_{jr} \leq W, \quad \forall r \in R \quad (6)$$

$$\sum_{j\in J} V.Q_{jr} \leq G, \quad \forall r \in R \quad (7)$$

$$\sum_{r\in R}\sum_{i\in J}\sum_{j\in J}\left(P_{ij} + \tau\right).X_{ijr} \leq L \quad (8)$$

$$X_{ijr} + X_{hir} \leq 1, \quad \forall i > 1, \forall j > 1, i \neq j, \forall h > 1, h \neq i, \forall r \in R, \quad (9)$$

$$\sum_{j\in J} X_{1jr} = 1, \quad \forall r \in R \quad (10)$$

$$\sum_{i\in J} X_{i1r} = 1, \quad \forall r \in R \quad (11)$$

$$\sum_{i\neq h} X_{ihr} = \sum_{h\neq j} X_{hjr}, \quad \forall i \in J, \forall j \in J, \forall r \in R \quad (12)$$

$$F_j = \frac{\sum_{r\in R} Q_{jr}}{D_j}, \quad \forall j \in J \quad (13)$$

$$A = \frac{\sum_{j\in J} F_j}{|J|}, \quad (14)$$

$$A - \delta \leq F_j \leq A + \delta, \quad \forall j \in J \quad (15)$$

$$X_{ijr} \in \{0,1\}, \quad \forall i \in J, \forall j \in J, \forall r \in R \quad (16)$$

$$Q_{jr} \geq 0, \quad \forall j \in J, \forall r \in R \quad (17)$$

where

D_j : demand of customer j
S : supply of goods in the logistic distribution center
B : unit weight of goods
W : weight capacity of distribution vehicle
V : unit volume of goods
G : volume capacity of distribution vehicle
P_{ij} : travel time from location i to location j
τ : loading and unloading time
L : volume capacity of distribution vehicle
F_j : service level of customer j
L : average service level
δ : service level balance factor

The objective function of this formulation, which is presented in equation (3), is to minimize total unmet demand for all customers. Constraint in equation (4) states that the amount of goods delivered to each customer is not exceeding its corresponding customer demand. Equation (5) states that the total amount of goods delivered to customers is not exceeding the supply of goods in the logistics distribution center. Constraint (6) and (7) explain the limitation of vehicle in term of weight and

IOP Conf. Series: Materials Science and Engineering **166** (2017) 012033 doi:10.1088/1757-899X/166/1/012033

volume, respectively. The equation (8) states the total service time limitation of vehicle including loading and unloading time. Equation (9) is defined to limit number of vehicle visit to each customer in each route, which is only single visit allowed. The logic of routes are described in constraints (10)-(12), in which the route starts and ends at the distribution center. Moreover, whenever a route enters a customer, it also departs from it. Constraints (13)-(15) are set for making the service level among customer are balanced. Constraints (16) and (17) limit the variables value.

4. Solution of the Case

Based on the data in Table 1, the calculation of equations 1 and 2 are resulting the location of logistic distribution center which is (-7.6402453, 110.4388979). Therefore, the travel time from this center to various village administrator offices and the travel time among office can be determined using common GPS service map, i.e. the Google Maps®. The travel times are summarized in Table 2, in which '1' denotes the logistic distribution center and '2' to '11' denote the customers, which are the village administrator office 2 to 11, respectively.

Table 2. Travel time from one location to other location (in minutes)

	1	2	3	4	5	6	7	8	9	10	11
1	0	27	29	21	19	13	6	7	8	7	17
2	27	0	3	7	11	18	22	26	30	24	33
3	29	3	0	9	13	21	24	28	31	26	35
4	21	7	9	0	5	13	16	21	23	18	28
5	19	11	13	5	0	17	15	19	22	17	26
6	13	18	21	13	17	0	14	16	10	5	25
7	6	22	24	16	15	14	0	5	10	9	13
8	7	26	28	21	19	16	5	0	8	11	9
9	8	30	31	23	22	10	10	8	0	5	17
10	7	24	26	18	17	5	9	11	5	0	20
11	17	33	35	28	26	25	13	9	17	20	0

4.1. Case 1

In this case, following problem parameters are considered S = 30000, B = 0.5, W = 5000, V = 1000, G = 8000000, τ = 30, L = 480, δ=0.05, and the demand each customer equal to the number of population in each village (detail in Table 1). Initially, the number of route is set to 5. It is noted that the total available supply in the distribution center are enough to supply all customers demand.

The mixed integer linear programming proposed above is coded and solving using LINGO program. The result can be summarized in Table 3 and 4, with the objective function is equal to 0. Since the service time availability for each vehicle is 480 minutes, therefore, 2 vehicles are required for this case, in which the first vehicle is serving route 1 – 3 with 400 minutes total travel time and the second vehicle is serving route 4 – 5 with 271 minutes total travel time.

Figure 1. Locations of logistic distribution center (yellow) and customers (blue) in the Google Maps®

Table 3. Route formed for Case 1

No	Route	Total travel time
1	1-7-10-1	112
2	1-2-3-1	149
3	1-4-8-1	139
4	1-5-6-1	139
5	1-9-11-1	132

Table 4. Quantity of goods delivered and service level each customers in Case 1

Customer j	D_j	Q_j	F_j
2	946	946	1
3	590	590	1
4	814	814	1
5	580	580	1
6	3912	3912	1
7	1826	1826	1
8	4254	4254	1
9	7104	7104	1
10	2040	2040	1
11	599	599	1

4.2. Case 2

Case 1 and 2 are similar except the total supply, $S = 20000$. It is noted that the total available supply in the distribution center are not enough to supply all customers demand. Using the same software, the result from the model can be obtained and summarized in Table 5 and 6. It is noted that the objective function is equal to 2665. Since the service time availability for each vehicle is 480 minutes, therefore, 2 vehicles are required for this case, in which the first vehicle is serving route 1 – 3 with 436 minutes total travel time and the second vehicle is serving route 4 – 5 with 240 minutes total travel time. From Table 6, it is implied that the average of service level A is 0.9061 with minimum service level is 0.8561 and maximum service level is 0.9559.

Table 5. Route formed for Case 2

No	Route	Total travel time
1	1-10-4-1	136
2	1-5-2-1	147
3	1-3-6-1	153
4	1-7-8-1	108
5	1-11-9-1	132

Table 6. Quantity of goods delivered and service level each customers in Case 2

Customer j	D_j	Q_j	F_j
2	946	904	0.9556
3	590	564	0.9559
4	814	697	0.8563
5	580	554	0.9552
6	3912	3349	0.8561
7	1826	1745	0.9556
8	4254	3644	0.8566
9	7104	6090	0.8573
10	2040	1936	0.9490
11	599	517	0.8631

5. Conclusions

This paper presented a mathematical model that can be used in the case situated in KRB II region whenever Mount Merapi is erupted and affected KRB III. The proposed model is able to determine the location of distribution center, vehicle route, and the amount of goods delivered to each customer. Two different problem setting are considered in the numerical examples, in which both result show the capability of this model. In the future, several situations that actually happened in the KRB II during disaster need to be considered in the model, for example: reliability of the distribution center and uncertain demand in each location.

References

[1] Guha-Sapir D, Hoyois P and Below R 2014 *Annual Disaster Statistical Review 2013: The numbers and trends* (Centre for Research on the Epidemiology of Disasters, Retrieved from http://www.cred.be/sites/default/files/ADSR_2010.pdf)

[2] Balcik B and Beamon BM 2008 Facility location in humanitarian relief *International Journal of Logistics Research and Applications* **11(2)** 101–121

[3] Pujawan IN, Kurniati N and Wessiani NA 2009 Supply Chain Management for Disaster Relief Operations: Principles and Case Studies *International Journal Logistics Systems and Management* **5(6)** 679–692
[4] Thomas AS and Kopczak LR 2005 *From Logistics to Supply Chain Management: The Path Forward In the Humanitarian Sector* (Fritz Institute, Retrieved from http://www.fritzinstitute.org/)
[5] Huang M Smilowitz K and Balcik B 2012 Models for relief routing: Equity, efficiency and efficacy *Transportation Research Part E: Logistics and Transportation Review* **48(1)** 2–18

Application of Holt exponential smoothing and ARIMA method for data population in West Java

A. Supriatna, D. Susanti, E. Hertini

Department of Mathematics, Faculty of Mathematics and Natural Science, Padjadjaran University, Bandung, West Java

E-mail: Asupriatna55@ymail.com

Abstract. One method of time series that is often used to predict data that contains trend is Holt. Holt method using different parameters used in the original data which aims to smooth the trend value. In addition to Holt, ARIMA method can be used on a wide variety of data including data pattern containing a pattern trend. Data actual of population from 1998-2015 contains the trends so can be solved by Holt and ARIMA method to obtain the prediction value of some periods. The best method is measured by looking at the smallest MAPE and MAE error. The result using Holt method is 47.205.749 populations in 2016, 47.535.324 populations in 2017, and 48.041.672 populations in 2018, with MAPE error is 0,469744 and MAE error is 189.731. While the result using ARIMA method is 46.964.682 populations in 2016, 47.342.189 in 2017, and 47.899.696 in 2018, with MAPE error is 0,4380 and MAE is 176.626.

1. Introduction

The population has a data pattern trend up because the population has increased. Increased happens will have an impact on many things, including the increasing need for food, house, or jobs. In addressing improvements, required a prediction of the number of people in order to define strategies. Prediction of total population can found using Holt and Autoregressive Integrated Moving Average (ARIMA) method. Holt is an exponential smoothing method by performing weighting on past observations and estimate trend in data. While ARIMA method is studied in depth by George Box and Gwilym Jenkins and consists of four stages: identification, estimation, diagnostic checking, and forecasting [4].

2. Methods and Materials

The data used is the population of West Java for 1998-2015 periods (Table 1, Figure 1).

2.1. Holt Exponential Smoothing

Holt is a double exponential smoothing method has two parameters. In the weighting parameters, the new value has a greater than previous observations. In addition, there is weighted to estimate the trend of data. Generally, Holt has two smoothing constants (the values between 0 and 1) and three equations [5]

Smoothing : $S_t = \alpha X_t + (1-\alpha)(S_{t-1} + b_{t-1})$

Smoothing of trend : $b_t = \gamma(S_t - S_{t-1}) + (1-\gamma)b_{t-1}$

Forecast : $F_{t+m} = S_t + b_t(m)$

Table 1. Data population for 1998-2015 [1]

Period	Data	Period	Data
1998	33.261.409	2008	42.194.809
1999	34.555.622	2009	42.693.951
2000	35.723.473	2010	43.053.732
2001	36.795.565	2011	43.826.775
2002	37.291.946	2012	44.548.431
2003	37.980.422	2013	45.340.799
2004	39.140.812	2014	46.029.668
2005	39.960.869	2015	46.497.175
2006	40.737.594		
2007	41.483.729		

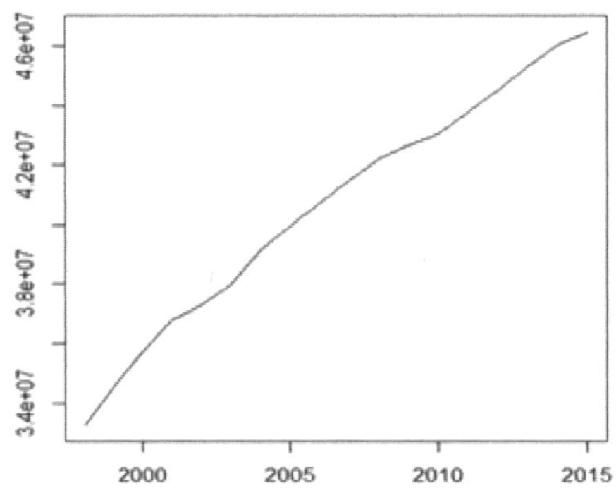

Figure 1. Plot of data population

2.2. Autoregressive Integrated Moving Average (ARIMA)
ARIMA is the short-term forecasting [2]. According Makridakis et al (1999), the stages of ARIMA are Identification, parameter estimation, diagnostic checking, and forecast. ARIMA models are generally expressed with (p,d,q) .
ARIMA Model with $(p,2,q)$ [4]:

$$(1-\mathrm{B})^d X_t = e_t$$

e_t : error,
X_t : Observation value for t (t= 1, 2, ..., n),
$(1-\mathrm{B})^d$: mathematics operation for differencing.

3. Results

3.1. Holt Exponential Smoothing
Holt method generally has two smoothing constants (with value between 0 and 1). To obtain the optimal parameters used by trial and error. Table 2 is the process of finding the value of α and γ. The value will selected with smallest MAPE error. Then the parameters to be used are α = 0.9 γ = 0.9.

Table 2. Selection Process for parameter α and γ

No	α	γ	MAPE
1	0,1	0,1	3,889468
2	0,1	0,5	2,131118
3	0,1	0,9	1,425721
4	0,2	0,1	2,556987
5	0,2	0,5	1,104799
6	0,2	0,9	0,714864
7	0,3	0,1	1,836226
8	0,3	0,5	0,74986
9	0,3	0,9	0,56141
10	0,4	0,1	1,41355
11	0,4	0,5	0,598116
12	0,4	0,9	0,591477
13	0,5	0,1	1,144885
14	0,5	0,5	0,51336
15	0,5	0,9	0,619235
16	0,6	0,1	0,961856
17	0,6	0,5	0,523453
18	0,6	0,9	0,592657
19	0,7	0,1	0,829935
20	0,7	0,5	0,520288
21	0,7	0,9	0,547798
22	0,8	0,1	0,730374
23	0,8	0,5	0,496619
24	0,8	0,9	0,496724
25	0,9	0,1	0,667488
26	0,9	0,5	0,484451
27	0,9	0,9	0,469744

Based on Holt method, with alpha 0.9 and gamma 0.9 is obtained prediction of the total population in 2016 is 47.028.975, populations in 2017 is 47.535.324, and populations in 2018 is 48.041.672. The value of MAPE error is 0,469744 and MAE error is 189.731.

3.2. Autoregressive Integrated Moving Average (ARIMA)

Time series analysis assumptions that must be fulfilled is stationary data. Based on the plot seen data contains trend patterns and not stationary in mean. In addition to viewing the plot, not stationary can viewable by ACF plot. If the ACF down slowly, then data is not stationary [3]. First, data must differencing because ACF plot down slowly (Figure 2). The result since first differencing (Figure 3), data still not stationary. Second differencing done to create data stationary (Figure 4).

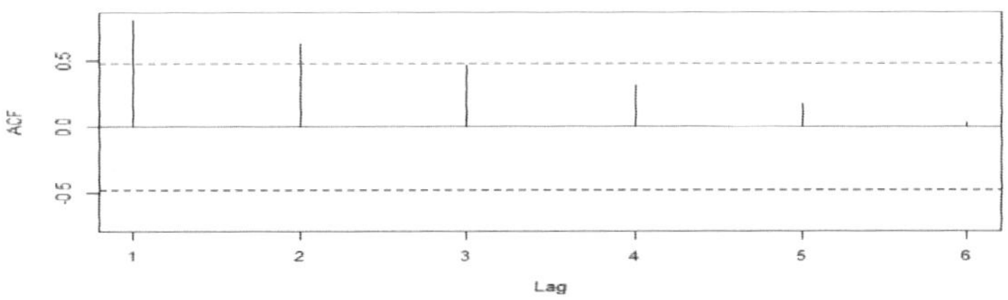

Figure 2. Plot ACF of data populations

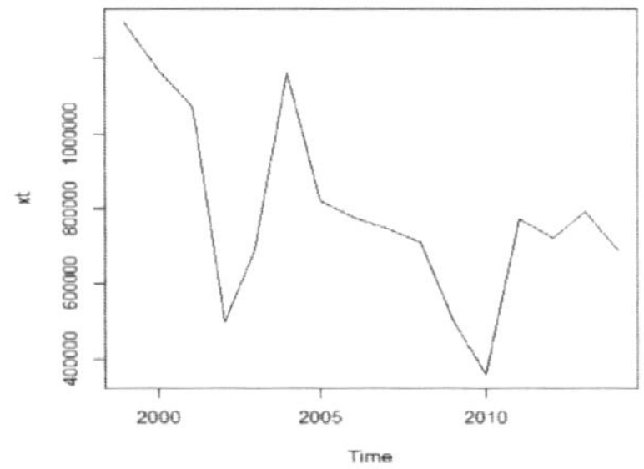

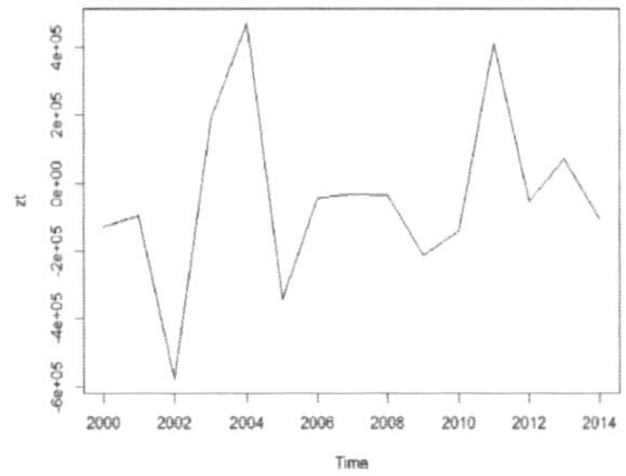

Figure 3. Result of first differencing **Figure 4.** Result of second differencing

Since data is stationary, plot ACF and PACF used to choose the model of ARIMA(p, d, q). ACF will determine the orde of Moving Average (q), order (d) is order differencing, and PACF will determine the orde of Autoregressive (p).

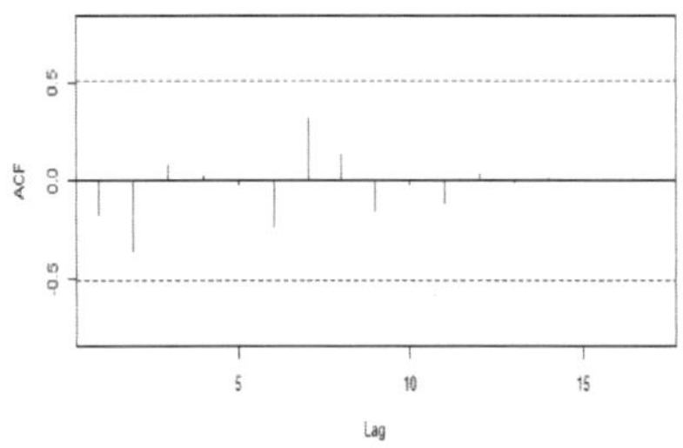

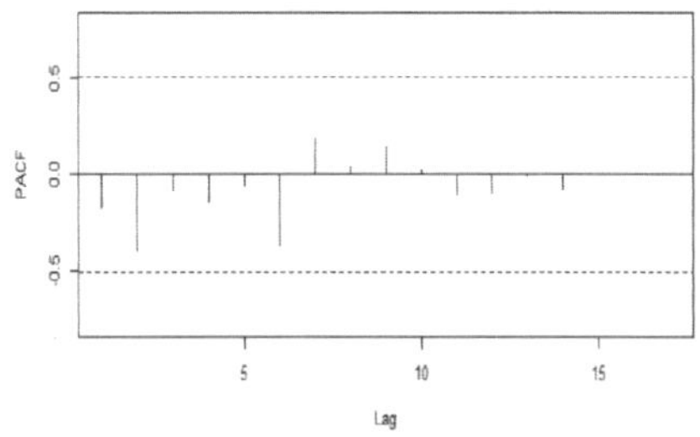

Figure 5. ACF after differencing **Figure 6.** PACF after differencing

Based on Figure 5 and Figure 6, ACF and PACF plot show no lag was out from Bartlett so that is probably is ARIMA (0,2,0). Furthermore, the suitable model used for forecast. The result of prediction are 46.964.682 populations in 2016, 47.432.189 populations in 2017, and 47.899.696 populations in 2018 with MAPE error is 0,4380 and MAE is 176.626.

4. Conclusions

The results of Holt method has prediction in 2016 is 47.028.975 populations, 2017 is 47.535.324 populations, and in 2018 is 48,041,672 population, with a MAPE error is 0.469744 and MAE is 189.731. While ARIMA method predictions in 2016 is 46.964.682 populations, 2017 is 47.432.189 populations, and in 2018 is 47.899.696 populations, with MAPE error is 0.4380 and MAE is 176,626. ARIMA method forecasting is better than Holt method because error values is smaller.

References

[1] Badan Pusat Statistik Provinsi Jawa Barat 2016 *Statistik Jumlah Penduduk Jawa Barat.*

[2] Ekananda, M 2014 *Analisis Time Series: Untuk Penelitian Ekonomi, Manajemen, dan Akutansi.* (Jakarta: PT. Mitra Wacana Media)

[3] Hyndman, J.R 2014 *Forecasting: Principles & Practice.* (Australia)

[4] Makridakis, S, et al 1999 *Metode dan Aplikasi Peramalan, Jilid 1.* (Jakarta: Binapura Aksara) [5] Noeryanti, et al 2012 *Aplikasi Pemulusan Eksponensial Dari Brown dan Holt Untuk Data yang Memuat Trend.* Prosiding Seminar Nasional Aplikasi Sains & Teknologi (SNAST) Periode III. Yogyakarta

An Optimization of Inventory Demand Forecasting in University Healthcare Centre

A T Bon[1] and T K Ng[2]

[1,2]Department Production and Operation Management
UniversitiTun Hussein Onn Malaysia 86400 Parit Raja, Johor, Malaysia
Email: talib@gmail.com

Abstract. Healthcare industry becomes an important field for human beings nowadays as it concerns about one's health. With that, forecasting demand for health services is an important step in managerial decision making for all healthcare organizations. Hence, a case study was conducted in University Health Centre to collect historical demand data of Panadol 650mg for 68 months from January 2009 until August 2014. The aim of the research is to optimize the overall inventory demand through forecasting techniques. Quantitative forecasting or time series forecasting model was used in the case study to forecast future data as a function of past data. Furthermore, the data pattern needs to be identified first before applying the forecasting techniques. Trend is the data pattern and then ten forecasting techniques are applied using Risk Simulator Software. Lastly, the best forecasting techniques will be find out with the least forecasting error. Among the ten forecasting techniques include single moving average, single exponential smoothing, double moving average, double exponential smoothing, regression , Holt-Winter's additive, Seasonal additive, Holt-Winter's multiplicative, seasonal multiplicative and Autoregressive Integrated Moving Average (ARIMA). According to the forecasting accuracy measurement, the best forecasting technique is regression analysis.

1. Introduction

Many companies do not know their demands and have to rely on demand forecasts to make decisions in production planning, sourcing and inventory management both in long and short term (Kerkkanen, 2010). According to Bon and Chong (2009), there are many factors affecting the frequent change of inventory's demand and those factors are trend, seasonality, and economic factors.Many organizations that utilized inventory optimization reduced inventory levels by up to 25 per cent in one year and enjoyed a discounted cash flow above 50 per cent in less than two years(Study of IDC Manufacturing Insight,2010). In fact, inventory optimization can better address demand volatility and supply variability and thus reducing the risk of both under stocks and overstocks for the organizations to optimize their inventory through demand forecasting.

Healthcare in Malaysia is mainly under the jurisdiction of the Ministry of Health Malaysia. Currently, the nation practices a dual healthcare system and has good access to a blend of government and private healthcare services through a network of hospitals and clinics.In this research, a case study of University Health Centre in University Tun Hussein Onn Malaysia (UTHM) will be conducted. Health services UTHM started back in May 2002, when the Student Health Centre was a part of The Students Affair Office. The University Health Centre provides outpatient services as well as services to emergency cases during office hours.

Under the 10th Malaysia Plan (2011–2015), the government has identified healthcare services as one of the 12 National Key Economic Areas (NKEA) to generate revenue for the country. Healthcare is getting important and important even has becomes necessity for publics in the world.Demand forecasting is one of the most crucial issues of inventory management as forecasts form the basis for the planning of production, transportation and inventory levels (Korpela and Tuominen, 1996). Due to large forecast errors usually negatively affect companies' operational performance, forecast accuracy is often considered as a necessity (Danese and Kalchschmidt, 2011).

Most of the time the healthcare provider will use their own judgement and experience to develop the decision on inventory demand. It means that there is yet have a proper and systematic way to forecast on the inventory demand. It is essential to apply forecasting methods in the aspect of inventory demand to avoid any wastage and lack of inventory situation happen. Besides that, since the actual demand is uncertainty the forecast also difficult to predict. Therefore, ongoing work should be carried out to improve the demand forecasting.

2. Literature Review

2.1 Forecasting

Forecasting means estimating a future event or condition which is outside an organization's control and provides a basis for managerial planning (Kerkkanen,2010). Everyone requires forecasts; the need for forecasts cuts across all functional lines as well as all types of organizations (Hanke and Wichern, 2009) because it is useful in making management decision. According to Panneerselvam (2005), in any organizations, all the functional managers will base their decision on the forecast value. Many companies rely forecasting heavily in decision making for production planning, sourcing and inventory management. Hence, forecasting is used by companies to determine how to allocate their budgets for an upcoming period of time. This is typically based on demand for the goods and services it offers, compared to the cost of producing them.

2.2 Healthcare Forecasting

Forecasting demand for health services is an important step in managerial decision making for all healthcare organizations. This task, which often is assumed by financial managers, first requires the compilation and examination of historical information. Although many quantitative forecasting methods exist, four common methods of forecasting are percent adjustment, 12-month moving average, trend line, and seasonalized forecast.Healthcare financial managers who want to project demand for healthcare services in their facility should understand the advantages and disadvantages of each method and then select the method that will best meet the organization's needs.

2.3 Demand Forecasting

Demand is the need for a particular product or component and it can come from a number of sources (e.g., customer order or producer's good) (Armstrong, 2000).Demand forecasting means estimate of most likely future demand for product under given conditions. Based on the study by Kerkkanen (2010), one of the fundamental managerial tasks is demand forecasting.Demand forecasting is the area of predictive analytics dedicated to understanding consumer demand for goods or services. That understanding is harnessed and used to forecast consumer demand. Knowledge of how demand will fluctuate enables the supplier to keep the right amount of stock on hand.

2.4 Forecasting Techniques

A forecasting technique is a procedure for computing forecasts from present and past values (Chatfield, 2000). According to Lawrence, Klimberg, and Lawrence (2009), forecasting techniques can be divided into two broad categories that are both qualitative and quantitative. Qualitativetechniquesareprojectionsbasedonjudgment,intuition,andinformed opinions,andtheyaresubjectivebynature(Kerkkanen ,2010). Examples of qualitative forecasting techniques include jury of executive opinion, Delphi method, sales force composite and consumer market survey.Quantitative forecasting techniques are used to forecast future data as a function of past data; they are appropriate when past data are available. Examples of quantitative forecasting techniques such as single moving average, single exponential smoothing, double moving average, double exponential smoothing, regression , Holt-Winter's additive, Seasonal additive, Holt-Winter's multiplicative, seasonal multiplicative and Autoregressive Integrated Moving Average (ARIMA).

3. Methodology

3.1 Research Design

According to Marican (2006), research design is like a map that clearly presents how a research will be carried out. Research design is the logic that links the data to be collected and the conclusions to be drawn to the initial question of study (Yin ,2009).The research design may be classified in three broad categories depending on objectives and types of research which are exploratory, descriptive and causal (Hair *et. al.*, 2003). This study was designed as a descriptive research. A descriptive research may be classified as both qualitative and quantitative.

3.2 Data Collection

This research will involve both qualitative and quantitative method for data collection. For qualitative process, the researcher will ask the problems existed in managing inventory demand in pharmacy, University Health Centre UTHM. For quantitative, the secondary data regarding inventory demand will be collected for data analysis.The most recent 5 years of high demand medicine like Panadol 650mg will be collected so that the forecasting techniques can be applied to analyze the data using risk simulator software and hence the best techniques can be selected to optimize the inventory demand.

4. Data Analysis

4.1 Data Analysis Background

Data analysis is one of the important elements of the research. After the data being collected, analysis data process will be started in order to determine and find out the result. The research data will be analyzed using Risk Simulator Software. Quantitative analysis was used for this study. Data analysis was committed once the historical data collected from University Health Centre.Before analyzing the data, the historical data were collected from University Health Centre, UTHM. The collected data was inventory demand regarding Panadol 650mg beginning from January 2009 until August 2014.Analyzing started by, firstly, forecast time horizon of collected data will be determined. It is an important preliminary step to selection of the forecasting methods to be used. The determined forecast time horizon was 1 month, a short range forecast. Then, time series plots of the data were constructed and visually inspected the data patterns. After the identification of the data pattern, the study proceeds to select and implement the appropriate forecasting methods based on the data pattern. Finally, the forecast accuracy measurements with the lowest values were selected as the best forecasting method.

4.2 Identification of Data Pattern

A time series are likely to consist one or more of the data patterns such as trend, cyclical, seasonal and horizontal component.

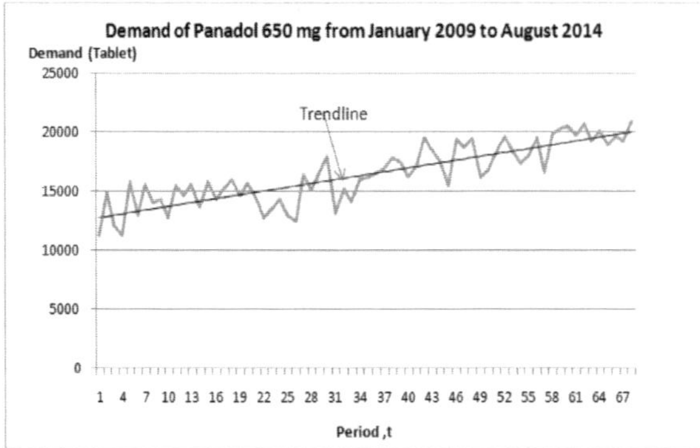

Figure 4.1: Time Series Plot for Panadol 650 mg from January 2009 to August 2014.

As from figure 4.1, it illustrates that there was a significant growth in the time series over an extended period of time. This means that there was a trend component in the time series plot.

4.3 Time-Series Forecasting Techniques

1. Single Moving Average
2. Single Exponential Smoothing
3. Double Moving Average
4. Double Exponential Smoothing
5. Regression Analysis
6. Holt-Winter's Additive
7. Seasonal Additive
8. Holt-Winter's Multiplicative
9. Seasonal Multiplicative
10. ARIMA(Autoregressive Integrated Moving Average)

4.3.1 Single Moving Average

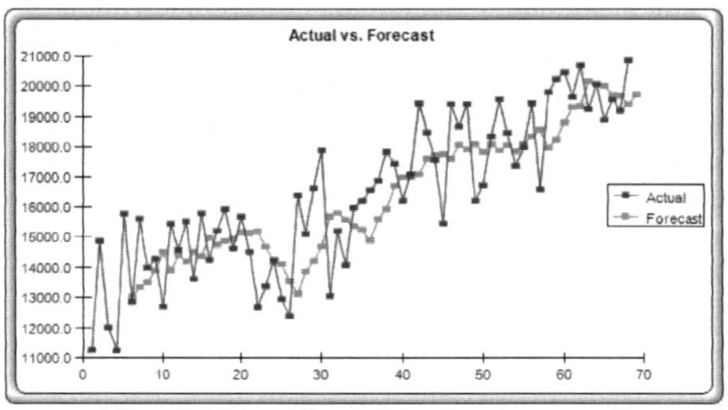

Figure 4.2: Forecasted Result Using Single Moving Average Forecasting Technique

For single moving average, the demand forecast for Panadol 650mg was 19729 tablets in September 2014 (t=69). The RMSE error measurement was 1406.8824.

4.3.2 Single Exponential Smoothing

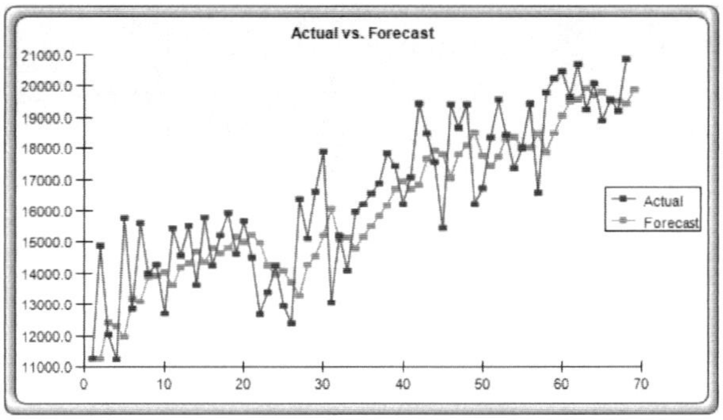

Figure 4.3: Forecasted Result Using Single Exponential Smoothing Forecasting Technique

For single exponential smoothing, the demand forecast for Panadol 650 mg during September 2014 (t= 69) with the value of 19898 tablets. The error measurement for RMSE was 1501.6296.

4.3.3 Double Moving Average

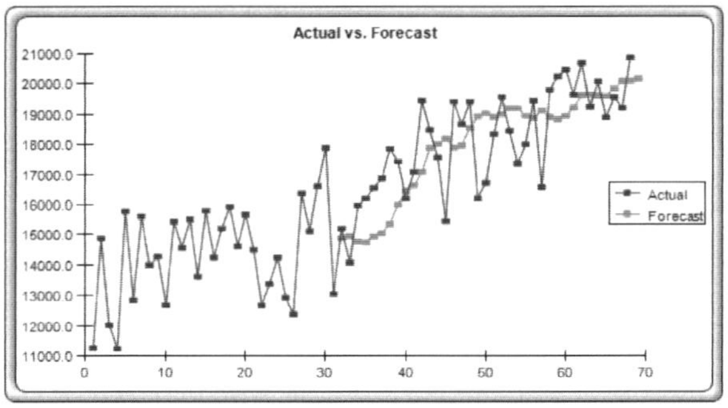

Figure 4.4: Forecasted Result Using Double Moving Average Forecasting Technique

For double moving average, the demand forecast for Panadol 650 mg in September 2014 (t=69) with the value of 20175 tablets. The RMSE error measurement was 1376.7254.

4.3.4 Double Exponential Smoothing

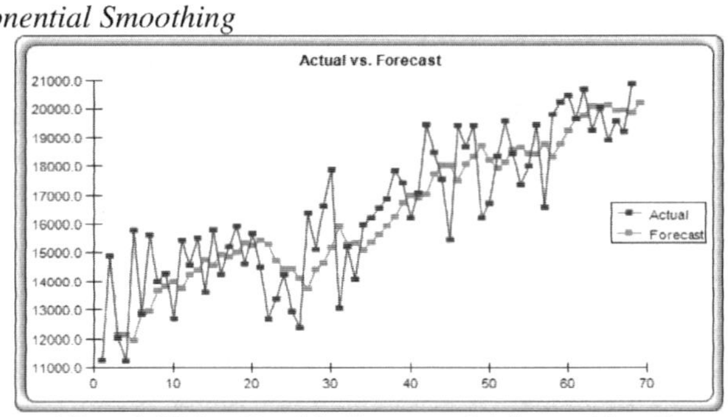

Figure4.5:Forecasted Result Using Double Exponential Smoothing Forecasting Technique

For double exponential smoothing, the demand forecast for Panadol 650 mg in September 2014 (t=69) was 20226 tablets. The error measurement for RMSE was 1413.5199.

4.3.5 Regression Analysis

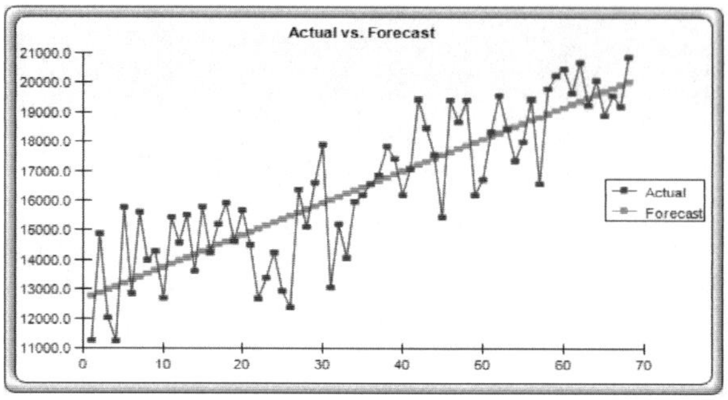

Figure4.6:Forecasted Result Using Regression Analysis

For regression analysis, the error measurement for RMSE was 1334.7793. The forecast using this technique will indicate continuous incremental pattern.

4.3.6 Holt-Winter's Additive

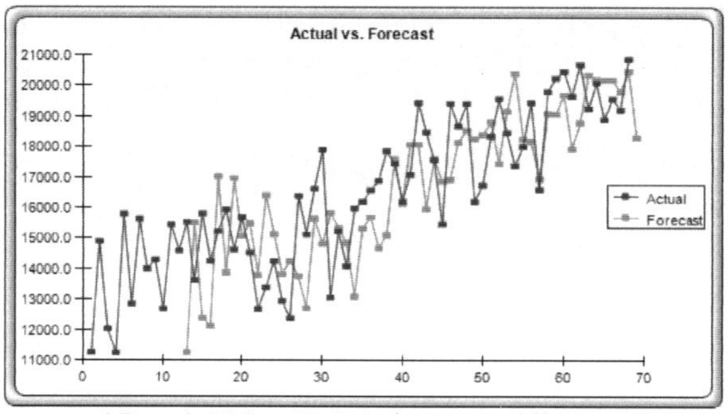

Figure 4.7: Forecasted Result Using Holt-Winter's Additive Forecasting Technique

For Holt-Winter's additive, the demand forecast for Panadol 650 mg in September 2014 (t=69) was 18301 tablets. The error measurement for RMSE was 1771.9117.

4.3.7 Seasonal Additive

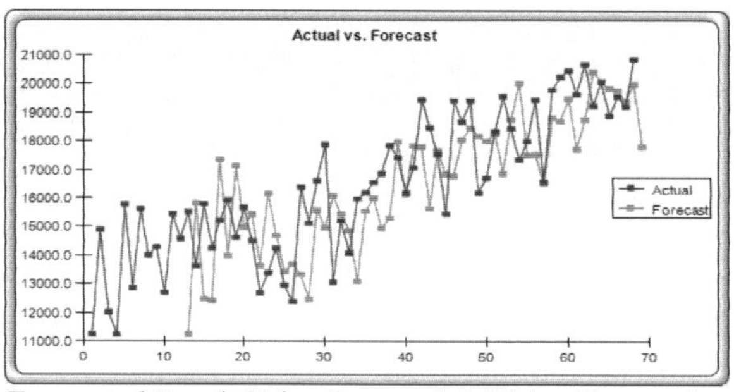

Figure 4.8: Forecasted Result Using Seasonal Additive Forecasting Technique

For seasonal additive, the demand forecast for Panadol 650 mg in September 2014 (t=69) was 17817 tablets. The error measurement for RMSE was 1793.8089.

4.3.8 Holt-Winter's Multiplicative

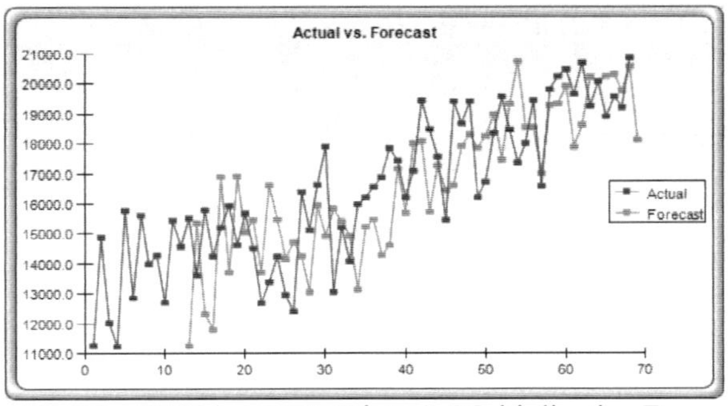

Figure 4.9: Forecasted Result Using Holt-Winter's Multiplicative Forecasting Technique

For Holt-Winter's multiplicative, the demand forecast for Panadol 650 mg in September 2014 (t=69) was 18114 tablets. The error measurement for RMSE was 1818.6557.

4.3.9 Seasonal Multiplicative

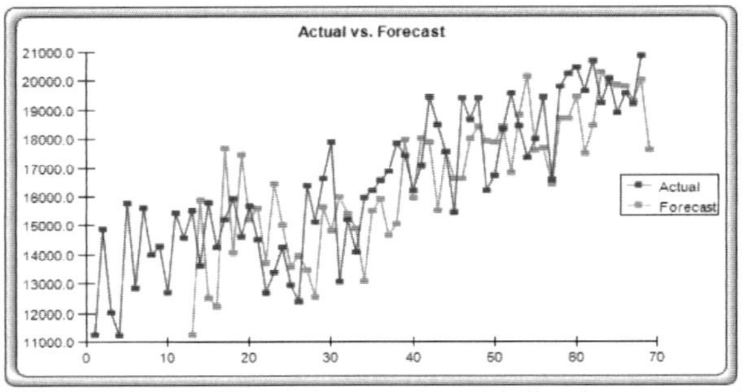

Figure 4.10: Forecasted Result Using Seasonal Multiplicative Forecasting Technique

For seasonal multiplicative, the demand forecast for Panadol 650 mg in September 2014 (t=69) was 17627 tablets. The error measurement for RMSE was 1849.2047.

4.3.10 Autoregressive integrated moving average (ARIMA)

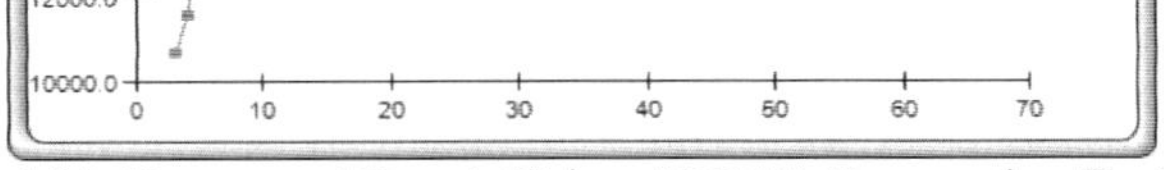

Figure 4.11: Forecasted Result Using ARIMA Forecasting Technique

For ARIMA, it showed that the actual value and the forecast value using ARIMA forecasting technique with parameters p = 2, d = 0, and q = 1 for Panadol 650mg.

4.4 Identification of best forecasting technique

The four forecast accuracy measurements used were RMSE, MSE, MAD, and MAPE. The smaller the forecast error is, the more accurate the forecasting methods.

Table 4.1: Forecasting Accuracy Measurement for 10 Forecasting Techniques

Forecasting Techniques	Ranking
Single Moving Average	3
Single Exponential Smoothing	5
Double Moving Average	2
Double Exponential Smoothing	4
Regression Analysis	1
Holt-Winter's Additive	7
Seasonal Additive	8
Holt-Winter's Multiplicative	9
Seasonal Multiplicative	10
ARIMA	6

Regression analysis technique with RMSE= 1334.7793 was in first ranking as it showed the lowest RMSE value. Next was double moving average technique with RMSE=1376.7254 in second ranking. The single moving average technique in third ranking with RMSE=1406.8824 followed by double exponential smoothing technique with RMSE=1413.5199 the forth ranking. The fifth rank was single exponential smoothing technique with RMSE=1501.6296. The sixth ranking was ARIMA technique with RMSE= 1666.1652. Furthermore, the seventh ranking and eighth ranking were Holt-Winter's additive and seasonal additive with RMSE=1771.9117 and RMSE=1793.8089 respectively. Holt-Winter's multiplicative with RMSE=1818.6557 was in ninth ranking whereas the last ranking was seasonal multiplicative with RMSE=1849.2047.It can be concluded that the best forecasting techniques to optimize the inventory demand is regression analysis.

5. Discussion and Conclusion

5.1 Discussion of Findings

The first research objective is to identify the possible forecasting methods to be applied for inventory demand. In order to achieve the objective, the data pattern needs to be identified first. There were 10 forecasting techniques have been selected for the Panadol 650mg as required for demand forecasting analysis.The second objective is to optimize the inventory demand using the best forecasting method. Once the forecasting techniques are completed to be used for demand forecast process, the forecasting accuracy measurements will be compared among the ten forecasting techniques. The best forecasting methods for the Panadol 650mg is regression analysis.

5.2 Contribution For University Health Centre

With the recommended forecasting method, the University Health Centre is able to produce more accurate forecast which can aid in making precise or reliable decision for the Panadol 650 mg inventory planning.The research can contribute to University Health Centre by assist the pharmacy using statistical forecasting techniques in a better inventory planning and avoid any overstocks or under stocks problems.

5.3 Recommendation For Future Research

The researcher recommends that more inventories should involve in demand forecasting process. This is a way to ensure that more accurate data of inventory demand quantities can be obtained to optimize the overall supply chain for the medicine.Further study should be carried out as well for other field like manufacturing industry. Manufacturing industry really need more accurate forecasting for their production to maximize the profit and ensure safety inventory always existed.

5.4 Conclusion

In a nutshell, the research objectives were achieved. The data pattern was trend and ten forecasting techniques were used. Among the ten forecasting techniques, the best technique which was regression analysis with least forecasting error provided. A better inventory management through forecasting techniques can be implemented and hence improve the quality and performance in a long term period for University Health Centre.

6. Reference

Abdel-Aal, R.E., A.M. Mangoud.*Modeling and forecasting monthly patient volume at a primary health care clinic using univariate time series analysis*. Computer Methods and Programs in Biomedicine.1998;56:235-47.http://dx.doi.org/10.1016/S0169-2607(98)00032-7

Anderson, D. R., Sweeney, D. J., Williams, T. A., Camm, J. D., Cochran, J. J., Fry, M. J. and Ohlmann, J. W. (2010). *Quantitative Methods for Business.*12th ed. South-Western: Cengage Learning.

Armstrong, J. S. and Collopy, F. (1998), "*Integrationof statistical methods and judgment for time series forecasting: Principles from empirical research*, "in G. Wright and P. Goodwin (Eds.), Forecasting with Judgment. Chichester: John Wiley.

Armstrong, J. S. (2000). *The Forecasting Dictionary.*Retrieved April 19, 2012, from ftp://ftp.cba.uri.edu/classes/jarrett/Briefcase/Forecasting%20Dicyionary.pdf

Armstrong, J. S. (2001). *Principles of Forecasting: A Handbook for Researchers and Practitioners.*

Armstrong J. S. (2009).Selecting Forecasting Methods.*Principles of Forecasting: A Handbook for Researchers and Practitioners, Norwell, MA: Kluwer Academic Publishers,* pp. 1 – 18.

Bartezzaghi E., Verganti R., Zotteri G., (1999). *A simulation framework for forecasting uncertain lumpy demand.* International Journal of Production Economics, 59 (1–3), pp. 499–510.

Bon, A. T. and Chong, Y. L. (2009).*The Fundamental on Demand Forecasting in Inventory Management,* 3(4), pp. 3937-3943.

Box G, Jenkins G. *Time series analysis:Forecasting and control.* San Francisco: Holden-Day; 2006.

Boylan, J. E., Syntetos, A. A., Karakostas, J.E., *"Classification for Forecasting and StockControl: A Case Study"*, Journal of the Operational Research Society, Vol 59, pp. 473-48, 2008.

Chatfield, C. (2000). *Time-series Forecasting.*London: Chapman and Hall/CRC.

Croston J.D., (2011). *Forecasting and stock control for intermittent demands.* Operational Research Quarterly 23(3), pp. 289–303.

Danese P. and Kalchschmidt M. (2011a).*The role of the forecasting process in improving forecast accuracy and operational performance.*Int. J. Production Economics 131 (2011), pp. 204 -214.

Fildes, R. & Hastings, R. (1994).*The Organization and improvement of Market Forecasting.Journal of the Operational Research Society*, 45(1), 1-16.

Ghobbar A.A., Friend C.H., (2003). *Evaluation of forecasting methods for intermittent parts demand in the field of aviation: a predictive model.*Computers & OR 30(14), pp. 2097-2114.

Hanke, J. E. and Wichern, D. W. (2009). *Business Forecasting.*9th ed. New Jersey: Pearson Education, Inc.

Hair, J. J., Bush, R., &Ortinau, D. (2003).*Marketing Research- Within A Changing Information Environment (2nd Edition).* New York: McGraw-Hil- Irwin.

Hoshmand, A. R. (2002). *Business and Economic Forecasting for the Information Age: A Practical Approach.* London: Greenwood Publishing Group, Inc.

Jones A J, Joy M P, *Pearson J: Forecasting demand of emergency care. Health Care Management Science* 2002; 5, 297 -305.

J. Scott Armstrong and Fred Collopy (1992). "*Error Measures For Generalizing About Forecasting Methods: Empirical Comparisons*". International Journal of Forecasting **8**: 69–80.

Kerkkanen, A. (2010). *Improving Demand Forecasting Practices in the Industrial Context.*Lappeenranta University of Technology: Ph. D. Thesis.

Korpela, J. and Tuominen, M. (1996).*Inventory Forecasting with a Multiple Criteria Decision Tool,* 45, pp. 159-168.

Lawrence, K. D., Klimberg, R. K. and Lawrence, S. M. (2009).*Fundamentals of Forecasting Using Excel.*New York: Industrial Press, Inc.

Makridakis, S., Andersen, A., Carbone, R., Fildes, R., Hibon, M., Lewandowski, R., Newton, J., Parzen, E. and Winkler, R. (1984), *The Forecasting Accuracy of Major Times-Series Methods.* Chichester: John Wiley.

Makridakis, S., Wheelwright, S. C., and Hyndman, R. J. (1998), *Forecasting Methods for Management*, Third edition. New York: John Wiley.

Marican, S. (2006).*PenyelidikanSains Social: PendekatanPragmatik.* Kuala Lumpur: PencetakanNazSdn. Bhd

Meade, N. and Islam, T. (2001), "*Forecasting the diffusion of innovations: Implications for time series extrapolation*," in J. S. Armstrong (Ed.) Principles of Forecasting. Norwell, MA: Kluwer Academic Publishers, pp. 4577-595.

Ozcan, Y. A., *"Quantitative Methods in Health Care Management: Techniques and Applications"*, San Francisco: Jossey Bass.2009.

Panneerselvam, R. (2005). *Production and Operations Management.*2nd ed. New Delhi: Prentice-Hall India (P) Ltd.

Prest, R., *Real Demand Forecasting*, http:// www. pharmamanufacturing.com/ articles/ 2007/178.html, last accessed: 02.12.2012.

Real Options Valuation Inc. (2005-2011). *Risk Simulator.* Retrieved April 21, 2012, from http://www.realoptionsvaluation.com/attachments/risksim.pdf

Scott Armstrong, Fred Collopy, Andreas Graefe and Kesten C. Green. "*Answers to Frequently Asked Questions*".Retrieved May 15, 2013.

Syntetos, A.A., Boylan, J.E., Croston, J.D., *"On the categorization of demand patterns"*, Journal of the Operational Research Society, Vol 56, pp. 495-503, 2005.

Varghese, V.M., Rossetti, M. D, *"Comparing Intermittent Demand Forecasters and Effects of Temporal Demand Aggregation"*,Industrial Engineering Research Conference,2011. Proceedings.

Winklehofer, H., Diamantopoulos, A. & Witt, S. F. (1996).*Forecasting Practice: A Review of the Empirical Literature and an Agenda for the Future Research.*International Journal of Forecasting, 12(2), 193-221.

Wittink, D. R, and Bergestuen, T. (2001), *"Forecasting with conjoint analysis,"* in J. S. Armstrong (Ed.) Principles of Forecasting:. Norwell MA: Kluwer Academic Publishers, pp. 147-167.

Yin, R. K. (2009).*Case Study Research: Design and Methods.* 5th ed. California: Sage Publication Inc.

Acknowledgments

This article is supported and sponsored by Registrar Office, UniversitiTun Hussein Onn Malaysia. Ministry of Higher Education, Malaysia

Student academic performance analysis using fuzzy *C*-means clustering

R. Rosadi[1], Akamal[1], R. Sudrajat[1], B. Kharismawan[2], Y.A. Hambali[3]

[1]Department of Computer Science, Padjadjaran University, Sumedang 45363, Indonesia
[2]Department of Mathematics, Padjadjaran University, Sumedang 45363, Indonesia
[3]Master Student of STEI, Institute of Technology Bandung, Bandung 40132, Indonesia

E-mail: r.rosadi@unpad.ac.id, akmal@unpad.ac.id, r.sudrajat@unpad.ac.id, baguskharismawan@unpad.ac.id, yudiahmadh@gmail.com

Abstract. Grade Point Average (GPA) is commonly used as an indicator of academic performance. Academic performance evaluations is a basic way to evaluate the progression of student performance, when evaluating student's academic performance, there are occasion where the student data is grouped especially when the amounts of data is large. Thus, the pattern of data relationship within and among groups can be revealed. Grouping data can be done by using clustering method, where one of the methods is the Fuzzy *C*-Means algorithm. Furthermore, this algorithm is then applied to a set of student data form the Faculty of Mathematics and Natural Sciences, Padjadjaran University.

1. Introduction

Evaluation of academic performance is one of the bases for monitoring the development of students' academic achievement in higher education level. In evaluating the student academic performance data, there are occasion that the student data is grouped especially when it involves a large amounts of data. Thus, the pattern of data relationship within and among groups can be revealed. With the help of clustering techniques, student data can be classified based on their academic achievement. Clustering is one method of data mining where the process of grouping data objects into different classes called clusters so that objects that are in a cluster have differences with objects in other clusters. One of the methods used in clustering is the so-called fuzzy *C*-means algorithm. This algorithm can be used to collect a set of objects into a desired cluster. The authors have done before researches relating to the academic data in 2015. It was classified the academic data through several methods, such as, *K*-Means, KNN, FCM, and others. A web-based mobile application has also been developed as a tool for information processing and presentation of results.

2. Method

2.1 Data Mining

Data Mining refers to a process or method for extracting or mining knowledge or patterns from large amounts of data. Data mining emerged due to the availability of large amounts of data and the need to transform the data into useful information or knowledge [1]. To perform the data mining process, the concept of KDD (Knowledge Discovery in Databases) is commonly needed, see Figure 1.

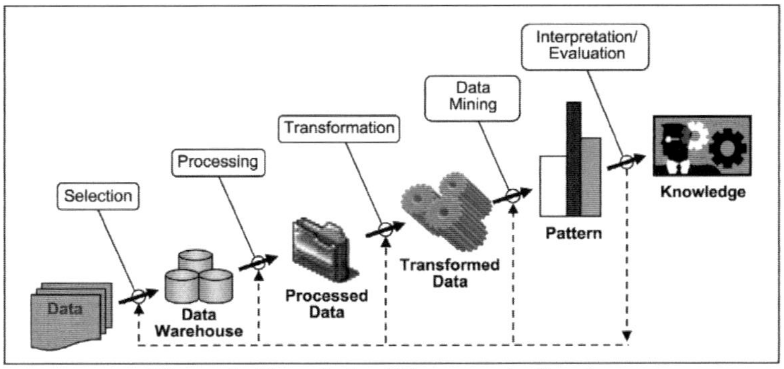

Figure 1. Knowledge Discovery in Database

2.2 Fuzzy Set

Fuzzy logic was first introduced by Lotfi A. Zadeh in 1965 (Zimmerman, 1991). Zadeh's definition of fuzzy sets is as follows: If X is a collection of objects denoted generically by x, then a fuzzy set A in X is defined as a set of ordered pairs:

$$A = \{(x, \mu_A(x)) \mid x \in X\}, \tag{1}$$

with $\mu_A(x)$ is called the membership function for the fuzzy set A. The membership function maps each element of X to a membership grade between 0 and 1 [2, 4].

2.3 Fuzzy C-Means Clustering

Fuzzy C-Means is an algorithm of data clustering, where each data becomes a member of a cluster with a degree defined by membership function [2]. Clustering process using Fuzzy C-Means algorithm has the following steps:

1. The data is stored X, a matrix of size $n \times m$, where n is the number of samples and m is the number of attributes.
2. Determine the number of clusters (c), rank (w), the maximum iteration (MaxIter), smallest error expected (ε), the initial objective function ($P_0 = 0$), and the initial iteration ($t = 1$).
3. Generate a random number μ_{ik} for $i = 1, 2, ..., n$ and $k = 1, 2, ..., c$ as an element of the matrix initial partition U. For $i = 1, 2, ..., n$, calculate

$$Q_i = \sum_{k=1}^{n} (\mu_{ik})^w \tag{2}$$

$$\mu_{ik} = \frac{\mu_{ik}}{Q_i} \tag{3}$$

4. Calculate the cluster center V_{kj}, with $k = 1, 2, ..., c$ and $j = 1, 2, ..., m$

$$V_{kj} = \frac{\sum_{i=1}^{n} (\mu_{ik})^w X_{ij}}{\sum_{i=1}^{n} (\mu_{ik})^w} \tag{4}$$

5. Calculate the objective function at iteration t

$$P_t = \sum_{i=1}^{n} \sum_{k=1}^{c} \sum_{j=1}^{m} (X_{ij} - V_{kj})^2 (\mu_{ik})^w \tag{5}$$

6. Calculate the partition matrix transition

$$\mu_{ik} = \frac{[\sum_{j=1}^{m}(X_{ij} - V_{kj})^2]^{\frac{-1}{w-1}}}{\sum_{k=1}^{c}[\sum_{j=1}^{m}(X_{ij} - V_{kj})^2]^{\frac{-1}{w-1}}} \quad (6)$$

7. If $|P_t - P_{t-1}| < \varepsilon$ or $t >$ MaxIter then stop, otherwise $t = t + 1$, repeat step 4.

Steps in Fuzzy *C*-Means algorithm can be illustrated by the following diagram (Figure 2).

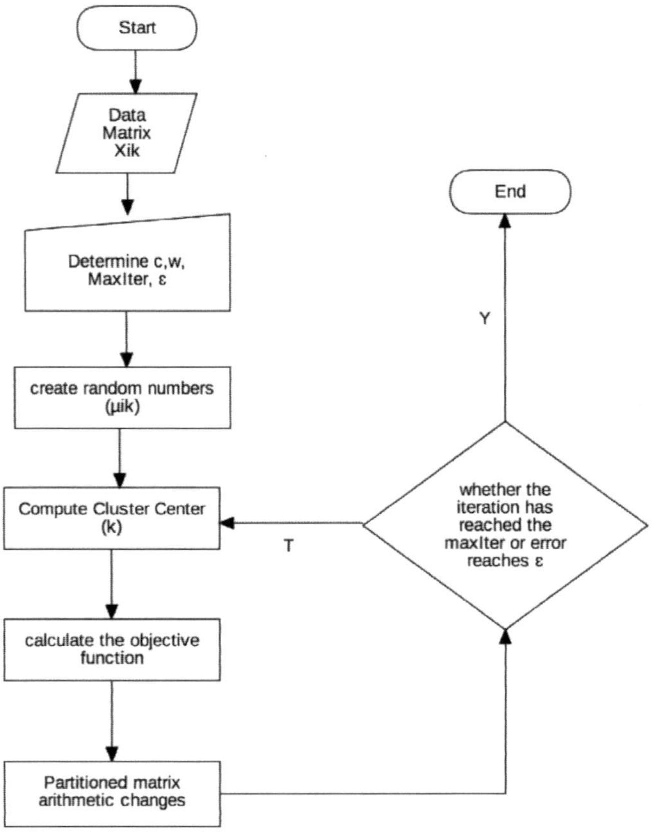

Figure 2. Diagram of Fuzzy *C*-Means Algorithm

2.4 Student Academic Performance

Academic performance is a result achieved in the academic field. Evaluation of students' academic performance can be seen from the GPA (grade point average) student. GPA is a number that shows achievement or academic performance of students cumulatively starting from the first semester [5].

3. Result and Discussion

Implementation of Fuzzy C-Means method can be seen in the trial grouping data using a web-based mobile application. In this study, the amount of data used is as many as 4255 records which to be grouped into four clusters, $w = 2$, and the minimum error is 0.0001. Once the data to be grouped has been stored into the database, then the clustering process can be done. The process and the results of clustering that has been done by the web-based mobile applications is provided in Figure 3 and the 4-clusters can be seen in Table 1.

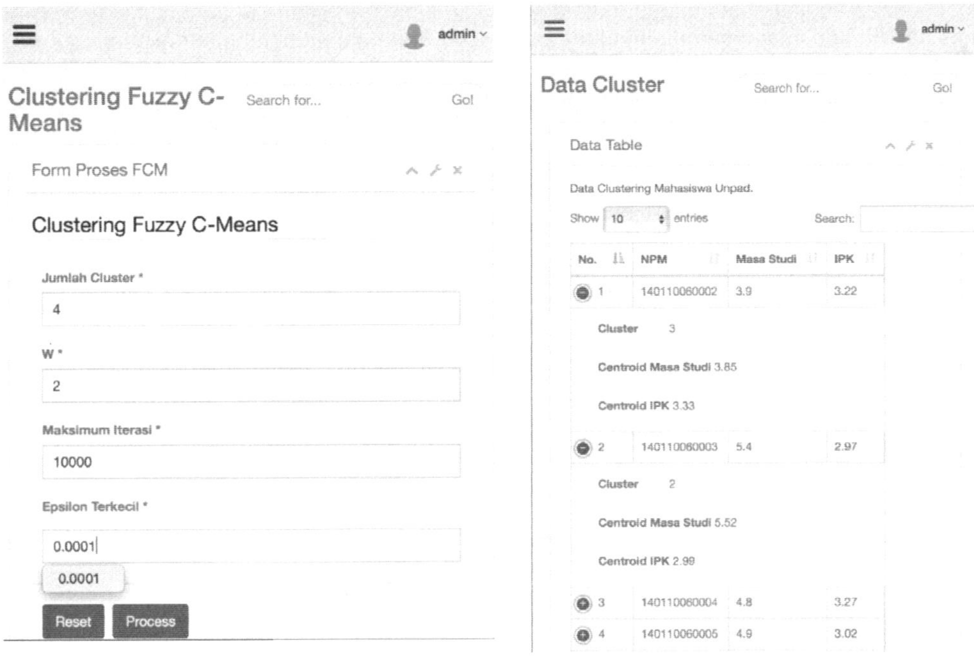

Figure 3. Mobile Web Application Process of Fuzzy C-Means Clustering

Table 1. Categories of Student Academic Performance Evaluation Process based Clustering

Cluster	Centroid of GPA	Centroid of Study duration	Category
1	2.85	7.05	Bad
2	2.99	5.52	Not Good
3	3.33	3.85	Very Good
4	3.13	4.57	Good

The member of each cluster can be presented by the application in the form of graphs, which is shown in Figure 4. Figures 5-9 show the clusters with respect to the origin of department.

4. Conclusion

Fuzzy *C*-Means method can be implemented into the program to classify student academic data. Based on the results of the clustering process of 4255 student academic data using applications. Four clusters suitable for use as a result of grouping data for each cluster of students do not have adjacent centroid that can have different categories. Cluster with the best academic performance categories, namely Cluster 3 totaling 1753 students, followed by a Cluster 4 totaling 1496 students, Cluster 2 totaling 676 students, and Cluster 1 totaling 330 students. From the results of the grouping that is associated with a department, the percentage of students that included in Cluster 1 (unfavorable category), the highest is the Department of Mathematics, Cluster 2 (category quite well) were the highest is Department of Biology, Cluster 3 (category Excellent) that the highest is Department of Statistics, Cluster 4 (both categories) were the highest is Department of Biology. Especially for the Department of Information Engineering and Geophysics, the condition of the current data that is processed at this time fewer new graduates, S1-Informatics 3 graduates, where all three go in Cluster 3 (excellent category) and S1 Engineering Geophysics 40 graduates, of which 31 graduates entered Cluster 3 (excellent category) and 9 graduates enter the Cluster 4 (both categories).

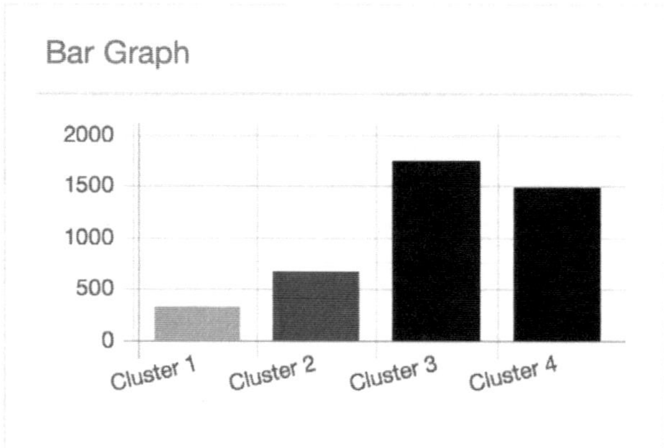

Figure 4. The resulting graph FCM Clustering

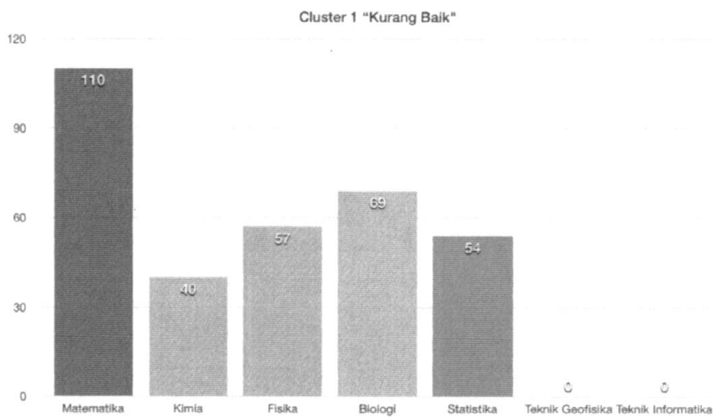

Figure 5. Cluster 1 based on department

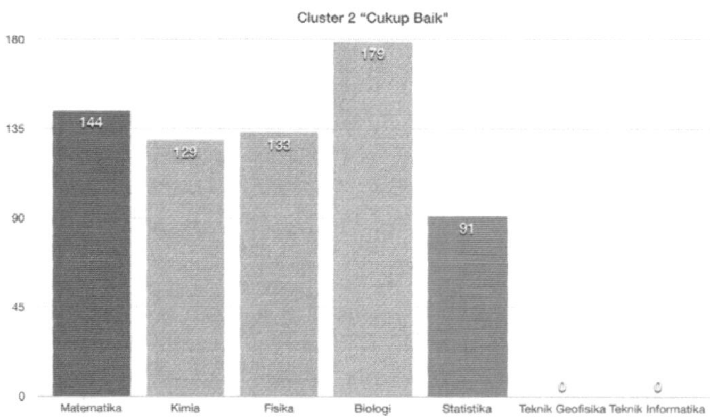

Figure 6. Cluster 2 based on department

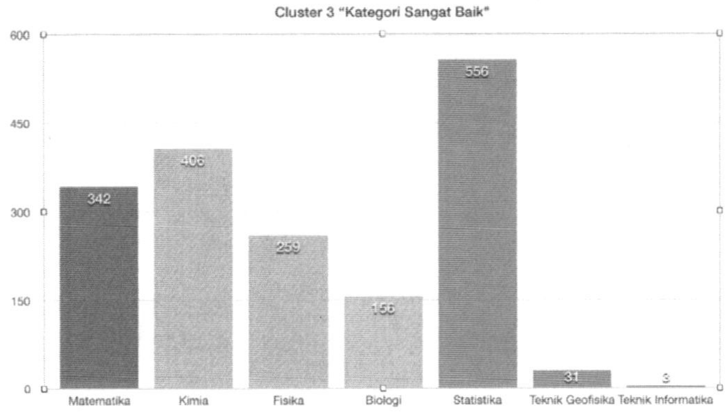

Figure 7. Cluster 3 based on department

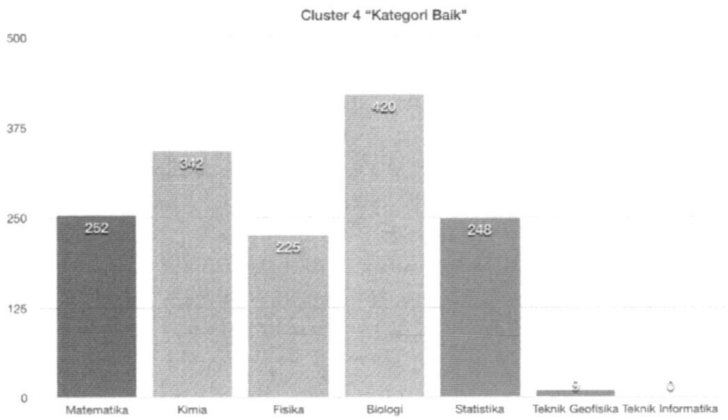

Figure 8. Cluster 4 based on department

Acknowledgment
Authors thank the organizers of DRPM Unpad for the HPKRD research grants, and FMPA Unpad as a facilitator. Author thank also other parties who have contributed a lot in research and completion of the seminar paper.

References

[1] Han, J, Kamber, M. 2006. Data Mining Concepts and Techniques. San Francisco: Morgan Kaufmann Publishers.
[2] Jang, JSR Sun, CT dan Mizutani, E. 2004. Neuro Fuzzy and Soft Computing. Singapore. Pearson Education.
[3] Kharismawan, Bagus. 2015, Aplikasi K-Means dan Fuzzy C-Means Clustering untuk Mengelompokkan Data Mahasiswa FMIPA Unpad: University of Padjadjaran.
[4] Kusrini, Luthfi. 2009, Algoritma Data Mining. Yogyakarta: Andi Offset.
[5] http://www.unpad.ac.id/pengajaran/evaluasi-hasil-belajar-dan-batas-waktu-studi/evaluasi-hasil-belajar/

Professional development model for science teachers based on scientific literacy

B. Rubini, D. Ardianto, I.D. Pursitasari, I. Permana

Science Education Study Program, Pakuan University, INDONESIA

Email: bibinrubini@unpak.ac.id

Abstract. Scientific literacy is considered as a benchmark of high and low quality of science education in a country. Teachers as a major component of learning at the forefront of building science literacy skills of students in the class. The primary purpose this study is development science teacher coaching model based on scientific literacy. In this article we describe about teacher science literacy and profile coaching model for science' teachers based on scientific literacy which a part of study conducted in first year. The instrument used in this study consisted of tests, observation sheet, interview guides. The finding showed that problem of low scientific literacy is not only happen the students, but science' teachers which is a major component in the learning process is still not satisfactory. Understanding science teacher is strongly associated with the background disciplinary. Science teacher was still weak when explaining scientific phenomena, mainly related to the material that relates to the concept of environmental. Coaching model generated from this study consisted of 8 stages by assuming the teacher is an independent learner, so the coaching is done with methods on and off, with time off for activities designed more.

1. Introduction

Mastery of Science and Technology is an important key to take on the challenges of the future. These challenges are related to the improvement of life quality, equitable development, and the ability to develop human resources. Therefore, science education as a part of education have an important role to prepare students who have science literacy, which are capable to think critically, creatively, logically, and have initiative to respond the issues in society that caused by the impact of science and technology development (Prayekti, 2006).

Science education is expected to become a vehicle for students to learn about themselves and the environment, as well as prospects for further development to applying it in everyday life (*National Department of Education*, Depdiknas, 2006). Toharudin *et al.*, (2011) argued that science literacy is an ability to understand science, communicate science (oral and written), and also it is knowledge of science to solve problems so the students have attitude and high sensitivity for themselves and their environment in making decisions based on considerations of science.

Developing science literacy is important because: (1) understanding of science offers personal satisfaction and pleasure that comes after understanding and studying nature; (2) everyone needs information and science thinking for decision making in their daily life; (3) everyone needs to use their abilities in public discourse and debate about important issues involving science and technology; and (4) science literacy is important in the world of work, as more and more jobs requiring high skills, thus requiring people to learn science, to reason, to think creatively, make decisions, and solve problems (National Research Council, 1996).

The importance of science literacy has been internationally recognized as the benchmark of high-low quality of education. This is responded by The Program for International Student Assessment (PISA) consisting of highly industrialized countries (*The Organization for Economic Cooperation and Development*, OECD). Results of PISA study in 2000, Indonesian students are in ranks 38th out of 41 countries participating in the science literacy test. In PISA 2003, PISA 2006, PISA 2009, and PISA 2012, Indonesian students' ability still weak on science literacy. The acquiring scores for Indonesian students on science literacy in 2000, 2003, 2006, 2009, 2012 are 393, 395, 393, 383, 382 respectively with an average score of all participating countries is 500 (*Indonesia Research and Development Board*, Balitbang, 2012). According to the analysis by OECD, science literacy scores in the range 335 $\leq$ 409 points are categorized as level 1 or lower than in the skill category. At this level, students have a limited science knowledge and its only be applied to several situations. Students at this level can only provide an easy science explanation and follows the evidence those were given explicitly (OECD, 2013). Science literacy of Indonesian students should be a concern to all parties. Several things were predicted as a cause of the condition. Low quality and quantity of human resources (teachers and staffs), the quality and quantity of educational facilities, the quality of teaching-learning process are some factors were raised. The results of further analysis of the data PISA for Indonesian students are listed below:

1. Students' literacy achievement is low, with an average 32% for the overall aspect, which consists of 29% content, 34% process, and 32% context applications.
2. Diversity between provinces is not significantly affected on the low level of science literacy learners Indonesia.
3. Diversity between provinces is not significantly affected on the low level of science literacy learners Indonesia.
4. The ability of Indonesian childrens to solve problems is very low.

These findings showed that most of Indonesian students were not able to associate their science knowledge with phenomenon that occur in the world, because they do not have experience to be linked. The comprehensive study on data of *Indonesia Research and Development Board* (Balitbang, 2012) revealed that the achievement of high school students in West Java (grades UN) showed deterioration compared to students of other developed regions such as East Java, Yogyakarta, Jakarta, and Bali. This result indicated that quality of education in West Java needs to be studied, analyzed, and immediately looked for the right solution, so that West Java can be immediately separated from the setback. The achievement setback can be assumed as a weakness in science literacy of students. To solve these problems, such efforts need to be done to rebuild relevance of the educational process with the aim. The other efforts those should be our concern are improving the learning process, strengthening human resources mainly teachers, and improving all the other needs of the educational process.

Study about students' science literacy in Bogor showed that achievement of students' science literacy has not shown satisfactory results. Achievements of students' science literacy is quite low, with an average 30% for the entire aspect, which consists of 29% for content, 30% for the process, and 31% for the context (Rubini & Ardianto, 2014). Rubini & Ardianto advanced idea for measure science literacy of science teacher. This study showed that average scientific literacy teachers was 63% to the overall aspect, which consists of 65% of content, 62% process and 62% applications of science.

Field observations with regard to the implementation of learning science is still dominated by stuffing science concepts to students. Science teachers often complain about classic problem such as lack of time and facilities to carry out science learning by applying science teaching strategies. Scientific practices regarded as additional activities that are allowed to do when they have time. Whereas this strategy that should be applied in science learning. The phenomenon of science teaching such as this will continue if the teacher had always assumed scientific practices in learning is not a requirement that must be met. To solve these problems, need to be made various efforts including rebuilding the relevance of the educational process with the goal. Such efforts can be packed in a pattern of training or coaching science teacher professionalism based scientific literacy.

2. Research Methodology

The primary purpose this study is development science teacher coaching model based on scienctific literacy. In this article we describe about teacher science literacy and profile coaching model for science' teachers based on scientific literacy which a part of study conducted in first year. This study involved 25 science' teachers in Bogor. 23 teachers who follow the teacher qualification test is S1 and the field of educational expertise Physics, Biology or Chemistry. Another science teachers qualified S1 with education background of computer and engineering fields (3 person). Instruments in this study consisted of scientific literacy test adopted from about PISA 2013 and guidelines for teacher interviews were used to explore the weaknesses of teachers in teaching science.

3. Results and Discussion

3.1 Teachers' Science Literacy (Contents Science)

The content aspect on PISA test includes acid rain, greenhouse effect, sport, pollution, cloning, biodiversity, energy, bacteria, photosynthesis, and pesticides. Figure represents the true answer of each item test of Science PISA test based on content.

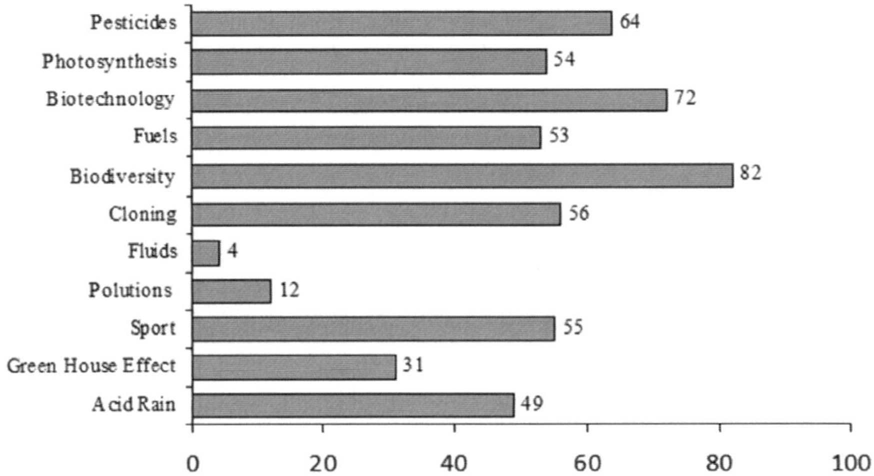

Figure 1. The Precentage of Teachers Achievement in Content Science Test

Figure 1 showed that teachers obtain the highest percentage of correct answers on biodiversity content is 82% and the lowest percentage correct answer on the fluid content (4%). This study proves that the educational background of teachers greatly affect the understanding of contents' science tested. The study involved 23 fields of science teachers with educational expertise Physics, Biology or Chemistry. A total of 15 teachers of them being teachers with educational expertise Biology field. Biodiversity tested more directed at Biology, so the impact on the high percentage of correct answers on this content. Teacher with Biology disciplinary would be easier to answer that question since the conceptual structure has been built up through experience everyday through their teaching activities. Vosniadoun and Ioannides (1998) also argues that the conceptual structure derived from everyday experiences and their constantly restructured to make the existing concept conceptual reached in accordance with the conceptual scientists. The results also showed that when the subjects were given about the material that is not in accordance with his expertise, the percentage of correct answers they were relatively low. This can be seen in the percentage of correct answers on contents fluid only reaches 4%, in other words, only 10 of the study subjects who responded well to the content.

Figure 1 also showed that, teacher's knowledge related to environmental issues is still low. Environmental problems represented by acid rain, pollution and the greenhouse effect. Achievement teacher in this content is still below 50%, meaning that only 12 teachers were able to answer correctly associated contents related to environmental issues. The findings are reinforced by the facts obtained through interviews with a number of teachers who stated that they were still confused and misconceptions when asked about environmental issues such as global warming, destruction of ozone, acid rain, radioactive pollution and others. The results of this study reinforced by research Spiropoulou *et al.* (2007) which states that 188 teachers in Greece is still limited knowledge when they were asked to discuss environmental issues. Another course of a study also revealed that students and teachers are less understanding about environmental issues and still have alternative conceptions about environmental issues. One of the main difficulties is when explaining the greenhouse effect and global warming (Acikalin, 2013). Some studies also support that position, students (Koulaidis and Christidou, 1999; Rebich and Gautier, 2005) and teacher pre-service (Groves and Pugh, 1999; Kahlid, 2001) still have a conception of the concept of the greenhouse effect.

The results of interviews with a number of teachers found some key issues that should be the concern of all parties, which are; background of qualified teachers is still very special cause difficulties in managing integrated learning. They generally dominate such fields as biology, but not for physics and chemistry. Currently, the government has made electronic school books labeled integrated. However, such alignment is still very little perceived primarily concerned with examples. Generally, teachers are still very low understanding of integrated learning models. Some training integration in science learning have ever they follow, but very few examples given. A lack of motivations from school to organize a unified science learning process, making it less motivating teachers to implement them consistently.

3.2 Profile Coaching Model Based on Science Literacy

Based on discussions with teachers and other stakeholders, and resource persons invited, basically there are some problems that can be addressed through training programs Science teacher in junior high school. The problem identification results are summarized in Table 1 below.

Table 1. Problems and Solutions through Training Programs

No	Problems	Solutions
1	Science teachers have difficulty when constructs integrated science	Training, design, packaging material & discussion integrated science; Surgical junior science curriculum
2	The weakness of teachers in understanding the integration in learning Science	Training, design models of integrated science teaching
3	The need of sharing experience of teachers in teaching science	Lesson activities among science teachers, which implements integrated science teaching Frequency and needs open lesson activities: held at least 4-6 times in one year (2-3 times per semester.
4	The need of teaching materials for the integrated science teaching	Teaching materials are made in groups of teachers Science, the research team made two-three learning package as a model
5	Science teacher rides to gather	Training packaged integrated, involving groups of science teachers, involves a commitment to the school, Head of Education Department (MOU)
6	Training mode typically spend, without a clear product	Mode of training needs to be designed on and off, more off the monitoring of researchers. More open lesson.

The results of the identification of the above problems, as well as various considerations /proposals submitted by FGD participants to complete, then the model is designed with professional training Science teacher in Junior High School as shown by the figure 2. as an outcome in this study.

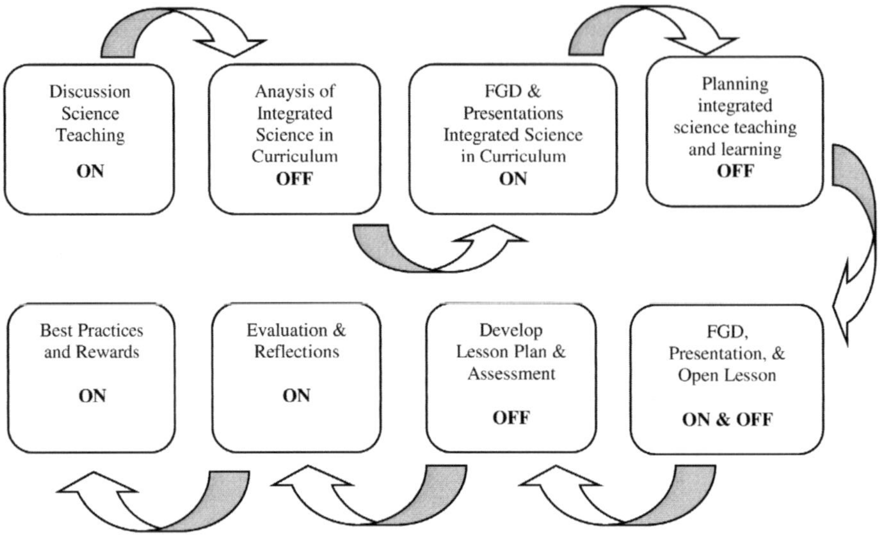

Figure 2. Coaching Model Science' Teachers

Coaching science teacher designed in cycles, and be completed during the first year. Coaching is designed in stages, with the assumption that teachers are independent learners, so the coaching carried with methods on and off, with time off for activities designed more. In addition, these considerations carried remember the teacher still remains bound by its duties and responsibilities in school, so the coaching model as expected not time-consuming, and in line with learning activities in schools.

Phase 1 (ON), in this stage discussions through MGMP, with guest speaker of researchers. The discussion aims to identify perceptions about learning science, including refreshing models of integrated science teaching. Material provided by the research team as a resource. Before, discussions, carried in advance of signing the commitment among teachers and resource persons. Phase 2 (OFF), at this stage the teacher in the group (class) to identify the science content in curriculum that can be taught with the concept of integration. The expectation of this activity is minimal resulting from each class (1, 2 and 3) four (4) integration of science content, so the total is expected to last 12 IPA integration in the content generated through self-employment. Independent work is scheduled for 1 month.

The next stage (Phase 3) is Focused Group Discussion (FGD) held face to face. In this activity, a group of teachers representing the results of an independent study related to integration in science content, which further improved and agreed upon by all participants of FGD. Involving FGD (Research Team). Designed a full day of activities. Once agreed, they then choose one of 12 learning material to be developed in lesson' plan. Selection was based on the possibility activities teaching and learning can be implemented in the month of July (the start of the new school year). Accordingly, this guidance will support real learning in school, and do not interfere with a predetermined schedule school. Phase 4 is carried out independently back. Each group developed a lesson plan with the same theme, complete with an evaluation tool based on its scientific literacy. This stage is carried out with less than 1 month. In addition, each group must make reports including reports for group presentations related to lesson plan have been made.

Phase 5, all the groups presented the results of the working group, giving feedback, and to set the final lesson plan that will be used as the base for an open lesson. In this activity, the group of teachers assign teachers model that will deliver real learning in the classroom. After completion of the planning, implementation of educational / open lesson conducted by the research team facilitator and moderator of school leadership to be a place open lesson. After the open lesson, practices are reflections and suggestions for improvements, and ends with the provision of rewards for teacher's models, as well as an active participant in the discussion. Follow-up of this activity is the distribution of groups/themes for each teacher / group to develop lesson plan (different theme, based on the themes that have been developed earlier, there are 11 themes). Phase 6, teachers making lesson plans based on the themes that have been agreed (11 theme) in the group. They work independently adjusted with the schedule which they have agreed.

Phase 7, at this stage all the groups presented the lesson plan has been designed, discussions, and provide feedback and improve lesson plans based on the feedback received. All teachers are getting all the lesson plan designed by all the groups, so that they acquire a minimum of 12 lesson plans of 12 science themes that can be taken home). In this activity evaluations carried by both facilitators and peers, giving rewards for the best group also carried out. Furthermore, the agreement for the participant teachers who will implement the model of learning in the open lesson activities (planned there 3 times open lesson on a different theme). The last stage includes activities of the lessons learned with the open lesson of the three themes that have been agreed on stage 7. Open lesson activities carried out on the day / week is different, and ends with reflection, evaluation, and giving rewards.

Basically the coaching model based on the principles of adult learning, more priority to independent work, as well as learning from the experience of teaching (lesson learned). In addition, to ensure that development does not disrupt the rhythm of work of teachers, the administration carried out in the mode "on" and "off". Mode "on" mostly reserved for perception and enhancing the knowledge of teachers, while the "off" is mostly done by the teachers in the group, with the tasks that have been agreed during the "On".

4. Conclusions

This study gives us that the problem of low scientific literacy is not only happen the students, but science' teachers which is a major component in the learning process is still not satisfactory. Understanding science teacher is strongly associated with the background disciplinary. Science teacher was still weak when explaining scientific phenomena, mainly related to the material that relates to the concept of environmental problems. The ability of teachers is very low in the indicator explain scientific phenomena, showing teachers are less accustomed to thinking critically and not accustomed to reason according to their knowledge.

The results also reveal that the problems in the perspective of learning science is a science teacher; 1. The science teacher is difficult to teach Integrated Science; 2. Science Teachers did not understand about integrated science learning models; 3. Motivation science teacher who still lacking in teaching integrated science consistently. Coaching model generated from this study consisted of 8 stages by assuming the teacher is an independent learner, so the coaching is done with methods on and off, with time off for activities designed more. The coaching model based on the principles of adult learning, more priority to independent work, as well as learning from the experience of teaching (lesson learned).

Acknowledgements

In this opportunity researchers thank Directorate of Research and Community Service through the Assignment Agreement Implementation of Competitive Grants Fiscal Year 2016 Number 03/LP-UP/KPHB/VII/2016.

References

[1] Bell, G., Warwick, J., Galbraith, P., "Higher Education Manajement and Operational Research", Sense Publischers, Rotterdam/Boston/Taipei, 2012.

[2] Acikalin, F. S. (2013). Middle school students' conceptions of environmental issues. International Journal of New Trends in Art, Sports & Science Education. 2(4): 23-27.

[3] Department of National education. (2006). *Integrated Science Learning Development Guide.* [online]. Available: *http:// www.puskur.net/inc/mdl/ 050_Model_IPA_Trpd.pdf.* [21 Juni 2007].

[4] Department of National education (2013). *Permendikbud No. 68 of 2013 on Curriculum 2013.* Jakarta: Ministry of Education.

[5] Groves, F.H dan Pugh, A.F. (1999). Elementary Pre-Service Teacher Perceptions of the Greenhouse Effect. *Journal of Science Education & Technology.* 9: 75-81.

[6] Kahlid, T. (2011). Pre-service Teacher's Misconceptoins Regarding Three Environmental

Issues. *Canadian Journal of Environmental Educations*. 6: 102-120

[7] Koulaidis, V & Christidou, V. (1999). Model of Students' Thinking Concerning the Greenhouse Effect and Teaching Implementations. *Science Education*. 83(5): 559-576.

[8] National Research Council. (1996). *National Science Education Standard.* Wahington, DC: National Academy Press.

[9] OECD. (2013). *PISA 2012 Assessment and Analytical Framework.* Paris: Organization for Economic Cooperation and Development (OECD).

[10] Organization for Economic Cooperation and Development. 2009. *Framework for Mathematics Assessment*. Paris: Organization for Economic Cooperation and Development (OECD).

[11] Prayekti. (2006). *Science, Technology and Society and Science learning.* [Online]. Available: *http:// www.duniaguru.com* [9 Januari 2008].

[12] Swaak, J., Jong, T,D., & Joolingen, W.R. (2004). The Effect of Discovery Learning and Expository Instruction on the Acquisition of Definitional and Intuitive Knowledge. Netherlands: *Journal of Computer Assisted Learning*. 20: 225-234.

[13] Spiropolou, D., Antonakaki, T., Kontaxaki, S & Bouras, S. (2007). Primary Teachers' Literacy and Attitudes on Education for Sustainable Development. *Journal of Science Education Technology*. 16: 443-450.

[14] Rebich, S dan Gautier, C. (2005). Concept Mapping to Reveal Prior Knowledge and Conceptual Change in a Mock Summit Course on Global Climate Change. *Journal of Geoscience Education*. 53(4): 355-365.

[15] Rubini, B. dan Ardianto, D. (2014). Analisis Literasi Sains Sisiwa di Kota Bogor. *Research Report*. Bogor: Pascasarjana Universitas Pakuan.

[16] Toharudin, U., Hendrawati, S. & Rustaman, A. (2011). *Building Science literacy for student.* Bandung: Humaniora.

Institute of Physics
Dirac House, Temple Back
Bristol BS1 6BE UK

CURRAN ASSOCIATES INC.
proceedings.com

ISSN: 1757-8981
ISBN 978-1-5108-3670-9

9 781510 836709